Microbiology for Nurses

V. Deepa Parvathi
Department of Human Genetics
College of Biomedical Sciences,
Technology and Research
Sri Ramachandra University
Porur, Chennai
Tamil Nadu

R. Sumitha
Department of Biomedical Sciences
College of Biomedical Sciences,
Technology and Research
Sri Ramachandra University
Porur, Chennai
Tamil Nadu

S. Smitha
C M Manguli Degree College
Sindagi, Bijapur
Karnataka

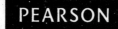

Chennai • Delhi

Assistant Editor—Acquisitions: R. Dheepika
Editor—Production: C. Purushothaman

Copyright © 2014 Dorling Kindersley (India) Pvt. Ltd

This book is sold subject to the condition that it shall not, by way of trade or otherwise, be lent, resold, hired out, or otherwise circulated without the publisher's prior written consent in any form of binding or cover other than that in which it is published and without a similar condition including this condition being imposed on the subsequent purchaser and without limiting the rights under copyright reserved above, no part of this publication may be reproduced, stored in or introduced into a retrieval system, or transmitted in any form or by any means (electronic, mechanical, photocopying, recording or otherwise), without the prior written permission of both the copyright owner and the publisher of this book.

ISBN 978-93-325-2527-6

First Impression

Published by Dorling Kindersley (India) Pvt. Ltd, licensees of Pearson Education in South Asia.

Head Office: 7th Floor, Knowledge Boulevard, A-8(A), Sector 62, Noida 201 309, UP, India.
Registered Office: 11 Community Centre, Panchsheel Park, New Delhi 110 017, India.

Compositor: Cameo Corporate Services Limited, Coimbatore.
Printer: Pushp Print Services

Dedicated to

**our beloved Parents
and Friends**

Brief Contents

Foreword   *xxi*
Preface   *xxiii*

UNIT 1: INTRODUCTION   1

1. Importance and Relevance of Microbiology to Nursing   3
2. Role of a Nurse in Microbiology   4
3. Historical Perspective   6

UNIT 2: GENERAL CHARACTERISTICS OF MICROBES   11

4. Structure and Classification of Microbes   13
5. Morphology—Size and Forms   28
6. Flagella and Motility—Hanging Drop Technique   46
7. Colonization, Nutrition, and Growth of Microbes   51
8. Culture Media   72
9. Laboratory Methods for the Identification of Microorganisms   82
10. Staining Techniques   97

UNIT 3: INFECTION CONTROL   111

11. Sources, Portals, and Transmission of Infections   113
12. Asepsis, Disinfection, and Sterilization—Types and Methods   119

13 Chemotherapy and Antibiotics	137
14 Standard Safety Measures and Biomedical Waste Management	152
15 Hospital Acquired Infection and Hospital Infection Control Programme	163

UNIT 4: PATHOGENIC ORGANISMS — 175

16 Bacteria	177
17 Viruses	231
18 Fungi	258
19 Parasites	275
20 Rodents and Vectors	285

UNIT 5: IMMUNOLOGY — 295

21 Immunity—Classification	297
22 Antigen and Antibody Reaction	304
23 Hypersensitivity Reaction	311
24 Serological Tests	322
25 Immunoprophylaxis	326
Glossary	*333*
Illustrations	*341*
Index	*373*

Contents

Foreword   xxi
Preface   xxiii

UNIT 1: INTRODUCTION   1

1 Importance and Relevance of Microbiology to Nursing   3

2 Role of a Nurse in Microbiology   4

3 Historical Perspective   6
    3.1 Immunity and Vaccination   8
       Multiple Choice Questions   9
       Short Notes   10
       Essay   10

UNIT 2: GENERAL CHARACTERISTICS OF MICROBES   11

4 Structure and Classification of Microbes   13
    4.1 Introduction   13
       4.1.1 Overview of Prokaryotic Cell Structure   14
       4.1.2 Overview of Eukaryotic Cell Structure   14
    4.2 Structure of Prokaryotic Cell   14
       4.2.1 Intracellular Structures   14
       4.2.2 Extracellular Structures   15
    4.3 Shapes of Microbes   16
       4.3.1 Advantages of Shape to the Cell   16
    4.4 Structures Involved in Attachment   16
    4.5 Classification of Microbes   17
       4.5.1 Scientific Nomenclature (Binomial Nomenclature)   17
       4.5.2 Taxonomic Hierarchy of *Escherichia coli*   17
       4.5.3 Two-Kingdom Classification   18

	4.5.4	Three-Kingdom System	18
	4.5.5	Four-Kingdom System	18
	4.5.6	Five-Kingdom System	18
	4.5.7	Carl Woese's Three-Domain System	20
4.6	Universal Tree of Life	22
4.7	Morphological Classification	23
4.8	Nutritional Classification	23
4.9	Biochemical Classification	24
4.10	Classification Based on Staining Reaction	24
4.11	Serological or Antigenic Classification	24
	Multiple Choice Questions	24
	Short Notes	26
	Essays	27

5 Morphology—Size and Forms   28

5.1	Introduction	28
5.2	Bacterial Morphology	28
	5.2.1 Size of Bacterial Cells	28
	5.2.2 Shape of Bacterial Cells	29
	5.2.3 Colony Morphology	30
5.3	Microscopy	31
	5.3.1 Types of Microscopes	31
5.4	Anatomy of Bacteria	34
	5.4.1 Architecture of a Bacterial Cell	35
	5.4.2 Cell Wall-Deficient Bacteria	43
	Multiple Choice Questions	43
	Short Notes	44
	Essays	45

6 Flagella and Motility—Hanging Drop Technique   46

6.1	Bacterial Motility	46
	6.1.1 Types of Bacterial Motility	46
6.2	Detection of Bacterial Motility	48
	6.2.1 Flagellar Staining	48
	6.2.2 Motility Test	48
	6.2.3 Direct Microscopic Examination	48
6.3	Importance of Bacterial Motility	49
	6.3.1 Chemotactic Behaviour	49
	6.3.2 Root Colonization	49
	6.3.3 Pathogenesis	49

		6.3.4 Twitching Motility	49
		Multiple Choice Questions	49
		Short Notes	50
		Essays	50

7 Colonization, Nutrition, and Growth of Microbes — 51

- 7.1 Colonization of Bacteria — 51
  - 7.1.1 Invasiveness — 51
  - 7.1.2 Toxigenesis — 51
  - 7.1.3 Adherence of Bacteria — 52
  - 7.1.4 Pathogenicity — 52
  - 7.1.5 Wound Colonization — 52
- 7.2 Microbial Nutrition — 53
  - 7.2.1 Autotrophic Bacteria — 53
  - 7.2.2 Heterotrophic Bacteria — 54
  - 7.2.3 Symbiotic Bacteria — 54
  - 7.2.4 Parasitic Bacteria — 54
  - 7.2.5 Major Elements — 55
- 7.3 Microbial Growth — 56
  - 7.3.1 Growth Factors — 57
  - 7.3.2 Uptake of Nutrients by Bacteria — 57
  - 7.3.3 Growth Curve — 60
  - 7.3.4 Continuous Culture — 62
  - 7.3.5 Synchronous Growth — 63
- 7.4 Influence of Environmental Factors on Microbial Growth — 64
  - 7.4.1 Water Acidity and Solutes — 64
  - 7.4.2 Temperature — 65
  - 7.4.3 Oxygen Requirement — 67
  - 7.4.4 pH — 68
  - 7.4.5 Pressure — 68
  - 7.4.6 Radiation — 68
- Multiple Choice Questions — 69
- Short Notes — 71
- Essays — 71

8 Culture Media — 72

- 8.1 Introduction — 72
- 8.2 History of Culture Media — 72
- 8.3 Importance of Culture Media — 73
- 8.4 Common Components Used in Culture Media — 73

		8.4.1	Agar	73

        8.4.1   Agar   73
        8.4.2   Peptones   73
        8.4.3   Water   73
        8.4.4   Extracts   74
        8.4.5   Body Fluids   74
   8.5 Classification of Culture Media   74
        8.5.1   Classification Based on Consistency   75
        8.5.2   Classification Based on Nutritional Components   76
        8.5.3   Classification Based on Functional Use   76
   8.6 Preparation and Storage of Culture Media   79
        Multiple Choice Questions   *80*
        Short Notes   *81*
        Essays   *81*

9 Laboratory Methods for the Identification of Microorganisms 82

   9.1 Good Laboratory Practices (GLP)   82
   9.2 Five I's in a Microbiology Laboratory   83
        9.2.1   Inoculation   83
        9.2.2   Incubation   83
        9.2.3   Isolation   83
        9.2.4   Inspection   86
        9.2.5   Identification   86
   9.3 Molecular Techniques and Typing   86
        9.3.1   Species Identification   86
        9.3.2   Typing   87
        9.3.3   Genotyping   88
   9.4 Biochemical Identification Techniques   89
        9.4.1   Biochemical Tests   89
        Multiple Choice Questions   *95*
        Short Notes   *95*
        Essays   *96*

10 Staining Techniques 97

   10.1 Definition and Protocol   97
        10.1.1 Uses of Stains   98
        10.1.2 Some Commonly Used Stains   98
        10.1.3 Staining Bacteria   100
   10.2 Simple Staining   100
        10.2.1 Staining of Bacteria from the Colony   100
        10.2.2 Staining of Bacteria from the Broth   100

10.3	Differential Staining	101
	10.3.1 Gram Staining	101
	10.3.2 Acid-Fast Staining	103
	10.3.3 Endospore Staining	104
	10.3.4 Capsule Staining	105
	10.3.5 Metachromatic Granule Staining (Albert's Staining)	106
	10.3.6 Flagella Staining	106
	10.3.7 Calcofluor White Staining	107
	Multiple Choice Questions	107
	Short Notes	109
	Essays	109

UNIT 3: INFECTION CONTROL   111

11 Sources, Portals, and Transmission of Infections   113

11.1	Introduction	113
11.2	Classification of Infections	113
11.3	Sources of Infections	114
	11.3.1 Humans	114
	11.3.2 Animals	114
	11.3.3 Insects	115
	11.3.4 Soil and Water	115
	11.3.5 Food	115
11.4	Portals of Entry and Exit	115
11.5	Modes of Transmission of Infections	116
	Multiple Choice Questions	117
	Short Notes	118
	Essays	118

12 Asepsis, Disinfection, and Sterilization—Types and Methods   119

12.1	Introduction	119
12.2	Asepsis	119
	12.2.1 Practice of Aseptic Techniques	120
12.3	Sterilization and Disinfection	120
12.4	Types of Sterilization	120
	12.4.1 Physical Sterilization	121
12.5	Types of Disinfection	131
	Conclusion	134
	Multiple Choice Questions	134

Short Notes — 136
Essays — 136

13 Chemotherapy and Antibiotics — 137
- 13.1 Introduction — 137
- 13.2 History of Antibiotics — 138
- 13.3 General Properties of an Ideal Antibiotic — 138
- 13.4 Classification of Antibiotics — 139
    - 13.4.1 Based on Target Organism — 139
    - 13.4.2 Based on Spectrum of Action — 140
    - 13.4.3 Based on Cidal and Static Activity — 140
    - 13.4.4 Based on Origin — 140
    - 13.4.5 Based on Mode of Action — 141
- 13.5 Antifungal Drugs — 145
- 13.6 Antiviral Drugs — 146
- 13.7 Antiprotozoal Drugs — 147
- 13.8 Antibiotic Resistance — 149
- Multiple Choice Questions — 150
- Short Notes — 151
- Essays — 151

14 Standard Safety Measures and Biomedical Waste Management — 152
- 14.1 Biomedical Waste Management — 152
    - 14.1.1 Collection and Segregation of Wastes — 153
    - 14.1.2 Containment and Labelling — 154
    - 14.1.3 Transportation — 155
    - 14.1.4 Treatment of Biomedical Waste — 155
    - 14.1.5 Disposal — 156
    - 14.1.6 Record Maintenance — 157
- 14.2 Risk Assessment — 157
    - 14.2.1 Pathogenicity — 157
    - 14.2.2 Route of Transmission — 157
    - 14.2.3 Infectious Agent Stability — 157
    - 14.2.4 Infectious Dose — 157
    - 14.2.5 Susceptibility of the Host — 157
    - 14.2.6 Concentration and Volume of the Pathogen — 158
- 14.3 Standard Safety Measures — 158
    - 14.3.1 Containment — 158
    - 14.3.2 Personal Protective Equipment — 158

15 Hospital Acquired Infection and Hospital Infection Control Programme — 163

    14.3.3 Biological Safety Cabinets — 159
    14.3.4 Facility as a Barrier — 159
    14.3.5 Biosafety Levels — 160
    Multiple Choice Questions — 161
    Short Notes — 161
    Essays — 162

- 15.1 Introduction — 163
  - 15.1.1 Occurrence of Infections — 163
  - 15.1.2 Microbial Causes — 165
- 15.2 Epidemiology of Nosocomial Infections — 165
  - 15.2.1 Nosocomial Infection Sites — 165
  - 15.2.2 Urinary Tract Infections — 165
  - 15.2.3 Surgical Site Infections — 166
  - 15.2.4 Nosocomial Pneumonia — 166
  - 15.2.5 Nosocomial Bloodstream Infections — 166
  - 15.2.6 Skin and Soft Tissue Infections — 167
  - 15.2.7 Other Nosocomial Infections — 167
- 15.3 Infection Control Programmes: Protocols — 167
  - 15.3.1 National Programmes — 167
  - 15.3.2 Hospital Programmes — 168
  - 15.3.3 Infection-Controlling Responsibility — 168
- 15.4 Nosocomial Infection Surveillance — 170
  - 15.4.1 Objectives — 170
- *Multiple Choice Questions* — 170
- *Short Notes* — 172
- *Essays* — 173

UNIT 4: PATHOGENIC ORGANISMS — 175

16 Bacteria — 177

- 16.1 *Staphylococcus* — 177
  - 16.1.1 General Properties — 177
  - 16.1.2 Cultural Characteristics — 177
  - 16.1.3 Biochemical Properties — 178
  - 16.1.4 Mode of Transmission — 178
  - 16.1.5 Virulence Factor — 178
  - 16.1.6 Pathogenesis — 179

xiv | Contents

	16.1.7 Laboratory Diagnosis	180
	16.1.8 Prevention and Treatment	181
16.2	*Streptococcus*	182
	16.2.1 Introduction	182
	16.2.2 General Properties	182
	16.2.3 Cultural Characteristics	182
	16.2.4 Biochemical Properties	182
	16.2.5 Mode of Transmission	182
	16.2.6 Pathogenesis	183
	16.2.7 Laboratory Diagnosis	183
	16.2.8 Treatment	183
16.3	*Neisseria*	184
	16.3.1 Introduction	184
	16.3.2 *Neisseria gonorrhoeae*	184
	16.3.3 *Neisseria meningitidis*	187
16.4	*Corynebacterium*	189
	16.4.1 General Properties	189
	16.4.2 Cultural Characteristics	189
	16.4.3 Biochemical Properties	189
	16.4.4 Mode of Transmission	189
	16.4.5 Pathogenesis	189
	16.4.6 Laboratory Diagnosis	190
	16.4.7 Treatment	190
16.5	*Enterobacteriaceae*	191
	16.5.1 Introduction	191
	16.5.2 *Escherichia*	191
	16.5.3 *Klebsiella*	194
	16.5.4 *Proteus*	196
	16.5.5 *Shigella*	198
	16.5.6 *Salmonella typhi*	199
16.6	*Mycobacterium*	201
	16.6.1 General Properties	201
	16.6.2 *Mycobacterium tuberculosis*	201
	16.6.3 *Mycobacterium leprae*	205
16.7	*Vibrio cholerae*	207
	16.7.1 General Properties	207
	16.7.2 Cultural Characteristics	208
	16.7.3 Biochemical Properties	208
	16.7.4 Mode of Transmission	208
	16.7.5 Pathogenesis	208

		16.7.6 Laboratory Diagnosis	209
		16.7.7 Prophylaxis and Treatment	209
16.8	*Spirochaetes*		209
		16.8.1 Introduction	209
		16.8.2 Size and Structure	210
		16.8.3 Habitat	211
16.9	*Mycoplasma*		211
		16.9.1 Introduction	211
		16.9.2 Morphology and General Characteristics	212
		16.9.3 Cultural Characteristics	213
		16.9.4 Biochemical Reactions	213
		16.9.5 Susceptibility to Physical and Chemical Agents	213
		16.9.6 Antigenic Properties	214
		16.9.7 Pathogenicity	214
		16.9.8 Laboratory Diagnosis	215
		16.9.9 Treatment and Prophylaxis	215
16.10	*Rickettsia*		216
		16.10.1 General Characteristics	216
		16.10.2 Cultural Characteristics	216
		16.10.3 Pathogenesis	216
		16.10.4 Laboratory Diagnosis	217
		16.10.5 Treatment	217
16.11	*Chlamydia*		217
		16.11.1 Classification	217
		16.11.2 General Characteristics	217
		16.11.3 Developmental Cycle	218
		16.11.4 Cultural Characteristics	219
		16.11.5 Pathogenesis	219
		16.11.6 Laboratory Diagnosis	220
		16.11.7 Treatment	220
	Multiple Choice Questions		*221*
	Short Notes		*227*
	Essays		*229*

17 Viruses — 231

17.1	General Properties of Viruses		231
	17.1.1 Introduction		231
	17.1.2 Structure of Viruses		231
	17.1.3 Replication of Viruses		232
	17.1.4 Classification of Viruses		236

17.2 Herpesviridae — 237
    17.2.1 Morphology — 237
    17.2.2 Herpes Simplex Virus — 238
    17.2.3 Varicella Zoster Virus — 239
    17.2.4 Epstein–Barr Virus — 240
    17.2.5 Cytomegalovirus — 241
17.3 Picornaviridae — 242
    17.3.1 Enteroviruses — 242
    17.3.2 Rhinoviruses — 244
    17.3.3 Hepatovirus (Hepatitis A Virus) — 245
17.4 Rhabdoviridae — 245
    17.4.1 Rabies Virus — 245
    17.4.2 Host Range and Growth Characteristics — 246
    17.4.3 Pathogenesis — 247
    17.4.4 Clinical Features — 247
    17.4.5 Immune Response — 247
    17.4.6 Laboratory Diagnosis — 247
    17.4.7 Post-exposure Treatment and Prophylaxis — 248
    17.4.8 Rabies Vaccine — 248
17.5 Retroviridae — 249
    17.5.1 Human Immunodeficiency Viruses — 249
    17.5.2 Laboratory Diagnosis — 252
    Multiple Choice Questions — *253*
    Short Notes — *256*
    Essays — *257*

18 Fungi — 258

18.1 Introduction — 258
18.2 Morphological Classification — 258
    18.2.1 Moulds — 258
    18.2.2 Yeasts — 259
    18.2.3 Dimorphic Fungi — 260
18.3 Taxonomical Classification — 261
18.4 Cell Wall of Fungi — 262
18.5 Fungal Reproduction — 262
18.6 Mycoses — 263
    18.6.1 Superficial Mycosis — 263
    18.6.2 Subcutaneous Mycosis — 266
    18.6.3 Deep Mycosis — 269
    18.6.4 Opportunistic Mycosis — 270

	Multiple Choice Questions	273
	Short Notes	274
	Essays	274

19 Parasites — 275
- 19.1 Introduction — 275
- 19.2 *Entamoeba histolytica* — 275
  - 19.2.1 Life Cycle — 276
  - 19.2.2 Pathogenesis — 277
  - 19.2.3 Laboratory Diagnosis — 277
  - 19.2.4 Treatment — 277
- 19.3 *Plasmodium* sp. — 278
  - 19.3.1 Life Cycle — 278
  - 19.3.2 Pathogenesis — 279
  - 19.3.3 Laboratory Diagnosis — 280
  - 19.3.4 Treatment — 280
- 19.4 Parasitic Helminths — 280
  - 19.4.1 *Taenia solium* — 280
  - 19.4.2 *Wuchereria bancrofti* — 282
- Multiple Choice Questions — 283
- Short Notes — 284
- Essays — 284

20 Rodents and Vectors — 285
- 20.1 Rodents — 285
  - 20.1.1 Introduction — 285
  - 20.1.2 *Yersinia pestis* — 285
- 20.2 Vectors — 287
  - 20.2.1 Introduction — 287
  - 20.2.2 Ticks — 288
  - 20.2.3 Lice — 290
  - 20.2.4 Mites — 291
- Multiple Choice Questions — 292
- Short Notes — 293
- Essays — 293

UNIT 5: IMMUNOLOGY — 295

21 Immunity—Classification — 297
- 21.1 Introduction — 297
  - 21.1.1 Recognition — 297

	21.1.2	Response	297
21.2	Types of Immunity		298
	21.2.1	Innate Immunity	298
	21.2.2	Adaptive Immunity	299
21.3	Cells of the Immune System		299
	21.3.1	B Lymphocytes	300
	21.3.2	T Lymphocytes	300
21.4	Humoral Immunity		301
21.5	Cell-Mediated Immunity		302
	Multiple Choice Questions		*302*
	Short Notes		*303*
	Essay		*303*

22 Antigen and Antibody Reaction — 304

22.1	Introduction		304
	22.1.1	Antigens	304
	22.1.2	Antibodies	304
22.2	Antigen–Antibody Interactions		305
22.3	Types of Antigen–Antibody Interactions		305
	22.3.1	Precipitation Reactions	305
	22.3.2	Agglutination Reactions	308
	22.3.3	Enzyme-Linked Immunosorbent Assay	308
	Multiple Choice Questions		*309*
	Short Notes		*310*
	Essays		*310*

23 Hypersensitivity Reaction — 311

23.1	Introduction		311
	23.1.1	Allergens	311
	23.1.2	Inclination to Allergic Reaction	311
23.2	Classification of Hypersensitivity Reaction		312
	23.2.1	Anaphylactic Hypersensitivity	312
	23.2.2	Type II Hypersensitivity Reaction	316
	23.2.3	Type III Hypersensitivity Reaction	317
	23.2.4	Cell-Mediated Hypersensitivity	319
	Multiple Choice Questions		*321*
	Short Notes		*321*
	Essay		*321*

24 Serological Tests — 322

- 24.1 Introduction — 322
- 24.2 Precipitation — 322
- 24.3 Agglutination — 323
  - 24.3.1 Widal Test — 323
  - 24.3.2 Anti-Streptolysin O Test — 324
  - *Multiple Choice Questions* — *325*
  - *Short Notes* — *325*
  - *Essay* — *325*

25 Immunoprophylaxis — 326

- 25.1 Introduction — 326
- 25.2 Active and Passive Immunity — 326
- 25.3 Adjuvants — 327
  - 25.3.1 Functions of Adjuvants — 327
- 25.4 Vaccines and Its Types — 328
  - 25.4.1 Live or Attenuated Vaccine — 328
  - 25.4.2 Killed Vaccines — 328
  - 25.4.3 Recombinant Subunit Vaccine — 329
  - 25.4.4 Conjugate Vaccine — 329
  - 25.4.5 Toxoid — 330
- 25.5 Vaccination Schedule — 330
- 25.6 Current Approaches in Vaccines — 331
  - *Multiple Choice Questions* — *331*
  - *Short Notes* — *332*
  - *Essay* — *332*

Glossary — *333*
Illustrations — *341*
Index — *373*

Foreword

As an academician who has taught biological sciences for forty years, few experiences are more gratifying than examining and evaluating scholarly write-ups of young authors. One such experience was when Deepa requested me to review her book *Microbiology for Nurses* jointly written with Ms R. Sumitha and Ms S. Smitha.

This book is different from the many books already available on the subject in its radical approach to the topic and the innate charm of its script. It showcases the authors' exemplary writing skills and subject knowledge with which they have, chapter by chapter, endeavored to present the elements of microbiology in a style that is comprehensible to students and health care professionals alike. The book will serve not only as a textbook for nursing students but also as a ready-reckoner to which they can turn to at any point in their career.

Divided into five units for the learner's convenience, the chapters of the book are succinctly illustrative and supported by a rich question bank. Unit 1 introduces microbiology from a historical perspective in three chapters; Unit 2 has seven chapters that illustrate the general characteristics of microbes; Unit 3 expounds on infection control in five chapters; Unit 4 focuses on pathogenic organisms in five chapters; Unit 5 is on immunology, which is covered in five chapters. A glossary of terms and acronyms used in the text has been added for the student's benefit. The book also presents selected images in colored plates, explicit in their detail thanks to the characteristic meticulousness of Pearson Education. These plates enable the learner to appreciate the structural intricacies laid out in the illustrations.

I strongly recommend *Microbiology for Nurses* by V. Deepa Parvathi, R. Sumitha and S. Smitha, not just for nursing students and the faculty who mentor them but for all health care professionals and students of medical microbiology, biotechnology and laboratory technicians. I am sure the approach the book has adopted as well as the method it espouses will help the reader to appreciate the relevance of microbiology to health care.

I congratulate the authors for their magnificent and sustained effort to lay down the essentials of the subject in a manner that is forthright in its attempt to reach the reader with clarity. The concerted effort of the team at Pearson Education in bringing out this book is indeed laudable.

SULTAN AHMED ISMAIL
Former Head, Department of Biotechnology
The New College, Chennai

Director
Ecoscience Research Foundation
Chennai

Preface

The science of microbiology has evolved and emerged as an important field in biomedicine and clinical practice. The emergence and re-emergence of various infectious agents, antimicrobial drug resistance and nosocomial infections have stirred remarkable advances in the understanding of concepts and hypotheses and kindled the development of various identification strategies and diagnostic tools.

It is important for biologists to understand the fundamentals, analyze theoretical concepts and explore new avenues through observation and scientific enquiry.

Microbiology for Nurses has been developed for students of nursing who study the subject as a part of their curriculum. It provides a simple, logical and comprehensive approach to learning the essentials of the subject. To this end, the objectives listed at the beginning of each chapter encapsulate the topics to be discussed and give the students an effective overview. An exhaustive glossary in simple sentences has been compiled to help students comprehend intricate biological terms and phrases. Every chapter concludes with pedagogical elements that consist of an array of multiple choice, short notes and essay questions designed to enable the students perform a progressive self-assessment of their understanding of the subject.

The content of this book has been based on the curriculum framed by the Nursing Council of India and classified into five distinct units. It builds on the science of microbiology from its introductory concepts, probing first into the history and roles of nurses, and progressing gradually to delineate the various methods used to classify and identify microorganisms. The book also expounds on the different techniques of sterilization, antibiotic therapy, biomedical waste management and nosocomial infections, which are of prime relevance in a clinical scenario. In addition, the book presents a detailed account of various pathogenic organisms and the role of immunology and vaccination in prophylaxis. It presents an assortment of diagrammatic illustrations, rendered conspicuous in color, to enhance clarity to the elucidations.

The chapters have been conceptualized and written by three authors, each contributing topics in their respective areas of expertise and confidence in a student friendly and teacher recommendable manner.

ACKNOWLEDGEMENTS

This book is indeed the culmination of our combined effort and contributions in our respective areas of specialization and confidence. Our mutual encouragements, constructive criticisms and suggestions were vital in designing the table of contents and bringing about an overall depth to the discussions in the book. We take this opportunity to thank a few people without whose support and motivation, we could not have completed this book.

We thank R. Dheepika, the acquisitions editor of the book, for her immense involvement and professionalism with which she followed up on the book's progress. Her suggestions and inputs at every stage contributed to its overall development. We also commend C. Purushothaman and his team of the production department for their keen eye for detail in the proofs and for developing our hand-drawn images booking an impressive layout.

We are obliged to Dr Gautham Annappa, Assistant Professor, Department of Studies & Research in Microbiology, Mangalore University, for his significant contributions to Unit 3.

We appreciate the efforts of Ms S. Smitha, who rendered the book's images as neat hand-drawn figures to accompany the manuscript.

We are indebted to all our teachers for imparting in us the confidence and for all their words of happiness and encouragement.

We are grateful to Sri. V. R. Venkataachalam, Chancellor of Sri. Ramachandra University, and the university's management for their support and encouragement.

We thank Dr Sultan Ismail for perusing the entire book and also for writing the foreword.

We thank Ms Betty Lincoln, our dear friend, for all her prayers and words of strength.

I express my gratitude to my dear father Mr A. Ravindran and my mother Dr Mallika Ravindran for being my pillars of strength.

I am thankful to my husband Mr A. Madhu for his encouraging words and support.

I am obliged to my children, M. Harini and M. Pragadesh, for their unconditional love. I specially thank Harini for helping us with a few hand-drawn images.

I thank Dr T. S. Lokeswari, Head of the Department of Biomedical Sciences, for instilling in me the confidence with her words of motivation and appreciation.

R. SUMITHA

I am indebted to my beloved parents, Dr B. R. Shankar and Smt. Devibai, for their unconditional love and support; and to my dear brothers, Chidanand and Gopal, for their help and encouragement.

I specially thank my soul-mate and sister Bhagyalakshmi and my brother-in-law Mr Sagar for their endless motivation.

I am indebted to my husband Dr K. Srinivas Naik for his undaunted support and help in all my endeavors.

Most importantly, I thank my sons for their beautiful smiles which kept my energy levels from sinking while I was working on this book.

S. SMITHA

I thank the Almighty for showering his blessings on us and helping us to bring this project to fruition. I take this opportunity to thank my beloved parents, Mr V. Venkatachalam and Dr Raji Venkatachalam, and my brother Vishnu, for their unconditional love and support in all my academic and personal endeavors.

I am indebted to my dear husband Mr Koushik, and my parents-in-law, Dr Ramesh Rao and Mrs Padmini Ramesh, for their encouraging words and without whose support I could not have completed this project.

My little son Ishaan is a source of happiness and energy in my life. He drives the best out of me by flashing his dimples!

I am grateful to Ms V. Priyanka for her contribution towards the chapter on sources and portals of infection.

My mentors, Dr Solomon FD Paul and Dr M. Ravi, have inculcated in me the art of scientific writing. I thank them for their support in all my professional accomplishments.

V. Deepa Parvathi

Unit 1

INTRODUCTION

Chapter 1 Importance and Relevance of Microbiology to Nursing   3
Chapter 2 Role of a Nurse in Microbiology   4
Chapter 3 Historical Perspective   6

1 Importance and Relevance of Microbiology to Nursing

Microbiology is the study of microscopic organisms and involves the handling, identification, and manipulation of organisms of microscopic size (not visible to the naked eye). These organisms possess a simple anatomic structure with differentiation of cells and tissues. The microbial world includes prokaryotic and eukaryotic cells, which fundamentally differ from each other in the organization of the nucleus. In addition, the microbial world includes acellular obligatory parasites—viruses—as well. Microorganisms are diverse and play an important role in ecosystem. They are also considered as one of the first living organisms to have evolved on earth. Many scientists across the globe have been working on the discovery of various microorganisms and the understanding of their implications on health, infections, and treatment modules. The important contributions of various scientists in the field of microbiology and the landmarks in microbiological research have been detailed in Chapter 3.

Apart from understanding the role of microorganisms with respect to their modes of entry, routes of transmission, pathogenesis, and treatment in various infectious diseases, much has gone into the development of various aseptic culture techniques for the isolation and characterization of microorganisms. The concepts of asepsis, sterilization, and disinfection provide an understanding to the control of infectious diseases, especially in a hospital set-up. Care must also be taken with respect to sanitary conditions to control infections.

Microbiology has utmost relevance in health care units, and it is important for health care professionals to understand the pathogenesis and control of infectious diseases. The mode of action of various antibiotics, the concept of antibiotic resistance, and multi-drug resistance are important apart from the methods of vaccination, types of vaccines, and schedules involved in vaccination. Understanding of the pathogenesis of virulent organisms, their epidemiology, host defense, and treatment strategies is an important aspect of medical microbiology. In addition, the role of the normal flora as opportunistic pathogens is critical.

Health care professionals need to understand that microorganisms do not limit their role to infectious diseases alone. They play a vital role in the ecosystem, in the degradation of organic materials, recycling, and bioremediation. Microorganisms have been employed as model organisms in genetics and recombinant DNA technology to engineer agents for pest control, and they have also been widely used in the fermentation industry. In addition, many bacteria and viruses have also been used in gene therapy strategies and drug development.

2 Role of a Nurse in Microbiology

This chapter deals with the role of a nurse in relation to the condition of the patient, disease stage, diagnosis, treatment module, and hospital environment in a microbiological perspective. The role of a nurse is varied, and it is important for a nurse to multitask intellectually with fundamental knowledge, decisive thinking, and application. Nurses dedicate a lot of their time and energy on the patients. This chapter focuses on evolving nurses with a conceptual clinical approach towards the following:

1. Asepsis, sterilization, and disinfection
2. Recognition of infection
3. Infection control—pathogenesis and transmission
4. Nosocomial infection
5. Immune system
6. Clinical thought process

It is important for nurses to learn the concepts and gain the knowledge and skills required to analyse and explore clinical situations. Their fundamental aim of providing quality patient care with professional ethics shall be strengthened if they think critically, plan strategies, and work with skill and effective communication. They need to understand the science of microbiology, from basic sciences to evidence-based nursing care.

The theoretical learning process of nurses shall include understanding the biology of microbes from their fundamental evidence of discovery (including various historical contributions of scientists across the globe and their experiments), their diversity in structure and pathogenesis, role of microbes in causing a disease, mechanisms involved in pathogenesis, routes of transmission, immune system of the host, infection control strategies adopted in hospitals, and various methods of asepsis and disinfection in terms of hospital equipment. The various tools and chemicals and their appropriate usage in asepsis, sterilization, and disinfection are of significance in patient care and hospital management. In addition, the different methods of sample collection, handling, storage, and transport of microbiological specimens and samples are considered vital in nursing education. Application towards the identification of microorganisms using laboratory tests (culture techniques and biochemical and serological tests) shall help a nurse read laboratory reports and understand the course of diagnosis and treatment strategies directed by the physician. Apart from this, it is important for nurses and other health care professionals (ward technicians, laboratory technicians, sanitary workers, etc.) to be educated adequately about self care and protective measures.

It is important for nurses to learn the pathogenesis of microorganisms towards understanding the anatomy of bacteria; the presence of capsules, fimbriae, and virulence factors; and cell wall anatomy. Knowledge of physical and chemical methods of controlling microbial infections through radiation and various chemical agents is essential to control nosocomial infections. Nurses also need to be familiar with the compositions and routes of administration of various vaccines and antibiotics along with their preparations and formulation data sheets. Mechanisms of action of antibiotics and allergic reactions apart from antibiotic resistance have to be critically understood. Moreover, knowledge of zoonoses, their reservoirs, and methods of transmission is vital in medical microbiology.

Understanding the host immune system in terms of types of immunities and various immune molecules such as interferons, cytokines, and complement proteins aids a nurse in deciphering the defence mechanism involved and generation of antibodies and establishment of immunity (active and passive). Apart from understanding clinical and medical microbiology, it is also an added advantage for nurses to realize the importance of microorganisms in food, agriculture, industries, and environment. Owing to the properties demonstrated and many advantages offered by microorganisms, they have been used as model organisms of research in genetics and recombinant DNA technology.

The flow of information in the forthcoming chapters has been designed in such a way as to help nurses understand fundamental microbiology with critical thinking and application as well as perform their role effectively.

3 Historical Perspective

CHAPTER OBJECTIVE

3.1 Immunity and Vaccination

Microbiology is the study of living organisms of microscopic size. The term microbiology was introduced by Louis Pasteur, who is referred to as the 'father of microbiology'. Microbiology has various sub-disciplines of which medical microbiology deals with the aetiology of infectious diseases in human beings and animals, their pathogenesis, methods of diagnosis, treatment modules available, and preventive measures involved.

Microorganisms of medical or clinical interest include bacteria, viruses, fungi, and protozoa. Most microorganisms can be observed with suitable magnification and staining techniques under light microscopes. The size of viruses, is generally less than 0.2 μm, and hence, they cannot be observed under conventional microscopes.

Infections and diseases have been a major challenge to the society and the clinical community as they cause high morbidity and health risks worldwide. Many serious infectious diseases like AIDS have posed a major threat to the health of individuals. It is here that the applications of microbiology have contributed significantly to the practice of medicine in terms of diagnosis and prevention and providing better health care to the society as a whole. The science of microbiology has evolved greatly with the advancement of various molecular and biochemical tools coupled with automation to increase the efficiency and accuracy of diagnosis. The technology in vaccine development has seen tremendous growth in terms of developing vaccines against a plethora of infectious diseases, with high success rates in preventing and controlling epidemics as well.

Microbiology has grown multifold because of simple observations and inventions made by early scientists. Anton van Leeuwenhoek, a Dutch lens maker, won the credit for devising a simple apparatus to aid the observation of microbial forms (e.g., shapes of bacteria—spherical, rod, and spiral). Though his effort was not initially recognized, two centuries later his contribution to the evolution of microscopy was acknowledged, and he is now referred to as the 'father of bacteriology'.

Louis Pasteur, in 1865, was consulted for the contamination of raw materials used for alcohol fermentation. His involvement in exploring the cause led him to focus on the role of microorganisms in the fermentation and spoilage of alcohol. Pasteur demonstrated that undesirable life forms could be destroyed by heating at 55°C–60°C for a short period of time. Subsequently, this process was modified and is now referred to as pasteurization.

John Tyndall (1820–1893) demonstrated that prolonged heating was required for completely eliminating life forms. He explained that bacteria existed in two forms: heat-stable and heat-sensitive. Prolonged or intermittent heating is required to destroy heat-stable forms. Tyndallization, is a process of intermittent heating which kills both forms.

Joseph Lister (1827–1912) appreciated the emerging germ theory of disease. He advocated the use of carbolic acid as an aerosol during surgeries and impregnation of dressings to reduce the risk of post-operative infections. He established the fundamental principles of antisepsis for good surgical practice and hence is known as the 'father of antiseptic surgery'.

The science of microbiology evolved with the introduction of sterilization techniques by Pasteur and perfection of bacteriological techniques, staining procedures, and cultures by Robert Koch. The causative agents of various diseases were reported by different investigators. The disease-causing organisms discovered by various scientists along with the year of discovery are provided in Table 3.1.

Table 3.1 Various Disease-causing Organisms Discovered by Scientists

S.no.	Scientist	Organism Discovered	Year
1	Robert Koch	Bacillus of anthrax	1876
		Bacillus of tuberculosis	1882
		Vibrio cholerae	1883
2	Hansen	Leprosy bacillus	1874
3	Neisser	*Gonococcus*	1879
4	Alexander Ongston	Staphylococci	1880
5	Eberth	Typhoid bacillus	1880
6	Klebs and Loeffler	*Corynebacterium diphtheriae*	1884
7	Rosenbach	*Staphylococcus*	1886
8	Weichselbaum	*Meningococcus*	1887
9	David Bruce	Malta fever - Brucella	1887
10	Schaudinn and Hoffman	Spirochaetes - *Treponema pallidum*	1905

The discovery of various microorganisms made it important to have a definite criterion to identify a microorganism as the causative agent of a disease. A microorganism is accepted as the causative agent of an infectious disease only if it follows Koch's postulates, which are as follows:

1. The organism must be present in the lesions in every case of the infectious disease.
2. It should be possible to isolate the organism in pure culture from the lesions.

3. Inoculation of the pure culture into suitable laboratory animals should produce a similar disease.
4. It should be possible to re-isolate the organism in pure culture from the lesions produced in experimental animals.
5. Specific antibodies to the organism should be demonstrable in the serum of the patient suffering from the disease.

3.1 IMMUNITY AND VACCINATION

Thucydides (464–404 BC), a Greek historian, observed during a plague epidemic in Athens that the sick and dying were able to recover because of the care provided by those who were already affected and knew that they would not contract the disease again. Immunization of healthy individuals with dried crusts of small pox was common in India and China. This would produce a mild form of small pox (variolation), and then the individual would recover and become resistant to the severe forms of the pox. Edward Jenner (1749–1823) tested the hypothesis that milkmaids who had been exposed to cowpox (vaccinia) from their herd never got infected with the dreaded small pox (variola) by inoculating them with the fluid from cow pox pustules. The vaccinated individuals reacted mildly and after recovery did not contract small pox when exposed to the virus.

Louis Pasteur discovered that certain bacteria lost their virulence after extensive culturing in the laboratory, and he suggested that such bacteria might be capable of offering immunity. He demonstrated experiments with attenuated bacilli of anthrax on sheep, goat, and cow and challenged them with virulent forms of anthrax. He observed that all the vaccinated animals survived whereas the controls did not. Pasteur's development of a vaccine against rabies was a significant breakthrough in vaccinations. He administered an attenuated suspension of rabies virus to Joseph Meister, a 13-year-old boy, who was bitten by a rabid dog, and saved him from the dreaded disease. The Pasteur Institute of Paris was built by public contribution during his lifetime for investigations on infectious diseases and preparation of vaccines.

The important landmarks in the field of microbiology are highlighted in Table 3.2.

Table 3.2 Important Landmarks in Microbiology

Year	Landmark
1676	Anton van Leewenhoek first described bacteria.
1765	Abbe Lazzaro Spallanzani conducted experiments to disprove the theory of spontaneous generation.
1796	Edward Jenner introduced the small pox vaccine.
1861	Louis Pasteur disproved the theory of spontaneous generation.
1863–1865	Pasteur devised pasteurization.
1865	Joseph Lister introduced antisepsis to treat wound infections.
1873	William Budd described the role of milk and water in the transmission of typhoid.

(continued)

Table 3.2 (continued)

Year	Landmark
1876	Robert Koch reported the isolation of pure culture of anthrax bacillus.
1880	Alphonse Laveran discovered the malarial parasite.
1881	Pasteur developed the anthrax vaccine.
1882	Robert Koch discovered *Mycobacterium tuberculosis*.
1884	Koch's postulates were published.
1884	Metchnikoff announced the phagocytic theory of immunity.
1884	Hans Christian Gram introduced Gram staining.
1885	Louis Pasteur offered the first vaccination against rabies.
1888	Roux and Yersin discovered the diphtheria toxin.
1889	Brieger discovered the tetanus toxin.
1890	Kitasato and Von Behring discovered the tetanus antitoxin.
1890	Robert Koch discovered tuberculin.
1892	Ziehl and Neelsen developed acid fast staining.
1896	Paul Ehrlich introduced the methods of standardizing antitoxins and toxins.
1898	Paul Ehrlich expounded the side chain theory of immunity.
1900	Karl Landsteiner discovered blood groups.
1901	Bordet and Gengou described complement fixation.
1928	Alexander Fleming discovered penicillin.
1940	Karl Landsteiner and Wiener discovered the Rh factor.
1941	Albert Coons developed the fluorescent antibody test.
1943	Selman Waksman discovered streptomycin.
1948–1951	Three types of polio viruses were established.
1953	Jonas Salk and team devised the polio vaccine.
1954–1963	Jon Enders and Thomas Peebles developed the measles vaccine.
1956	Sabin developed the first oral polio vaccine.

MULTIPLE CHOICE QUESTIONS

1. The early microscope was invented by _____
   (a) Louis Pasteur
   (b) Anton van Leeuwenhoek
   (c) Robert Koch
   (d) Edward Jenner

   Ans. b

2. Tyndallization was named in honour of _____.
   (a) Robert Koch
   (b) Edward Jenner
   (c) Joseph Lister
   (d) John Tyndall

   Ans. d

3. Who is referred to as the father of antiseptic surgery?
   (a) Robert Koch	(b) Edward Jenner
   (c) Joseph Lister	(d) John Tyndall
   Ans. c

4. Koch's postulates were proposed by _____.
   (a) Louis Pasteur
   (b) Anton van Leeuwenhoek
   (c) Robert Koch
   (d) Edward Jenner
   Ans. c

5. Tuberculin was discovered by _____.
   (a) Louis Pasteur
   (b) Anton van Leeuwenhoek
   (c) Robert Koch
   (d) Edward Jenner
   Ans. c

6. Penicillin was discovered by _____.
   (a) Alexander Fleming
   (b) Paul Ehrlich
   (c) Karl Landsteiner
   (d) Ziehl and Neelsen
   Ans. a

7. Blood groups were discovered by _____.
   (a) Alexander Fleming
   (b) Paul Ehrlich
   (c) Karl Landsteiner
   (d) Ziehl and Neelsen
   Ans. c

8. Research on pox vaccination was done by _____.
   (a) Edward Jenner
   (b) Louis Pasteur
   (c) Alexander Fleming
   (d) Paul Ehrlich
   Ans. a

9. Vaccination against rabies was offered by _____.
   (a) Edward Jenner
   (b) Louis Pasteur
   (c) Alexander Fleming
   (d) Paul Ehrlich
   Ans. b

10. Sabin and Salk are vaccines against _____.
    (a) Tuberculosis	(b) Polio
    (c) Rabies	(d) Cholera
    Ans. b

SHORT NOTES

1. Mention 10 important landmarks in microbiological research.
2. What are Koch's postulates?
3. Why is Louis Pasteur known as the father of microbiology?
4. What is the role of a nurse in microbiology?

ESSAY

1. Give an overview of the history of microbiology. Add a note on the important landmarks and contributions of Louis Pasteur.

Unit 2

GENERAL CHARACTERISTICS OF MICROBES

Chapter 4	Structure and Classification of Microbes	13
Chapter 5	Morphology—Size and Forms	28
Chapter 6	Flagella and Motility – Hanging Drop Technique	46
Chapter 7	Colonization, Nutrition and Growth of Microbes	51
Chapter 8	Culture Media	72
Chapter 9	Laboratory Methods for the Identification of Microorganisms	82
Chapter 10	Staining Techniques	97

Unit 2

GENERAL CHARACTERISTICS OF MICROBES

4 Structure and Classification of Microbes

CHAPTER OBJECTIVES

4.1 Introduction
4.2 Structure of Prokaryotic Cell
4.3 Shapes of Microbes
4.4 Structures Involved in Attachment
4.5 Classification of Microbes
4.6 Universal Tree of Life
4.7 Morphological Classification
4.8 Nutritional Classification
4.9 Biochemical Classification
4.10 Classification Based on Staining Reaction
4.11 Serological or Antigenic Classification

4.1 INTRODUCTION

Microorganisms are those that scale to micro level in terms of measurement and dimensions. They are made up of one cell (unicellular) or a few cells (multicellular). They are diverse, exist abundantly on Earth, and can survive in the most extreme environmental conditions. Microorganisms can be divided into several categories based on their distinct characteristics.

The studies on microorganisms focus on their beneficial effects and also their pathogenic impact. In addition, microbes play an important role in maintaining the ecological balance in ways of decomposing organic wastes, in agriculture, and in industries. The application of microbes in research is diverse and challenging.

This chapter focuses primarily on the structure and classification of microbes in terms of their morphology and then proceeds towards varied classification methods.

Cells are divided into two basic types, which are as follows:

1. Prokaryotic cells
2. Eukaryotic cells

4.1.1 Overview of Prokaryotic Cell Structure (Figure 4.1)

Prokaryotic cells are smaller when compared with eukaryotic cells. In these cells, the cytoplasm is enclosed by the cell membrane, and the cell organelles freely float, without any membrane, in this cytoplasm. Nucleus is called nucleoid as it lacks the membrane surrounding it; that is, it contains a naked DNA molecule. Blue green algae and bacteria are examples of prokaryotes.

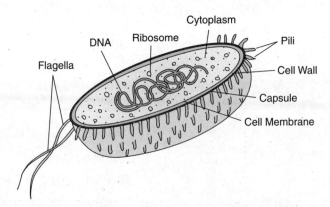

Figure 4.1 Structure and Contents of a Typical Prokaryotic (Bacterial) Cell (See page 341 for the colour image)

4.1.2 Overview of Eukaryotic Cell Structure

Eukaryotic cells have a true nucleus, that is, the nucleus is membrane bound. The DNA is divided among many chromosomes and resides in the nucleolus. These cells are larger and are well defined with a central membrane-bound nucleus, and a variety of intracellular structures and membrane-bound organelles in the cytoplasm. This cytoplasm is surrounded by the cell membrane. In case of plants, a cell wall is present protecting the membrane that lies next to it. Protists, fungi, animals, and higher plants are examples of eukaryotes.

4.2 STRUCTURE OF PROKARYOTIC CELL

4.2.1 Intracellular Structures

The bacterial cell is surrounded by the cell membrane, which is also called the plasma membrane. This membrane acts as a protecting shield for the cell; it encloses all the essential belongings of the cell within it. The cell comprises the cytoplasm, cell organelles, chromosome, essential nutrients, proteins, and other necessary components of cytoplasm. Generally, prokaryotes do not have membrane-bound organelles and a membrane-bound nucleus; rather, they have few intracellular structures. They do not have mitochondrion and chloroplasts. Bacterial hyperstructures are nothing but the cytoskeletal structures seen in prokaryotes. Protein-bound bacterial organelles are called carboxysomes. Nucleoid, which is a single circular chromosome suspended in the cytoplasm in an irregular fashion, can be seen in prokaryotes. The chromosome is associated with protein and RNA. The ribosomes are different from that of Archaea and Eubacteria. Intracellular nutrient storage granules are constructed by the bacteria for storing essential nutrients, which can be used by the bacteria when needed. Glycogen, polyphosphate, sulphur, and polyhydroxyalkanoates are some of the storehouse materials of the cell. Some photosynthetic bacteria like *Cyanobacteria* synthesize internal gas vesicles. These gas vesicles help them to float by providing buoyancy to the cell.

4.2.2 Extracellular Structures

Cell Envelope

In some bacteria, the cell wall can be seen lying just outside the cytoplasmic membrane. The cell wall is mainly made up of murein (also called peptidoglycan), which is a polysaccharide chain cross-linked by peptides containing D-amino acids. Cell wall content will differ with organism; for example, bacterial cell wall is made up of peptidoglycan, fungi cell wall of chitin, and plant cell wall of cellulose. Archaea differ from bacteria by lacking peptidoglycan in their cell wall. In bacteria, two different types of cell wall composition can be seen, one in Gram-positive bacteria and the other in Gram-negative bacteria (Figure 4.2).

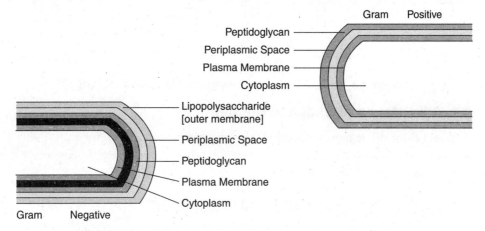

Figure 4.2 Cell Wall Differences between Gram-Positive and Gram-Negative Bacteria
(See page 341 for the colour image)

Gram-positive bacteria possess a thick cell wall that contains many layers of peptidoglycan and teichoic acids. Gram-negative bacteria have a relatively thin cell wall that is made up of a few layers of peptidoglycan surrounded by a second lipid membrane containing lipopolysaccharides and lipoproteins.

S Layer

In many bacteria, an S layer covers the outside of the cell. This S layer is made up of a crystalline protein. It provides chemical and physical shield for the cell surface and acts as a macromolecular diffusion barrier. The S layer has diverse effects but its functions are poorly understood.

4.3 SHAPES OF MICROBES

Microbes exhibit different shapes (Figure 4.3), and they are as follows:

1. Spherical: Cells that are spherical are called cocci. If the cell divides once in one axis, it produces diplococci. If it divides more than once to produce a chain, then it is called streptococci. If it divides regularly in two planes at right angles or at different angles to produce a cuboidal packet, then it is called staphylococci.
2. Rod: Cells that are cylindrical are called rods or bacilli.
3. Spiral: Cells that are coiled are called spiral or spirilla.

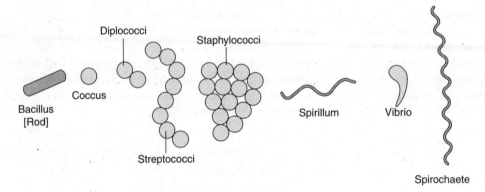

Figure 4.3 Shapes of Microbes (See page 342 for the colour image)

4.3.1 Advantages of Shape to the Cell

1. Cocci shape offers more resistance to drying when compared with bacilli.
2. Bacilli shape helps the intake of dilute nutrients into the cell from the environment as the surface area of rods is higher.

4.4 STRUCTURES INVOLVED IN ATTACHMENT

1. Glycocalyx: Glycocalyx forms multicellular aggregates by binding the cells together. Few bacterial cells adhere to solid surfaces using this structure; for example, some aquatic bacteria adhere to rocks. Some bacteria are involved in plaque formation leading to dental caries.
2. Fimbriae: Possession of fimbriae is an inherited trait. Fimbriae arise from the cytoplasmic membrane or sometimes from just below the membrane. Fimbriae are not involved in motility. Fimbriae are numerous and shorter when compared with flagella.
3. Pili: Pili are similar to fimbriae but are fewer (sometimes only one per cell) and longer. There are three functional types of bacterial pili, which are as follows
   (a) Pili that act as receptor sites for attachment of some phages
   (b) Pili that act as sex pili for bacterial conjugation processes
   (c) Pili that help in the attachment of pathogenic bacteria to human tissues

4.5 CLASSIFICATION OF MICROBES

Taxonomy is the science of classification, identification, and nomenclature. Organisms are generally organized into subspecies, species, genera, families, and higher orders. Classification of bacteria is the orderly arrangement of bacteria in groups. There are innumerable species of bacteria on Earth, many of which have not been identified. When attempting to classify a bacterium, a variety of characteristics are used, including visual uniqueness and laboratory tests. It is essential to classify and identify the bacteria in the environment because they are economically very important. Bacteria have important applications in industries, agriculture, and medicine and also play a major role in maintaining the ecological balance.

Some bacteria can be identified through simple visual inspection. Identification begins with evaluating the appearance of the bacterial colony followed by examination under the microscope for considering their shape, groupings, and features such as the number and location of flagella. If visual examination is not sufficient to identify the organism, various laboratory methods can be used. Staining techniques are employed to mediate identification in addition to various culture techniques such as the use of differential special culture medium. Many other biochemical tests are employed to identify bacterial by-products, and sophisticated tests (Polymerase Chain Reaction based) are available to analyse the DNA of the bacteria.

4.5.1 Scientific Nomenclature (Binomial Nomenclature)

Carolus Linnaeus is the founder of 'modern taxonomy'. Linnaeus was a Swedish physician and also a botanist. Binomial nomenclature originated from him; one can name living organisms and group similar organisms into categories using this system. In binomial nomenclature, each organism is given two names, a generic name and a specific name. The generic name is given for the genera, and the specific name is for the species. Both the names together give the scientific name of the organism. Scientific nomenclature has been accepted and used all over the world; it is the universal system of naming organisms. Naming through this system allows the scientists around the world to speak similar while talking about living organisms. This helps to overcome the confusion of multiple common names, which are used in different regions of different countries for a particular plant, animal, or any organism. A scientific name should always be italicized or underlined separately. The first letter of the genus name should always be capitalized and that of the species name should be in lower case. *Vibrio cholerae*, *Staphylococcus epidermidis*, and *Saccharomyces cerevisiae* are examples of scientific names of some organisms.

4.5.2 Taxonomic Hierarchy of *Escherichia coli*

Domain: Bacteria
Kingdom: Monera
Phylum or division: Proteobacteria
Class: Gammaproteobacteria
Order: Enterobacteriales
Family: Enterobacteriaceae
Genus: *Escherichia*
Species: *coli*

4.5.3 Two-Kingdom Classification

In the mid-17th century, Carolus Linnaeus classified all living organisms into two kingdoms, namely, kingdom Plantae and kingdom Animalia.

4.5.4 Three-Kingdom System

Haeckel (1866), a Swiss naturalist, was the first to construct a natural kingdom for microbes, which had been discovered by Antony van Leeuwenhoek nearly two centuries ago. Haeckel placed all unicellular (microscopic) organisms in a new kingdom 'Protista' and gave the three-kingdom system consisting of kingdom Protista, kingdom Plantae, and kingdom Animalia.

4.5.5 Four-Kingdom System

After the development of the electron microscope (circa 1950), it was found that Haeckel's kingdom Protista had some organisms with membrane-bound nucleus and some organisms that lacked intracellular compartments. The latter organisms were then shifted to and grouped under a new kingdom Monera.

4.5.6 Five-Kingdom System

R. H. Whittaker, a botanist at the University of California, polished the system into five kingdoms in 1967 (Figure 4.4). Whittaker grouped fungi under a separate kingdom. He made a basic contribution to the clarification of relationships among lower organisms. It reflects the evolutionary concepts remarkably well.

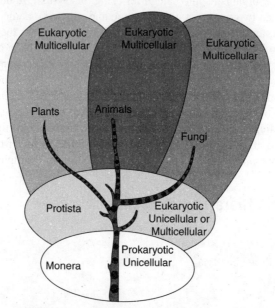

Figure 4.4 Whittaker's Five-Kingdom Classification
(See page 342 for the colour image)

Whittaker's five-kingdom classification consists of the following kingdoms:

1. Kingdom Monera (prokaryotic organisms)
2. Kingdom Protista (primitive eukaryotic organisms)
3. Kingdom Mycota (exclusively fungi)
4. Kingdom Metaphyta (advanced eukaryotic plants)
5. Kingdom Metazoa (all multicellular animals)

The features of five-kingdom classification are summarized in Table 4.1.

Table 4.1 Features of Five-kingdom Classification

Kingdom	Organisms	Approximate Time of Evolution (Million Years Ago)	Major Characteristics	Major Selective Factor in the Environment
Monera	All prokaryotes (e.g., bacteria and blue green algae)	Early middle Precambrian (3000–1000)	UV photoprotection, photosynthesis, and aerobiosis	Solar radiation and increasing atmospheric oxygen concentration
Protista	All eukaryotic algae (e.g., green, yellow–green, red, brown, and golden brown), all protozoans, slime moulds, and so on	Late Precambrian to early Paleozoic (1500–500)	Mitosis and meiosis, obligate recombination, and more efficient nutrition	Depletion of organic nutrients
Mycota	Conjugation fungi, sac fungi, club fungi, yeasts	Phanerozoic (700 on)	Tissue development, dikaryotic, advanced mycelia development, and absorptive nutrition	Nature of nutrient source and transitions from aquatic to terrestrial environments
Metaphyta	All green plants	Phanerozoic (700 on)	Tissue development for autotrophic specializations and photosynthetic nutrition	Transitions from aquatic to terrestrial environments
Metazoa	All animals developing from blastulas	Phanerozoic (700 on)	Tissue development for heterotrophic specializations and ingestive nutrition	Transitions from aquatic to terrestrial and aerial environments

4.5.7 Carl Woese's Three-Domain System

Carl Woese, an American microbiologist and physicist, classified life on Earth into three domains (Figure 4.5), which are as follows:

1. Archaea
2. Bacteria
3. Eukaryotes

Earlier to the three-domain concept, life on Earth was grouped into two kingdoms, namely, prokaryotes and eukaryotes. Carl Woese divided kingdom prokaryotes into two groups, namely, Archaea and bacteria, on the basis of differences in 16S rRNA genes. Thus, the concept of three domains of life came into existence. Each of these (i.e., Archaea, bacteria, and eukaryotes) arose separately from their ancestor with inadequately developed genetic apparatus, on the basis of differences in 16S rRNA, which is called a progenote. To reflect these primary lines of descent, Carl designated each group as domain and divided it into several different kingdoms.

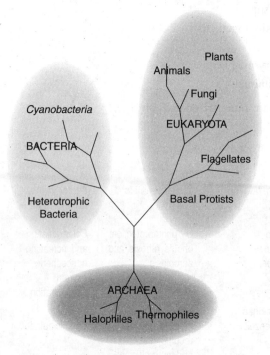

Figure 4.5 Three-Domain Concept (See page 343 for the colour image)

Based on the comparison of nucleotide sequences of the smaller subunit ribosomal RNA (ssrRNA), Carl Woese began the phylogenetic analysis of all forms of cellular life. Woese settled down for the ssrRNA rather than considering other molecules in the cells such as the DNA, protein, and so on because of the following reasons:

1. rRNA is found in every cell.
2. Isolation of rRNA from cells is quite easy.
3. Thousands of copies of rRNA are present in cells.
4. rRNA can be analysed to determine the exact sequence of nucleotide bases in it.
5. As the RNA sequence of base pairs is a complimentary copy of the DNA, it can be used to encode the gene.
6. By computer analyses and statistical methods, the base sequences in different rRNA molecules can be compared. This gives precise similarities and dissimilarities in cellular genomes.
7. rRNA molecules always carry out the same function, and their structure changes very little over time. This is why rRNA comparison is especially useful.
8. The rRNA nucleotide sequence differences between organisms indicate their relation to each other.

This three-domain system adds a level of classification 'above' the kingdoms present in the three- or five-kingdom system. The present system has the following kingdoms in the three domains.

Archaea Domain

Archaea are prokaryotic cells that are typically characterized by membranes that are branched hydrocarbon chains connected to glycerol by ether linkages. The ability of tolerating extreme temperatures and highly acidic conditions is strengthened by the existence of these ether-containing linkages in Archaea.

Kingdom Archaebacteria: The ability to thrive in extreme environments is the most striking characteristic of this kingdom. Few examples are stated below.

1. Halophiles: Organisms that live in places where the concentration of salt is high
2. Thermoacidophiles: Organisms that live in acidic conditions with high temperatures
3. Methanogens: Organisms that metabolize hydrogen and carbon dioxide into methane

Bacteria Domain

Bacteria, like Archaea, are prokaryotic cells. But they are kept in a separate domain because they do not have ether-containing linkages like Archaea. Bacterial membranes are made up of unbranched fatty acid chains linked to glycerol by ester linkages. It is nearly impossible to determine all the bacterial species existing on this planet, but vast diversity can be seen in this domain. The best examples of this domain are *Cyanobacteria* and *Mycoplasma*.

Kingdom Eubacteria: Most of the known pathogenic prokaryotic organisms come under this kingdom. Few important examples of pathogens are given below.

1. Spirochaetes: These are Gram-negative bacteria and are responsible for causing syphilis and Lyme disease.
2. Firmicutes: These are Gram-positive bacteria. They include *Bifidobacterium animalis*, which resides in the human intestine.
3. *Cyanobacteria*: These are photosynthesizing bacteria.

Eukaryota Domain

As the name suggests, eukaryotes are cells that have membranes similar to that of bacteria. Eukaryotes are further grouped into the following:

1. Kingdom Protista (algae, protozoans, and so on)
2. Kingdom Fungi (yeasts, moulds, and so on)
3. Kingdom Plantae (flowering plants, ferns, and so on)
4. Kingdom Animalia (insects, vertebrates, and so on)

The cell wall in eukaryotic cells does not contain peptidoglycan as in bacteria. Some eukaryotic cells do not even possess a cell wall. In kingdom Plantae and kingdom Animalia, cells are organized into well-developed tissues. Cell wall is present only in the members of kingdom Plantae. The basis of three-domain system is the rRNA content of the cell, which is so unique that these three domains of life are recognized by the scientists even today. Domains Archaea and Bacteria are similar in not having the membrane-bound nucleus but domain Eukarya differs from both by having a well-organized membrane-bound nucleus. Archaea and Bacteria domains differ from each other in having distinct biochemistry and RNA markers. The main purposes of the three-domain system are as follows:

1. The three domains of cellular organisms found in nature can be compared.
2. It can give correct knowledge to differentiate a prokaryotic cell from a eukaryotic cell.

The eukaryotes are sub-divided into four kingdoms, which are discussed below.

Kingdom Protista: Organisms that fall under kingdom Protista are simple, and most of them are unicellular eukaryotic organisms (e.g., euglenoids, protozoans, slime moulds, and algae).

Kingdom Fungi: This kingdom comprises both unicellular and multicellular eukaryotic organisms. Cells do possess cell wall that is made up of chitin, but the cells are not well organized into tissues. The organisms are saprophytes as they obtain nutrients through absorption (They do not carry out photosynthesis.). Sac fungi, club fungi, yeasts, and moulds are a few examples.

Kingdom Plantae: This kingdom includes plants, which are multicellular organisms made of eukaryotic cells. The cells have cell walls and are organized into tissues. They obtain nutrients by absorption and photosynthesis. Mosses, ferns, conifers, and flowering plants are a few examples.

Kingdom Animalia: This kingdom is made up of animals, which are multicellular organisms composed of eukaryotic cells. The cells are organized into tissues, and the cell wall is absent. They obtain nutrients primarily by ingestion. Examples include insects, worms, reptiles, birds, and mammals.

4.6 UNIVERSAL TREE OF LIFE

Microorganisms can alter their virulence capabilities in order to adapt to new environments. This change is a very slow process occurring within an organism primarily through mutations, chromosomal rearrangements, gene deletions, and gene duplications. Natural selection occurs as these new inherited changes pass on to new progenies. This kind of gene transfer from a parent to its offspring is called vertical gene transmission. It is now known that microbial genes are transferred not only through vertical transmission but also through horizontal gene transmission, which is the transfer of genes to relatives that are only distantly related through mechanisms such as transformation, transduction, and conjugation or even through genetic elements such as plasmids, transposons, integrons, and chromosomal DNA. As a result, the old three-branched 'tree of life' is now regarded as the 'universal tree of life' (Figure 4.6).

There are several other approaches also to classify bacteria based on different aspects, which are discussed in the following sections.

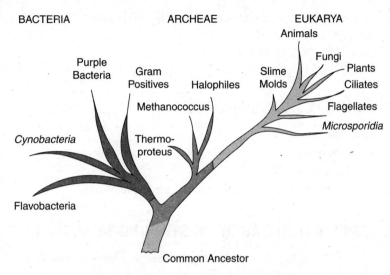

Figure 4.6 Universal Tree of Life (See page 343 for the colour image)

4.7 MORPHOLOGICAL CLASSIFICATION

Based on their morphology, the bacteria can be divided into the following groups:

1. Filamentous or higher bacteria (actinomycetes)
2. True or lower bacteria
   (a) Gram-positive bacilli
      (i) Aerobes (e.g., *Corynebacterium* and *Bacillus*)
      (ii) Anaerobes (e.g., *Clostridium* and *Lactobacillus*)
   (b) Gram-positive cocci
   (c) Gram-negative cocci
   (d) Gram-negative bacilli

4.8 NUTRITIONAL CLASSIFICATION

1. Based on their energy requirements, the bacteria can be classified into the following:
   (a) Phototrophs: They derive energy from sunlight.
   (b) Chemotrophs: They use chemicals to derive energy.
2. Based on their ability to synthesize essential metabolites, the bacteria can be classified into the following:
   (a) Autotrophs: They synthesize organic compounds from carbon dioxide and nitrogen.
   (b) Heterotrophs: They depend on preformed organic compounds (e.g., pathogens).
3. Based on their oxygen requirements, the bacteria can be classified into the following:
   (a) Aerobes
   (b) Anaerobes

4. Based on temperature, the bacteria can be classified into the following:
   (a) Psychrophiles (below 20°C)
   (b) Mesophiles: (25°C to 40°C)
   (c) Thermophiles: (55°C to 80°C)

4.9 BIOCHEMICAL CLASSIFICATION

1. Lactose fermenting
2. Non-lactose fermenting
3. Late lactose fermenting
4. Production of indole, H_2S, catalase, oxidase, urease, and so on

4.10 CLASSIFICATION BASED ON STAINING REACTION

1. Based on Gram staining, the bacteria can be classified into the following:
   (a) Gram-positive bacteria
   (b) Gram-negative bacteria
2. Based on acid-fast staining, the bacteria can be classified into the following:
   (a) Acid-fast bacilli
   (b) Non-acid-fast bacilli

4.11 SEROLOGICAL OR ANTIGENIC CLASSIFICATION

1. Serovars or serotypes
2. Serogroups

MULTIPLE CHOICE QUESTIONS

1. A slippery outer covering in some bacteria that protects them from phagocytosis by host cells is _____.
   (a) Capsule   (b) Cell wall
   (c) Flagellum   (d) Peptidoglycan

   Ans. a

2. When flagella are distributed all around a bacterial cell, the arrangement is called _____.
   (a) Polar   (b) Random
   (c) Peritrichous   (d) Encapsulated

   Ans. c

3. A bacterial cell wall does all of the following except _____.
   (a) Giving shape and rigidity to the cell
   (b) Being associated with some symptoms of a disease
   (c) Being the site of action for some antibiotics
   (d) Protecting the cell from phagocytosis

   Ans. d

4. Which of the following contains polysaccharide?

(a) Gram-negative cell wall
(b) Pili
(c) Capsule
(d) Plasmids

Ans. a

5. Differential staining of bacteria on Gram staining is due to the _____.
   (a) Difference in the cell wall layer components of Gram-positive and Gram-negative bacteria
   (b) Difference in the cell structure of Gram-positive and Gram-negative bacteria
   (c) Difference in the mode of nutrition of Gram-positive and Gram-negative bacteria
   (d) None of the above

Ans. a

6. Surface appendages of bacteria meant for cell–cell attachment during conjugation are _____.
   (a) Pili   (b) Flagella
   (c) Spinae   (d) Cilia

Ans. a

7. The optimum temperature for an organism is the one at which _____.
   (a) It grows with the strongest generation time
   (b) It has the longest time between cell divisions
   (c) It is near one extreme of its range of tolerated temperatures
   (d) Its enzymes begin to denature

Ans. a

8. Identify the only taxonomic category that has a real existence from the following _____.
   (a) Phylum   (b) Species
   (c) Genus   (d) Kingdom

Ans. b

9. In the five-kingdom system, the main basis of classification is _____.
   (a) Structure of cell wall
   (b) Nutrition
   (c) Structure of nucleus
   (d) Asexual reproduction

Ans. b

10. If the five-kingdom system of classification is used, in which kingdom would you classify the Archaea and nitrogen-fixing organism?
    (a) Protista   (b) Fungi
    (c) Plantae   (d) Monera

Ans. d

11. The phylogenetic system of classification was put forth by _____.
    (a) Theophrastus
    (b) George Bentham and Joseph Dalton Hooker
    (c) Carolus Linnaeus
    (d) Adolf Engler and Karl Prantl

Ans. d

12. The classification of organisms based on evolutionary as well as genetic relationships is called _____.
    (a) Numerical taxonomy
    (b) Phenetics
    (c) Biosystematics
    (d) Cladistics

Ans. d

13. The practical purpose of taxonomy or classification is _____.
    (a) To know the evolutionary history
    (b) To explain the origin of organisms
    (c) To facilitate the identification of unknown species
    (d) To identify medicinal plants

Ans. c

14. Single-celled eukaryotes are included in _____.

(a) Fungi	(b) Protista
(c) Monera	(d) Archaea

Ans. b

15. Arrange the following in the ascending order of Linnaean hierarchy _____.

    (a) Kingdom–Phylum–Class–Order–Family–Genus–Species
    (b) Kingdom–Family–Genus–Species–Class–Phylum–Order
    (c) Kingdom–Order–Species–Genus–Class–Family–Phylum
    (d) Species–Genus–Family–Order–Class–Phylum–Kingdom

Ans. d

16. Taxonomic hierarchy refers to _____.

    (a) The list of botanists or zoologists who have worked on the taxonomy of a species or group
    (b) The classification of a species based on fossil records
    (c) The step-wise arrangement of all categories for the classification of plants and animals
    (d) The group of senior taxonomists who decide the nomenclature of plants and animals

Ans. c

17. The kingdom of multicellular eukaryotic heterotrophs whose cells do not have cell walls comes under _____.

    (a) Fungi	(b) Animalia
    (c) Protista	(d) Plantae

Ans. b

18. The kingdom of multicellular photosynthetic autotrophs that have cell walls containing cellulose is _____.

    (a) Animalia	(b) Plantae
    (c) Protista	(d) Fungi

Ans. b

19. The most inclusive taxonomic category that is larger than a kingdom is _____.

    (a) Phylum	(b) Kingdom
    (c) Class	(d) Domain

Ans. d

20. The classification system in which each species is assigned a two-part scientific name is _____.

    (a) Taxon
    (b) Taxonomy
    (c) Binomial nomenclature
    (d) Derived character

Ans. c

SHORT NOTES

1. Differences between Gram-positive and Gram-negative bacteria
2. Function of the structure involved in attachment
3. Bacterial classification based on morphology with relevant diagrams
4. Bacterial classification based on staining
5. Three-kingdom classification system
6. Bacterial classification based on nutritional requirements

ESSAYS

1. Explain the three-domain system of classification with relevant comparisons and diagrams.
2. Explain in detail the five-kingdom classification with illustrations.
3. Explain the structure of a prokaryotic cell (both intracellular and extracellular).

5 Morphology—Size and Forms

CHAPTER OBJECTIVES

5.1 Introduction

5.2 Bacterial Morphology

5.3 Microscopy

5.4 Anatomy of Bacteria

5.1 INTRODUCTION

Around 1650 species of bacteria have been identified; among them, 900 constitute eubacteria, and the rest constitute 'higher bacteria'. The most common and simplest one is the true bacteria. These bacteria are spherical, ovoid, or rod-shaped. Most of them are responsible for causing diseases in human beings. Higher bacteria are grouped into five orders, which are as follows:

1. Actinomycetales or fungus-like bacteria
2. Chlamydobacteriales or alga-like bacteria
3. Myxobacteriales or slime bacteria
4. Spirochaetales or protozoa-like bacteria
5. Rickettsiales or minute and rod-shaped bacteria

5.2 BACTERIAL MORPHOLOGY

Bacteria are microscopic living organisms that are structurally very simple. They are strictly unicellular and thrive solitarily. However, some bacteria are found living in groups (For example, some groups of bacterial cells are embedded in a mucilage layer.).

5.2.1 Size of Bacterial Cells

Bacteria range from 1 mm in diameter (largest end of the scale) to 200 nm in length (smallest end of the scale). Large bacteria (e.g., *Thiomargarita namibiensis* and *Epulopiscium fishelsoni*) can be observed without the help of a microscope. The smallest known bacteria are so minute that they were once considered to be viruses (e.g., *Mycoplasma*). Commonly found bacteria range between 5 and 0.5 µm in length. Bacilli are the largest bacteria, and they range up to 5–8 µm in length. The smaller ones are usually cocci. The size of the smallest bacteria is about 100–200 nm in diameter, which is approximately the size of the largest viruses (poxviruses). Some spirochaetes infrequently grow up to 500 µm in length, and the *Cyanobacterium Oscillatoria* reaches up to 7 µm in diameter. Recently, a huge bacterium, *Acanthurus nigrofuscus*, has been discovered in the intestine of brown surgeonfish. *Epulopiscium fishelsoni* (an example of a very big prokaryote) is a bacterium that is as big as an average eukaryotic cell.

5.2.2 Shape of Bacterial Cells

Bacterial cells vary in shape and fundamentally are of four groups, which are as mentioned below.

Spherical Type

In spherical type, cells are circular in shape. This type of cells are termed as 'cocci'. Cocci are of different forms as mentioned below:

1. Micrococcus (if the coccus is single)
2. Diplococcus (if the coccus lives in pairs)
3. Streptococcus (if the coccus exists in chains)
4. Staphylococcus (if the coccus occurs in clusters, i.e., grape-like)
5. Sarcina (if the colony exists in cubical packets of eight or more)

All the above-mentioned forms of association are due to the difference in the plane of cell division and non-partition of daughter cells.

Bacillus Type

In bacillus type, the cells are elongated and rod-like, and such cells are called 'bacilli'. Bacilli may again be of different types. They might be single, in pairs (diplobacilli), or in groups to form a chain (streptobacilli).

Spirillum Type

In spirillum type, the cells are spirally coiled and hence referred to as spirillum. Some of the important genera that fall under this type are *Spirillum*, *Microspira*, *Rhodospirillum*, and *Leptospira*.

Vibrio Type

In this type, the cells are comma-shaped. The rod-shaped cell is curved at one end and hence appears like a 'comma'. *Vibrio cholerae* is an important species of this type.

Some bacteria tend to change their shape according to environmental changes. Such bacteria fall under two categories, which are as follows:

1. **Monomorphic:** Here, the bacterium exhibits only a single shape. Even if the environmental or physiological condition gets altered, the bacterium tends to show the same shape.

2. **Polymorphic or pleomorphic:** Here, the bacterium exhibits different shapes as the environmental or physiological condition gets changed. Examples of polymorphic bacteria are *Corynebacterium diphtheriae* and *Mycoplasma*.

5.2.3 Colony Morphology

Colony morphology refers to the observable characteristics of a bacterial colony. As bacteria are cultured primarily on solid media, appearance of bacterial colonies is an important tool for microbiologists.

Principles of Microbiology

The principles of microbiology are as follows:

1. To examine the culture plates
2. To examine the rapidity of colony development
3. To determine the characteristics of different colonies
4. To determine the number and forms of colonies (quantitative analysis)

Colony morphology can be described by means of surface features of the colony such as shape, margin, edge, and colour. The appearance of a colony results from the characteristics of the individual bacterium observed collectively (Figure 5.1).

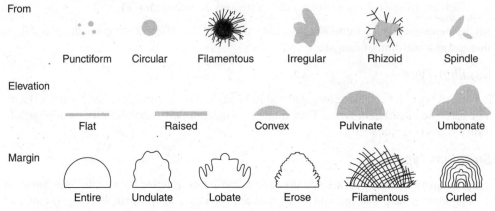

Figure 5.1 Illustration of the Various Forms, Elevations, and Margins of Bacterial Colonies

An important tool in the description and identification of microorganisms is determining the morphology of the single colony growing on the surface of a culture plate. This might be utilized at the superficial level to distinguish between different types of microorganisms.

Bacterial Identification

The appearance of colonies differs with bacterial strain, species, or genus. Hence, colony morphology cannot be used as an appropriate way for identifying bacteria. The major drawback of colony morphology is that many different bacteria when grown on culture media show similar colony morphologies.

5.3 MICROSCOPY

Microscope is an important invention in the field of microbiology. Anton van Leeuwenhoek (1632–1723), a Dutch tradesman and scientist, invented the magical instrument 'microscope'. Microorganisms are very minute and cannot be seen through naked eyes. Hence, microscopes are required to observe microorganisms. They can magnify the image 1000 times larger than the original size of the microorganism, which is helpful in the detailed study of microorganisms.

5.3.1 Types of Microscopes (Figure 5.2)

There are different types of microscope based on their application. Microscopes are generally made up of the lenses (one or more), to magnify the specimen; disk, where the objects are placed; and light source. Based on their source of radiation, resolution, and magnification, microscopes are broadly classified into two types: light microscope and electron microscope (EM).

Light Microscope

The light microscope, also called optical microscope, uses natural sunlight or bulb light as the source of light. It has low resolution power when compared with EM. It is used to observe micro-objects such as cells. Light microscope is further classified into the following configurations:

Simple Microscope: The simple microscope is considered as a 'primitive microscope' as for the magnification of the specimen. It utilizes only a single convex lens.

Compound Microscope: The compound microscope is the most commonly used microscope. It uses two optical lenses, that is, the ocular lens and the objective lens, to magnify the object. It provides about x2000 magnification of the specimen. On comparison with simple microscope,

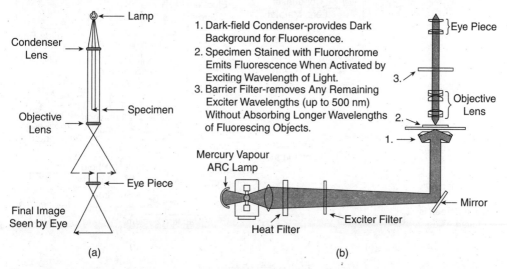

Figure 5.2 Ray Diagram of Principle of (a) Light Microscope, (b) Fluorescence Microscopy, (c) Dark-field Microscopy, (d) Phase-contrast Microscopy, (e) Transmission Electron Microscope, and (f) Scanning Electron Microscope (See page 344 for the colour image)

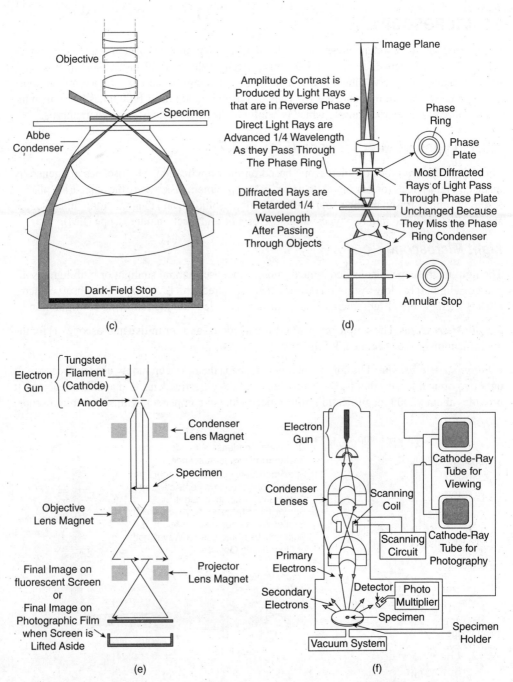

Figure 5.2 (Continued)

compound microscope is larger, heavier, and uses two lenses rather than only one (to focus the light into the eye and to form an image). The compound microscope helps us to observe bacteria, algae, protozoa as well as plant and animal cells. The compound microscope is further divided into different types, which are briefly discussed below.

1. **Standard compound light microscope:** This microscope consists of an eyepiece with the rotating nose piece lying in the same line. The rotating nose piece can be fitted with two or more objective lenses. The following magnifications can be achieved depending on the objective lens used: 4x, 10x, 40x, or 100x. The mechanism behind is as follows: the light passes from the stage through a hole into the specimen and from the specimen into the lens resulting in the formation of a magnified image of the specimen to the observer.
2. **Inverted microscope:** Using this microscope, the specimen can be viewed from an overturned position and an upright image of the specimen can be observed. This microscope is an upside-down microscope and can be used to view large and thick specimens and also liquid cell cultures.
3. **Stereo microscope:** It is also known as dissection microscope. Magnification power of this microscope is very low. With the aid of stereo microscope, larger-sized specimens can be observed. It helps the user to see the specimen in a three-dimensional view as it contains two optical paths that are at different angles. The advantage of stereo microscope is that it can be used to view live samples. It is used in anatomy and physiology for microsurgery, dissection, fine repair, and sorting and in forensics.
4. **Fluorescence microscope:** This microscope utilizes high-energy short-wavelength light that excites the electrons of certain molecules present within the sample. This microscope causes the electrons to move into a higher orbit, and when they return back to their original energy levels, they emit low-energy and long-wavelength light. This emitted light comes under visible spectrum that helps in the construction of an image.
5. **Digital microscope:** The digital microscope consists of Closed Circuit Cameras (CCD) or Complementary Metal Oxide Semiconductor (CMOS) sensors and optical lenses. Magnification of this microscope is 1000x. Digital microscopes with a 15-inch monitor and 2-million pixel camera are used more commonly; here, a digital CCD camera is fixed to the microscope, which in turn is connected to an LCD computer. This microscope is used to obtain high-quality recorded image of the specimen.

Based on the lighting technique, microscopes are categorized as follows:

1. **Dark-field microscope:** In dark-field microscope, for the scattering of light, a special type of condenser lens is used. Due to this, the light gets reflected off the specimen at an angle so that any light object can be seen against a dark background. This microscope is used to observe live spirochaetes.
2. **Phase contrast microscope:** In this microscope, the light passes into the specimen at different speeds. Phase contrast microscope also uses a special condenser lens, which aids in tossing the light out of phase. This microscope is used to view unstained specimens and also to study cells and cell organelles such as mitochondria, lysosomes, and Golgi bodies.

Electron Microscope

One of the most advanced microscopes used today is the electron microscope (EM). This microscope employs a beam of very short-wavelength electrons. The resolution of this microscope gets amplified by the electrons that hit the specimen that comes in their path. This microscope is used to study both minute viral cells and larger molecules. Different kinds of EMs are briefly explained below.

Transmission Electron Microscope (TEM): The electron beam is deflected on the densely coated specimen, and the image is observed on a dark and light background. This microscope is used to study cells. Ultrathin sections of organisms such as viruses are placed on a wire grid, the cells are then stained with palladium or gold and viewed under a TEM.

Scanning Electron Microscope (SEM): A type of EM that has lower magnification power than a TEM is SEM. This microscope is used to view 3D images of microorganisms and other specimens. In order to stain the specimens mounted on a SEM, gold and palladium are used.

Reflection Electron Microscope (REM): REM detects elastically scattered electrons, instead of following the principle of TEM or SEM. This microscope uses the electron beam that is incident on a surface.

5.4 ANATOMY OF BACTERIA

The bacterial cell structure (Figure 5.3) can be organized into three categories, which are as follows:

1. Appendages and coverings
2. Cell envelope
3. Internal structures

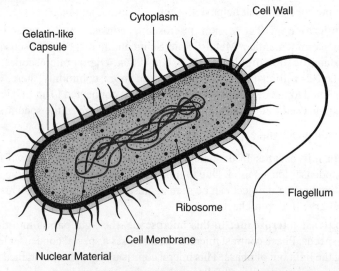

Figure 5.3 Bacterial Anatomy (See page 345 for the colour image)

5.4.1 Architecture of a Bacterial Cell

The structure of prokaryotes is relatively simple in construction when compared with that of eukaryotes. As individual units, the prokaryotes carry out all cellular functions. A prokaryotic cell comprises the essential structural components such as nucleoid (genetic material), ribosomes, cell membrane, cell wall, and certain surface layer.

Cell Wall

The cell wall of bacteria is composed of unique components and has got very important functions, which are mentioned below:

1. The cell wall is responsible for immunological distinction and immunological variations.
2. The cell wall serves as the most important site for the attack by antibiotics.
3. The cell wall facilitates ligands as receptor sites for the attachment of drugs or viruses.

A prokaryotic cell has a rigid cell wall, which protects the cell's protoplast from mechanical damage or lysis. Usually, the concentration of cell cytoplasm greatly varies from the concentration in the outside environment due to the accumulation of solutes. The cell membrane is a very delicate elastic structure. The cell wall is made up of porous and rigid material with high tensile strength. Murein is the ubiquitous component of a bacterial cell that is mainly responsible for the rigidity and shape of the cell.

Gram-Positive, Gram-Negative, and Acid-Fast Bacteria: Bacterial cell wall consists of murein, also known as peptidoglycan or mucopeptide. The cell wall differs in its composition in two groups of bacteria, namely, Gram-negative and Gram-positive. Bacterial murein is a unique type of peptidoglycan. Peptidoglycan is a polymer of sugars (a glycan) cross-linked by short chains of amino acids (peptide). The definitive component of murein is N-acetylmuramic acid of peptidoglycans. However, in the case of Archaea, the cell wall is made up of protein, polysaccharides, or peptidoglycan-like molecule; murein is completely lacking. This serves as one of the important differentiating features between Bacteria and Archaea. In Gram-positive bacteria, the cell wall is thick (15–80 nm), consisting of several layers of peptidoglycan. Peptidoglycan layers lie at right angles to a group of molecules, that is, teichoic acids. It is unique to the cell wall of Gram-positive bacteria. There are numerous different peptide arrangements among peptidoglycans.

In Gram-negative bacteria, the cell wall is relatively thin (10 nm) and made up of a single layer of peptidoglycan enclosed by a membranous structure called the outer membrane. This outer membrane is made up of a unique component, lipopolysaccharide (LPS). LPSs serve as endotoxins as they are toxic to animals. The outer membrane is usually considered as a part of the cell wall in Gram-negative bacteria. The most important difference between Gram-negative and Gram-positive bacterial cell wall is the interpeptide bridge of amino acids connecting nearby side chains to each other. Gram-positive bacteria are more sensitive to penicillin than Gram-negative bacteria because peptidoglycan does not have a shield of outer membrane and is a more abundant molecule. The differentiating features between Gram-positive and Gram-negative bacteria are given in Table 5.1. Cell wall differences between Gram-positive and Gram-negative bacteria are given in Figure 5.3.

It is difficult for microbiologists to distinguish acid-fast organisms using a standard technique such as Gram staining. Acid-fast organisms have a small amount of peptidoglycan and mycolic

Table 5.1 Comparative Characteristics of Gram-Positive and Gram-Negative Bacteria

Characteristic	Gram-Positive Bacteria	Gram-Negative Bacteria
Gram reaction	Retain crystal violet dye and stain blue or purple	Can be decolourized to accept counterstain (safranin) and stain pink or red
Peptidoglycan layer	Multilayered (thick)	Single-layered (thin)
Teichoic acids	Present (many)	Absent
Outer membrane	Absent	Present
Periplasmic space	Absent	Present
LPS content	Virtually nil	High
Lipid and lipoprotein content	Low (acid-fast bacteria have lipids linked to peptidoglycan)	High (because of the presence of outer membrane)
Flagellar structure	Two rings in the basal body	Four rings in the basal body
Toxins produced	Exotoxins	Endotoxins and exotoxins
Resistance to physical disruption	High	Low
Cell wall disruption by lysozyme	High	Low (requires pretreatment to destabilize outer membrane)
Susceptibility to penicillin and sulphonamide	High	Low
Susceptibility to streptomycin, chloramphenicol, and tetracycline	Low	High
Inhibition by basic dyes	High	Low
Susceptibility to anionic detergents	High	Low
Resistance to sodium azide	High	Low
Resistance to drying	High	Low

acid, which is a waxy substance. These bacteria are highly resistant to staining and treatment due to this waxy mycolic acid. In order to stain such bacteria, one must use concentrated dyes and the staining process should be combined with heat treatment. An example of a bacteria with acid-fast cell wall is *Mycobacterium tuberculosis*, the causative agent of tuberculosis. To stain the bacteria in laboratory, they must be heat-fixed and treated with acid–alcohol.

Cytoplasmic Membrane

The cytoplasmic membrane is a delicate structure enclosing the cytoplasm. The transportation of substances between the environment and the cell is specifically facilitated by the cytoplasmic membrane. It is composed of phospholipid (40%) and protein (60%). The phospholipids have a glycerol head, which is the water-soluble (hydrophilic head) region, attached to the fatty acid (hydrophobic tail) region. Hence, phospholipids are called amphipathic molecules. As the cytoplasm contains water, the molecules naturally form a bilipid layer. Various structural and enzymatic proteins that carry out the most important functions of the membrane are dispersed throughout the bilayer membrane. This typical arrangement of proteins and phospholipids is termed as 'fluid mosaic membrane' (Figure 5.4).

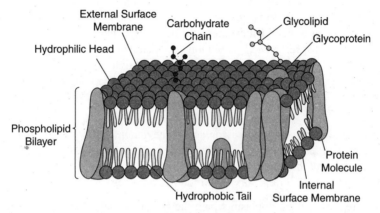

Figure 5.4 Fluid Mosaic Model of Cell Membrane (See page 346 for the colour image)

Through specific interactions with the phospholipid molecules, within the membrane, membrane proteins are located in various positions. Proteins associated with the membrane fall under three main groups, which are as follows:

1. Integral protein
2. Outer-surface protein
3. Inner-surface protein

Integral proteins are firmly embedded in the membrane. They help in the transportation of substances across the cytoplasmic membrane. Uniport, symport, and antiport transport mechanisms are facilitated by these proteins. Overall, membrane proteins play a characteristic role in cellular activities.

Cytoplasm

The gel-like matrix where the functions for cell growth, metabolism, and replication take place is known as cytoplasm or protoplasm. It is composed of water, nutrients, wastes, enzymes, and gases. All the cell organelles, that is, ribosomes, mitochondrion, chromosome, and plasmid, are suspended in the cytoplasm. The membrane-bound nucleus is absent in prokaryotes unlike eukaryotes. Here, the chromosome is a single continuous strand of DNA suspended in the cytoplasm and is termed as nucleoid. All these scattered cellular components throughout the cytoplasm are sheathed by the cell envelope.

Ribosome

Ribosomes consist of RNA and protein. They are granular-shaped organelles responsible for protein synthesis. The bacterial cell contains 70S ribosomes, and the eukaryotic cell contains 80S ribosomes. Several ribosomes float freely in the cytoplasm. They are responsible for reading the instructions on the DNA and directing the production of bacterial protein of interest. After the protein is obtained, the ribosomes detach away from the DNA and again float freely in the cytoplasm.

Components of 70S Ribosome: It has 30S and 50S subunits. The 30S subunit is made up of 16S rRNA and 21 different proteins, and the 50S subunit is made up of 23S rRNA and 5S rRNA and 34 different kinds of protein. Ribosomes are active only when the two subunits, that is, 30S and 50S, are combined together. Magnesium ion and chemical energy are required for attaching the two subunits to make the ribosome functional. The activity of 70S ribosome can even be blocked by the usage of antibiotics such as erythromycin and streptomycin.

Bacterial Genome

The most important component in the cytoplasm is the genome, which is located in the central region of the cell and is called nucleoid. Nucleoid is a DNA molecule freely floating in the cytoplasm. It controls all the functions of the bacterial cell and aids in the production of desired proteins (which are needed for the survival of the bacterial cell).

Nucleoid: The region of the cytoplasm where the chromosomal DNA lies is called nucleoid. The nucleoid is not membrane-bound. Most of the bacteria have a single, circular auxiliary DNA strand, called plasmid, suspended in the cytoplasm. It is responsible for replication.

Chromosome: There are two kinds of DNA: bacterial chromosome and plasmid. The bacterial chromosome is freely suspended in the cytoplasm without any membrane enclosing it. Plasmids are circular DNA. Both kinds perform the same function, that is, store genetic information in the sequence of DNA. The properties of a bacterial chromosome are stated below:

1. Bacterial chromosomes are located within the nucleoid region without being surrounded by any envelope.
2. Each cell has only one chromosome.
3. The size of the chromosome varies with cell. In *Escherichia coli*, the chromosome is 4640 kbp in size.
4. The components of the chromosome also vary from cell to cell. Some cells have single-stranded DNA, some double-stranded DNA, and some circular DNA.
5. DNA does not interact with histone protein.
6. DNA replication, transcription, and gene regulations take place mainly because of DNA but with the aid of RNA and proteins.
7. Plasmids are much smaller than chromosomes. Plasmids contain drug-resistant genes as well as heavy metal-resistant genes.

Several other specific and important functions are carried out by plasmids. Plasmids exhibit special properties such as antibiotic drug resistance, heavy metal resistance, and possession of certain factors essential for the infection of animals and plants, which are transmitted to other bacterial cells during reproduction. Through the above-mentioned properties, the plasmids provide protection to bacteria.

Inclusions

Most part of the cytoplasm is occupied by some distinct granules called inclusions. Some granules are provided with certain areas in which nutrients or other energy-producing substances are stored, and these are termed as storage granules. The substances that are stored as reserve food are glycogen (carbohydrate energy source), lipids (fats), polyphosphate (a stabilizer that helps to hold water), or sulphur or nitrogen, in some cases. For example, glycogen and polybetahydroxybutyric acid granules might be stored as reserves for carbon and energy source. Polyphosphate inclusions are reserves of phosphate to provide energy when needed. Elemental sulphur also serves as a reserve and is mainly stored by some phototrophic organisms. Some inclusions are in the form of membranous vesicles, which comprise photosynthetic pigments and enzymes.

External Structures

Capsule: The capsule is made up of polysaccharides and serves as a protective covering for bacteria. It is found only in some species. The most important role of the capsule is to keep the bacterium protected from phagocytosis by larger microorganisms. It also helps to protect the cell from desiccation. In the major disease-causing bacteria such as *E. coli* and *Streptococcus pneumoniae*, capsule is the chief virulence factor. Non-capsulated mutants of the above organisms are avirulent; that is, they cannot cause the disease.

Flagella: Flagella are hair-like propellers that are composed of helically coiled protein subunits called flagellin. Here, the flagellin units are fixed inside the bacterial membrane by means of hook and basal body, which are driven by membrane potential. The chief function of flagella is anchoring and locomotion. The position of flagella may vary with species of bacteria. Bacteria may possess one or more flagella on their surface; and the flagella might be anchored at different parts of the cell (Figure 5.5).

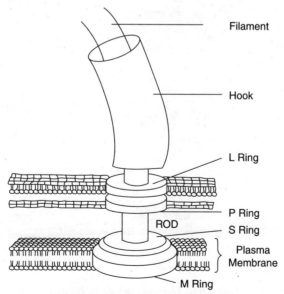

Figure 5.5 Anatomy of Flagellum

Flagella are responsible for the cell's movement towards food and away from harmful substances. Bacteria swim to approach food by moving straight or by tumbling in a new direction. As the concentration of chemo-attractant increases, the swimming period of bacteria also increases. By determining whether the bacteria swims or tumbles, we can determine the direction of flagellar spinning. Antigenic and strain determinants are also expressed by flagella.

Structure of flagella: Flagella are long, filamentous hair-like appendages. The structure of flagella can be divided into the filament, hook, and basal body. The filament is made up of flagellin, a protein subunit. The hook connects the filament to the basal body; it is made up of a single type of protein. The basal body is responsible for the rotation of flagella by obtaining energy from the activity of proton pumps. The arrangement of the basal body in the cell will slightly differ in Gram-positive bacteria when compared with Gram-negative bacteria. Gram-positive bacteria contain a rod and only the inner pairs of rings in the basal body, whereas Gram-negative bacteria have a rod and several rings in the basal body.

The flagella are classified into the following types based on their distribution (Figure 5.6):

1. **Monotrichous flagella:** Only one flagellum is present at only one end of the cell.
2. **Lophotrichous flagella:** A cluster of flagella are present at only one end of the cell.
3. **Amphitrichous flagella:** Flagella are distributed at both the ends of the cell.
4. **Peritrichous flagella:** Flagella are distributed all around the cell.

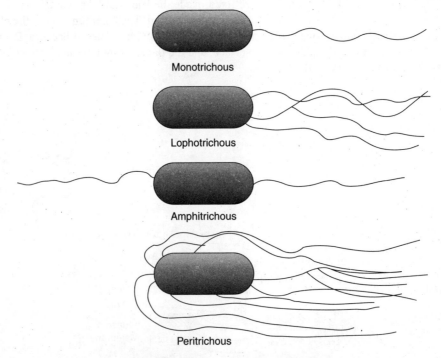

Figure 5.6 Types of Flagella Distribution

Properties of flagella: The properties of flagella are mentioned below:

1. **Chemotaxis:** The movement of bacteria is dependent on chemical stimuli: the bacteria move towards or away from chemical stimuli.
2. **Magnetotaxis:** The movement of bacteria is due to the Earth's magnetic field. This movement can be seen only in magnetotactic bacteria that have magnetosomes including iron.
3. **Phototaxis:** The bacterial movement is dependent on the differences in light density.

Fimbriae: Fimbriae or pili are the hair-like structures seen on the outer surface of bacteria. Fimbriae are made up of protein subunits called pilin. The diameter of pili ranges from 3–25 nm. Fimbriae differ from flagella in their small size and uncoiled structures. Several hundreds of pili are distributed uniformly over the entire surface of the bacterial cell.

Pili help in the adherence of bacteria to the host or other bacterial cells. Fimbriae act as adhesins (adherence factor), lectins, evasins, and aggressins. Pili are an important virulence factor in *E. coli*, which colonizes and infects the urinary tract, *Neisseria gonorrhoeae*, and other bacteria. The tips of pili might contain lectins (a kind of protein), which are responsible for binding to specific sugars. The larger segments of bacterial chromosomes can be transferred between bacteria with the aid of sex pili, which are encoded by the plasmid (F). If bacterial cells lack pili, many disease-causing bacteria might lose their ability to infect as they will fail to attach to the host cell. During conjugation, specialized pili are utilized to exchange the fragments of plasmid DNA.

Endospore

When the environment becomes unfavourable due to the lack of nutrients, microorganisms should adapt to the changes. Few bacteria that are motile might move in search for nutrients, or they may produce enzymes for using alternative resources. One of the most important examples of survival strategy adapted by certain low Guanine & Cytosine Gram-positive bacteria is endospore formation. Endospore formation is a complex developmental process that an organism undergoes when nutrient scarcity occurs. In times of extreme stress, the bacterium tends to construct a dormant and highly resistant cell for protecting its genetic material. Endospores can withstand any extreme condition, in which normally no bacterium would be able to survive. Some examples of extreme conditions are chemical damage, high UV irradiation, enzymatic destruction, high temperature, and desiccation. Endospores are not easily destroyed by antimicrobial treatments. This extraordinary property of resistance of endospores makes them incredibly important. Many microbes form certain cysts or spores, but the low G+C Gram-positive bacteria form endospores, which is 'the most' resistant to harsh conditions. Endospores germinate to form vegetative cells when the environment becomes favourable.

Endospore Structure: Endospore structure comprises mainly the exosporium, spore coat, cortex, and core. The outermost layer made up of protein is exosporium. Spore coat is made up of several layers of spore-specific proteins. Cortex consists of loosely cross-linked peptidoglycan. Core wall, cytoplasmic membrane, cytoplasm, ribosome, nucleoid, and other cellular compartments constitute the core. In addition to all the above components, dipicolinic acid–calcium complex is present inside the core; it maintains dehydrated conditions inside the spore and helps to stabilise DNA against heat denaturation.

Endospore Formation: When environmental conditions turn harsh, for example, when there is a shortage of essential nutrients, the vegetative cells of endospore-forming bacteria detect the harsh conditions and develop the endospores. The process to form one endospore requires 8–10 hours (Figure 5.7).

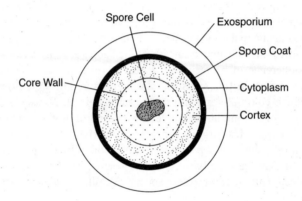

Figure 5.7 Endospore Structure

Sporulation: Sporulation of the endospore occurs through the following steps:

1. Firstly, the bacterial DNA gets replicated resulting in the protrusion of the plasma membrane between the replicated chromosomes and forms a forespore.
2. Calcium and dipicolinic acid formed in the cortex enclose the forespore resulting in the formation of second membrane.
3. At last, the endospore gets enclosed in the external spore coat, which lies between the inner and outer membrane.
4. Once the endospore is formed, the bacterial cell is said to be dormant.
5. Like the seed of a plant, the bacterium does not metabolize food or reproduce; instead, it exists in an inert form during the dormant stage.
6. The endospore germinates into a new bacterium when conditions turn favourable.

Endospore Staining: Endospore staining involves the following steps:

1. A smear of bacteria (capable of forming endospores) is prepared.
2. The smear is then subjected to staining with malachite green.
3. The slide is heated on a water bath.
4. The stained endospore-forming cells are observed under a microscope.

5.4.2 Cell Wall-Deficient Bacteria

The bacteria that are able to grow as spheroplasts or protoplasts and that usually grow with a conventional cell wall are said to be cell wall-deficient bacteria or bacterial L-forms. For a better understanding of the functional and structural organization of the cytoplasmic membrane and cell division, the stable protoplast-type L-forms are utilized. L-forms having a single bilayer membrane represent a unique expression system for the production of recombinant proteins. L-forms are used to form intracellular associations with plant cells and can extract induced disease-resistance species thereby providing a novel method for plant protection.

MULTIPLE CHOICE QUESTIONS

1. A light microscope has an objective lens with a magnification of 40x and an ocular lens with a magnification of 10x. What will be the total magnification of the image?
   (a) 40x
   (b) 50x
   (c) 400x
   (d) 450x
   Ans. c

2. The cocci that mostly occur in pairs are _____.
   (a) Streptococci
   (b) Diplococci
   (c) Tetracocci
   (d) None of the above
   Ans. b

3. The hook and filaments of flagella are composed of a protein called _____.
   (a) Keratin
   (b) Gelatin
   (c) Flagellin
   (d) Casein
   Ans. c

4. Which one of the following components is not found in the cytoplasm of majority of prokaryotes?
   (a) Capsule
   (b) Ribosomes
   (c) Granules
   (d) Nucleoid
   Ans. a

5. The term that refers to the fine hair-like bristles that aid in adhesion in prokaryotic organisms is _____.
   (a) Flagella
   (b) Pili
   (c) Glycocalyx
   (d) Fimbriae
   Ans. d

6. How are the flagella distributed over an organism when they are peritrichously arranged?
   (a) Multiple flagella are attached to one end of the cell.
   (b) A single flagellum is attached to both ends of the cell.
   (c) Multiple flagella are randomly dispersed over the cell surface.
   (d) Multiple flagella are attached to both ends of the cell.
   Ans. c

7. Unstained living cells and organisms can be observed best using _____.
   (a) Transmission electron microscope
   (b) Scanning electron microscope
   (c) Fluorescent microscope
   (d) Phase contrast microscope
   Ans. d

8. Pili, elongated tubular structures composed of pilin, are found only on which type of organisms?
   (a) Gram-positive bacteria
   (b) Gram-negative bacteria
   (c) Fungi
   (d) Spirochaetes
   Ans. b

9. Why are encapsulated bacteria generally more pathogenic than unencapsulated strains?
   (a) Because the capsule stimulates a potent immune response in the host
   (b) Because phagocytes do not recognize the capsule as foreign
   (c) Because the capsule helps prevent phagocyte attachment to the organism
   (d) Because the capsule causes the phagocyte to mutate
   Ans. c

10. Which statement describes why Gram-negative organisms are more sensitive to lysis than Gram-positive organisms?
    (a) Gram-negative organisms have a well-developed periplasmic space.
    (b) Gram-negative organisms have endotoxin.
    (c) Gram-negative organisms have a cell wall that is one-layer thick.
    (d) Gram-negative organisms have an outer membrane.
    Ans. c

11. The acid-fast stain is used to identify organisms containing which of the following chemicals in their cell walls?
    (a) Lipoteichoic acids
    (b) Porin proteins
    (c) Mycolic acid
    (d) Peptidoglycan
    Ans. c

12. How many chromosomes do bacteria possess?
    (a) One   (b) Six
    (c) Ten   (d) Zero
    Ans. a

13. Under what conditions are endospores formed?
    (a) When the cell reproduces
    (b) When the cell forms inclusion bodies
    (c) When carbon and nitrogen sources become depleted
    (d) When the cell germinates
    Ans. c

14. Which type of bacteria naturally lack the cell wall?
    (a) Archaea
    (b) *Mycoplasma*
    (c) Photosynthetic bacteria
    (d) Appendaged bacteria
    Ans. b

15. Which type of microscope is used to study ultrathin sections of organisms stained with gold and palladium?
    (a) Inverted microscope
    (b) Transmission electron microscope
    (c) Florescence microscope
    (d) Dark-field microscope
    Ans. b

SHORT NOTES

1. Electron microscopy
2. Polymorphism
3. Bacterial classification based on morphology
4. Intracellular organelles in bacteria

5. Various types of bacterial colony
6. Cytoplasm of bacterial cell

ESSAYS

1. What are endospores? How are they formed?
2. Describe the structure, function, and types of flagella.
3. Describe briefly the different kinds of microscope and their uses.

6 Flagella and Motility—Hanging Drop Technique

CHAPTER OBJECTIVES

6.1 Bacterial Motility
6.2 Detection of Bacterial Motility
6.3 Importance of Bacterial Motility

6.1 BACTERIAL MOTILITY

The ability of an organism to move by itself is called motility. Motility or locomotion is an important characteristic of bacteria. The terms motility, movement, and locomotion are used synonymously. Many but not all bacteria show motility, that is, self-propelled motion, under suitable conditions. Motion can be achieved by any one of the three mechanisms discussed below.

6.1.1 Types of Bacterial Motility

Flagellar Motility

Flagellum is the locomotory organ of a bacterium. Flagella are responsible for the motility of bacteria in aqueous environments. Not all bacteria contain flagella: only few cocci possess flagella and 50% of the rod-shaped bacteria and almost all spirilla have flagella. Because spherical cells do not have a good geometry, unlike linear bacteria, directional movement by flagella is not so convenient. Hence, most cocci are non-motile. Flagella are thin with a diameter of about 20 μm. They are semi-rigid propellers that are held to the bacterial cell at one end and free at the other end. Unless they are stained with some suitable dye, flagella are not clearly visible as their diameter is very small. Flagellar number varies with cell; it may be one to several. Distribution of the flagella also differs from one cell to another: flagella might be arranged in clumps or distributed all over the cell surface or even a single flagellum may be present.

Chemical Structure of Flagella: The flagellum is fixed to the cell by a complex protein structure, which is termed as hook, and the basal body. Typically, flagella are made up of a protein called flagellin. These proteins are bound in a long chain that enfolds itself to form a helical structure. Several flagellin protein subunits are needed to make a single turn of the helix. When the basal body rotates, the flagellum shows movement, which in turn helps the bacteria in locomotion. This happens when one ring in the basal body rotates in correlation with another ring. In order to rotate the ring, energy is derived from proton-motive force. Energy is released when protons are translocated from outside the cell to inside the membrane.

Few examples of flagellar motility are as follows:

1. Motility using flagella can be seen in *Pseudomonas, Vibrio, Spirillum, Azospirillum, Klebsiella, Salmonella, Proteus,* and so on.
2. *Escherichia coli* moves all over a solution in the clockwise direction with the turns of flagella taken inside the solution. This type of movement can be stated as smooth swimming. The bacterium can even overturn the direction of flagellar rotation. The bacterium moves in the opposite direction when the flagella turn clockwise.
3. Bacteria with peritrichously arranged flagella show a tumbling movement as the flagella tend to move the cell in all directions.

The average speed of bacteria is 50 µm/second. Locomotion of bacteria generally occurs in water. If the flagellum gets damaged and cut off, the bacterium has the ability to re-synthesize it to a suitable length. Re-synthesis of flagellum starts from the tip; flagellin monomers tend to build the new flagellum until it reaches a suitable length.

Spirochaetal Movement

The 'spirochaetes' includes the most important genera such as *Spirochaeta, Cristispira, Treponema, Borrelia,* and *Leptospira*. Spirochaetal movements can be seen in all these genera. The spirochaetal movement is facilitated by the flagella-like axial filaments lying in between the inner and the outer membrane of the cell wall. Two or more fibrils or axial filaments are inserted into the inner membrane, which acts as the basal body (otherwise called motor). Several types of movement such as creeping, spinning, swimming, or flexing can be seen in spirochaetes. To understand the spirochaetal movement, one can imagine a flexible helical rod in the air.

Gliding Movement

Gliding movements can be seen in bacteria that lack the flagellar structure. Instead, they secrete a slimy substance and use it for locomotion by gliding on surfaces. The exact mechanisms behind gliding movements are not yet known. Several scientists have made an effort to study about gliding movements, and few suggestions made by them for such movements are as follows:

1. Some scientists believe that fimbriae-like outgrowth arising at the poles of glider cells might be the reason behind gliding movements.
2. According to some scientists, contractile waves or pushing by secreted slime is responsible for gliding movements.
3. Few assume that such movements might be possible because of surface tension.

The 'gliding bacteria' include special bacteria that show gliding motility. Some of the important genera that fall under this category are *Myxococcus, Cyanobacterium, Achroonema, Alysiella, Leucothrix, Thioploca, Saprospira, Oscillatoria, Beggiatoa, Stigmatella, Polyangium, Nannocystis, Chondromyces, Cystobacter, Melittangium, Flexithrix, Herpetosiphon,* and *Archangium*.

6.2 DETECTION OF BACTERIAL MOTILITY

Detection of bacterial motility helps in the diagnosis and identification of bacteria to some extent. There are several techniques for the detection of bacterial motility, which are discussed below.

6.2.1 Flagellar Staining

Flagella are stained by Leifson's method. In this method, the dye used will precipitate the protein filament, making the flagella visible under microscope as it is coloured due to precipitation. It also demonstrates the arrangement of flagellin. Very occasionally, flagellar distribution is used to distinguish morphologically related bacteria. For example, in Gram-negative rods, which are motile, one can distinguish between enteric bacteria and Pseudomonads by flagellar distribution. In enteric bacteria, peritrichous flagella can be seen, and Pseudomonads have polar flagella.

6.2.2 Motility Test

Motility test is done to determine whether the bacterium is motile or non-motile. For this test, test tubes with semi-solid medium are taken. In one test tube, the bacterium that is suspected to be non-motile is inoculated in a straight-line stab with a sterile needle. For the other sample, which is suspected to be motile, inoculation should be done in the same way. After incubation, turbidity will be seen only along the straight line where the inoculum was present (as the bacteria grow along the inoculation line), proving that the bacteria are non-motile. Whereas in the other test tube, the organism tends to go away from the stab making the semi-solid medium turbid (as the bacteria swim away from the line of inoculation), proving the motility of the bacteria (Figure 6.1).

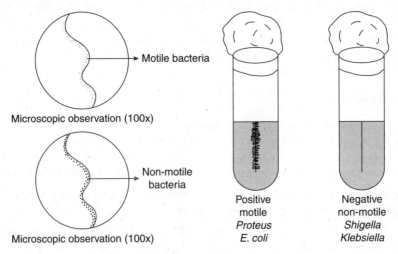

Figure 6.1 Motility Test

6.2.3 Direct Microscopic Examination

By this method, one can observe the movement of living bacteria. For this, a drop of water from a pond should be taken and placed onto a clean cover slip. The cover slip should then be placed on the cavity slide in inverted position without disturbing the drop of water. The slide must then be observed under x400 or x1000 objective lens. This process is commonly called wet mount. Brownian movement of the bacteria will be seen under the microscope. It is not true motility but just due to the random collisions between the bacteria and water molecule. If the bacterial cells tend to swim from one side of the microscopic field to another, it is true motility.

6.3 IMPORTANCE OF BACTERIAL MOTILITY

6.3.1 Chemotactic Behaviour

The tendency of bacteria to move towards chemical attractants is known as chemotactic behaviour. During unfavourable conditions, for example, if nutrients get depleted in the region where the bacteria normally grow, chemotactic bacteria move towards nutrients. Therefore, chemotactic behaviour helps bacteria survive by selecting a suitable environment.

6.3.2 Root Colonization

Root colonization is the establishment of bacteria in the rhizosphere region. As motile bacteria can swim faster towards exudates or any nutrients, they are effective root colonizers when compared with non-motile bacteria. For example, Pseudomonads and Azospirilla are effective root colonizers as they are motile.

6.3.3 Pathogenesis

To infect a host and colonize all over the host, it is important for an organism to be motile. When motile, the organism can move faster, get attached to the host tissue, and start to colonize. For example, a pathogen after colonization of the cell wall of say intestine, it can move to colonize other vital organs and succeed in spreading the infection. *Campylobacter*, *Salmonella*, *Escherichia*, and *Vibrio* are examples of motile pathogens.

6.3.4 Twitching Motility

Even though some bacteria do not have flagella, they show a kind of jumping type of motility. This is known as twitching motility. It is observed in *Acinetobacter* spp. This type of movement is seen especially on semi-solid media. The mechanism behind such movement is unclear, but some scientists assume that twitching motility is due to piliated cell surface.

MULTIPLE CHOICE QUESTIONS

1. Bacterial locomotion is mainly due to _____.
   (a) Capsule
   (b) Flagella
   (c) Cytoskeleton
   (d) Both (a) and (b)

   Ans. b

2. Which one of the following bacteria is motile?
   (a) *Bacillus anthracis*
   (b) *Klebsiella pneumoniae*
   (c) *Salmonella typhi*
   (d) *Shigella flexneri*

   Ans. c

3. For the rotation of bacterial flagella, energy is _____.
   (a) ATP driven
   (b) Proton driven
   (c) None of the above
   (d) Both (a) and (b)

   Ans. b

4. Cilia are different from flagella because _____.
   (a) Cilia are shorter than flagella.
   (b) Cilia are distributed all over the cell surface.
   (c) Cilia are more in number than flagella.
   (d) All the above are correct.

   Ans. d

5. Bacterial flagella are typically made up of a protein called _____.
   (a) Vimentin
   (b) Flagellin
   (c) Tubulin
   (d) Dynein

   Ans. b

SHORT NOTES

1. Different types of bacterial motility
2. Chemical structure of flagella
3. Gliding movement

ESSAYS

1. Describe the techniques used to detect bacterial motility.
2. How is motility important to bacteria? Explain.

7 Colonization, Nutrition, and Growth of Microbes

CHAPTER OBJECTIVES

7.1 Colonization of Bacteria

7.2 Microbial Nutrition

7.3 Microbial Growth

7.4 Influence of Environmental Factors on Microbial Growth

7.1 COLONIZATION OF BACTERIA

The establishment of pathogenic microorganisms at the convenient entry into the host tissue is called colonization. The host tissue that is in contact with the outer environment is usually prone to be colonized by most of the pathogens. The digestive tract, respiratory tract, conjunctiva, and urogenital tract are the most common entry sites in the human body. The adherence mechanisms developed by the pathogens are responsible for infections. Pathogens should also develop the capability to withstand the constant pressure exerted by the immune system of the host at the region of infections. Invasiveness and toxigenesis are the most important properties of pathogens that are responsible for causing diseases in any host.

7.1.1 Invasiveness

The ability of the pathogen to invade and colonize the host tissue is invasiveness. Pathogens must have certain characteristics such as adherence and initial multiplication in order to colonize. The production of extracellular substances such as 'invasin' aids in the immediate invasion of the host tissue. The actual invasive process is facilitated by overcoming the host defence mechanisms.

7.1.2 Toxigenesis

The ability of the pathogens to produce toxic substances is toxigenesis. These toxic substances produced might be carried out all over the host's body by means of blood and lymph causing cytotoxic effects. These toxic substances are both soluble and cell-associated.

7.1.3 Adherence of Bacteria

A specific receptor and a ligand are required to facilitate the adherence of a bacterium. On the surface of eukaryotic cells, the receptors are generally residues of peptides or specific carbohydrates. Bacterial ligands are also called 'adhesins', which are typically macromolecular components present on the surface of bacteria. Adhesins and receptors usually interact in a complementary fashion, same as that of antigen–antibody reactions and enzyme–substrate relationships.

7.1.4 Pathogenicity

Infection is nothing but colonization of the pathogen into the host tissue. It begins with the entry of the pathogen into the body of the host where it flourishingly grows and develops. Most pathogens gain entry into the host through the mucosa of oral cavity, nasal openings, eyes, genitalia, anus, or any open wounds. Few pathogens can colonize at the initial site of entry but the majority of them tend to migrate deeper causing systemic infections in different organs. Some pathogens are intracellular (i.e., grow within the host cell), whereas some are extracellular (i.e., grow freely in bodily fluids). The pathogens that are virulent produce a special kind of protein that helps them to colonize in different parts of the host tissue. For example, *Helicobacter pylori* produces the enzyme urease that enables it to survive in the acidic environment of the stomach of a man. When this bacterium starts to colonize on the stomach lining, it might lead to gastric ulcer and even cancer.

7.1.5 Wound Colonization

Wounds are certain damaged areas on the skin such as cuts and burns. When microorganisms infect and colonize such areas, it is termed as wound colonization. When specific conditions are offered, even non-pathogenic microorganisms turn into pathogenic microorganisms. In order to cause an infection, certain conditions are required even for the most virulent pathogens. Some colonizing bacteria such as *Corynebacterium* sp. and streptococci, which live in symbiotic relationship with their host, prevent the adhesion and colonization of pathogenic bacteria. Thus, wounds can heal quickly due to such beneficial organisms.

Wound is an oxygen-deficient area that generally gets infected and colonized by anaerobic pathogens. Most of the time, wounds are colonized by extrinsic organisms. However, if proper care is not taken, then the opportunistic organisms that reside within the host will cause infections. For example, anaerobic bacterial species colonize the mammalian colon. Various species of *Staphylococcus* on the skin of human beings, which are in either a commensal or a mutualistic relationship with the host, never colonize. However, if they enter the wound without resistance, it generates a trouble for the host. Several criteria are necessary for a wound colonization to commence; few are as follows:

1. The path of entry of the pathogen should be through the wound.
2. There must be acceptance of the pathogen by the host.
3. The intrinsic organisms should have the ability to infect and colonize the wound.
4. The quantity of the initial inoculum of the pathogen should be high.
5. The immunity of the host should be low.

7.2 MICROBIAL NUTRITION

In order to get energy and to maintain cellular biosynthesis, every organism must be provided with the essential substances needed for growth from its environment. These essential substances required for bacterial growth are referred to as 'nutrients'. Most bacteria can be grown away from their natural habitats in laboratories if suitable nutrients are provided in the form of culture media. Bacteria can be easily grown on culture media because they contain all the essential nutrients that are needed for bacterial growth. However, all bacteria cannot be grown in laboratories as some are symbionts and some are obligate intracellular parasites of other cells. The host cells must satisfy the nutritional requirements of mutualists and parasites as these are the residents of the host.

Based on their nutrition, the bacteria are classified into the following:

7.2.1 Autotrophic Bacteria

The bacteria that can synthesize organic food from inorganic substances are called autotrophs. For example, in autotrophs, the need for carbon is fulfilled by the utilization of carbon dioxide. And to reduce this carbon, hydrogen sulphide, ammonia, or hydrogen is used as the source of hydrogen. Autotrophic bacteria are of two types, which are mentioned below.

Photoautotrophic Bacteria

These bacteria contain photosynthetic pigments in thylakoids that utilize the solar energy to synthesize food. Bacterial photosynthesis is completely different from that of green plants; here, water is not used as the hydrogen donor. Hence, oxygen is not released as the by-product of photosynthesis. As a result, this process is known as anoxygenic photosynthesis.

$$CO_2 + H_2S + Sunlight \rightarrow Sugar + Sulphur + Water$$

Chemoautotrophic Bacteria

Chemoautotrophs construct organic compounds from inorganic substances. Here, while the oxidation of inorganic substances occurs, energy gets liberated, which in turn is used to construct food, that is, organic compounds. Some of the common chemoautotrophic bacteria are stated below.

Nitrifying Bacteria: These bacteria derive energy by oxidizing ammonia into nitrates (e.g., *Nitrosomonas* and *Nitrobacter*).

$$NH_4^+ + 2O_2 \rightarrow NO_2 + 2H_2O + Energy$$

Sulphur Bacteria: These bacteria derive energy by oxidizing hydrogen sulphide to sulphur (e.g., *Thiobacillus* and *Beggiatoa*).

$$2H_2S + 2O_2 \rightarrow 2S + 2H_2O + Energy$$

Iron Bacteria: These bacteria derive energy by oxidizing ferrous ions into ferric ions (e.g., *Ferrobacillus* and *Gallionella*).

$$4FeCO_3 + 6H_2O + O_2 \rightarrow 4Fe(OH)_3 + 2CO_2 + Energy$$

7.2.2 Heterotrophic Bacteria

Heterotrophic bacteria cannot synthesize their own food; they are always dependent on external sources. The heterotrophs are of different forms, which are mentioned below:

1. Bacteria that obtain their nutritional requirements from dead and decaying matter are called saprophytic organisms. By secreting certain exogenous enzymes, saprophytes break down complex organic matter into a simple soluble form. During the uptake of nutrients, they release energy.
2. Decomposition is carried out by certain heterotrophs. During it, aerobic breakdown of organic matter takes place.
3. Fermentation is also accomplished by heterotrophic organisms. Here, anaerobic breakdown of organic matter occurs. Fermentation reactions are incomplete and always release foul gases.
4. Putrefaction, which is the breakdown of protein molecules, is also done by heterotrophs.

Decomposition of organic compounds is employed in numerous industrial areas such as retting of fibres, curing of tobacco, and ripening of cheese.

7.2.3 Symbiotic Bacteria

Some bacteria can live with other organisms in such a way that both are not harmed by each other but rather are benefitted by one another. This mutually beneficial relationship is called symbiosis. And the bacteria that live in such a relationship are called symbiotic bacteria. These bacteria take in essential nutrients from their hosts and facilitate their hosts through some of their biological activities. The most well-known examples of symbiotic bacteria are as follows:

1. Bacteria such as *Pseudomonas* and *Rhizobium* live in the root nodules of leguminous plants. These bacteria directly reduce atmospheric nitrogen to ammonia and provide the source of nitrogen for host plants. In return, plants offer nutrition and protection for these bacteria.
2. *Escherichia coli* are the non-pathogenic bacteria that reside in the human alimentary canal. These bacteria get shelter and food from the human host and provide protection against certain harmful putrefying bacteria. In addition, these bacteria release vitamin K and vitamin B_{12}, which are very essential for blood components.
3. The cellulose-digesting bacteria that live in the alimentary canal of ruminant mammals such as cows and goats are another example. The relationship observed here is the same mutual relationship as can be seen in *E. coli* residing in the human alimentary canal.

7.2.4 Parasitic Bacteria

Parasitic bacteria also cannot synthesize their own food and always need a host for their survival. These bacteria obtain their food from their hosts such as animals and plants. The majority of parasitic bacteria are pathogens and are responsible for several chronic diseases in the host by exploiting them. Parasitic bacteria also secrete toxins, which are lethal for the host.

7.2.5 Major Elements

The nutritional requirements of bacteria are at an elementary level. The cell's elementary composition consists of C, H, O, N, S, P, K, Fe, and Ca and traces of Zn, Ni, Mn, Co, Cu, and Mo. Bacteria uptake these elements as water, as some inorganic ions, microelements, and macroelements. These elements serve as either functional or structural units of the bacterial cell.

Macroelements

Macroelements include C, H, O, P, K, N, S, Ca, Fe, and Mg. These elements are required in large quantities by the bacterial cell, hence called macroelements. Among the macroelements, C, H, and O are responsible for the cellular structures. Phospholipids, ATP, nucleic acids, and teichoic acid require phosphorous as a macroelement. A primary regulatory of osmolarity is potassium. Nitrogen is required for sugars, amino acids, nucleic acids, and many vitamins. Calcium, magnesium, and sulphur are required for carbohydrates, vitamins and amino acids. The macroelement iron is an important co-factor for enzymes.

Microelements

The elements that are required in very small quantities are known as microelements or trace elements. Trace elements include Zn, Ni, Co, Mn, Cu, and Mo. Microelements are responsible for the maintenance of cell and enzyme functions. The source, quantity, and functions of major elements are outlined in Table 7.1.

Table 7.1 Major Elements, Their Sources, and Functions in Bacterial Cells

Major Element	Source	Quantity (% of Dry Weight)	Functions
Carbon	Organic compounds (carbon dioxide)	50	Cellular materials are chiefly made up of this element.
Oxygen	Organic compounds (water, carbon dioxide, and oxygen)	20	This element is the electron acceptor in aerobic respiration. It is also a constituent of cellular materials and cell water.
Nitrogen	Organic compounds (ammonia, nitrate, and nitrogen)	14	It is the main element of nucleotides, nucleic acids, co-enzymes, and amino acids.
Hydrogen	Organic compounds (water and hydrogen)	8	It is the major constituent of cell water.
Phosphorus	Inorganic phosphates	3	it is the major constituent of phospholipids, teichoic acid, lipopolysaccharide (LPS), nucleotides, and nucleic acids.
Sulphur	Organic sulphur compounds (SO_4, SO, and H_2S)	1	Several co-enzymes, methionine, glutathione, and cysteine contain sulphur.

(continued)

Table 7.1 (continued)

Major Element	Source	Quantity (% of Dry Weight)	Functions
Potassium	Potassium salts	1	Potassium is the chief co-factor for few enzymes and a constituent of cellular inorganic cation.
Magnesium	Magnesium salts	0.5	It is a constituent of cellular inorganic cation and a co-factor in certain enzymatic reactions.
Calcium	Calcium salts	0.5	It has the same function as that of Mg. In addition, it is a component of endospores.
Iron	Iron salts	0.2	It is a constituent of certain non-haem iron proteins, a component of cytochromes, and a co-factor for some enzymatic reactions.

In the above table, trace elements have been ignored as they are required in very small quantities. Sometimes, they are represented as contaminants of media components or of water. Trace elements are metal ions needed in small amounts. However, trace elements, as metal ions, usually act as co-factors for maximum enzymatic reactions that are essential for the cell. The trace elements that get to qualify as cations in bacterial nutrition are Mn, Co, Zn, Cu, and Mo.

7.3 MICROBIAL GROWTH

In the case of bacteria, growth is not the increase in size, rather it is the increase in cell number. Microbial growth completely depends upon the capability of the cell to construct new protoplasm from the nutrients present in its environment. Bacterial reproduction mainly occurs by binary fission. Every cell becomes double the size and gets divided into two cells. Binary fission proceeds in a very systematic way as follows:

1. Increase in cellular structures and components
2. Doubling of the DNA by replication to make two sets of DNA
3. Segregation of the doubled bacterial DNA
4. Formation of the cross wall, which facilitates the cell to divide into two progeny cells

After replication, the DNA molecules are said to be connected to a point on the cell membrane. Between the two points, the new membrane is synthesized by mesosomes. Both DNA molecules move in the opposite direction, and the cross wall is laid down between the two chromosomal compartments. The cell gets divided into two progeny cells as soon as the cell wall formation gets completed. The time required for one cell to become two by means of binary fission is known as generation time. The generation time varies with bacterial species: it might be as short as 15 minutes or as long as several days.

7.3.1 Growth Factors

Organisms sometimes fail to synthesize certain essential substances from the nutrients available. Such essential substances are called growth factors. Growth factors accomplish some important roles in biosynthesis, but they are required in very less amounts by the cell. A blocked or missing metabolic pathway in the bacterial cell results in the need for a growth factor. Growth factors are grouped as follows:

1. Amino acids (needed for protein synthesis)
2. Purines and pyrimidines (needed for nucleic acid synthesis)
3. Vitamins (needed as co-enzymes and functional groups of certain enzymes)

Few bacteria do not require any growth factor because they can synthesize all the essential substances (e.g., *E. coli*). *Escherichia coli* synthesizes purines, pyrimidines, vitamins, and amino acids as part of its intermediary metabolism. Few other bacteria such as *Lactobacillus* need several amino acids, purines, pyrimidines, and vitamins for growth. For such bacteria, growth factors should be supplied in advance through the culture media used for their growth. Growth factors are incorporated into the bacterial cell to execute their specific roles in metabolism and not as a source of carbon or energy. When growth factors are required by the mutant strains of bacteria but the same are not required by the wild types, they are termed as auxotrophs.

7.3.2 Uptake of Nutrients by Bacteria

For the uptake of the nutrients required for growth, specific mechanisms are required by the bacteria. The mechanism should be capable of carrying the required essential substances against the concentration gradient. The mechanism should be able to pass the nutrients through a selectively permeable membrane.

Following are some mechanisms commonly used for the uptake of nutrients:

Facilitated Diffusion

In this mechanism, the molecules move about freely from the region of higher concentration to the region of lower concentration. It is passive diffusion where the molecules move freely due to accidental thermal energy. The rate of diffusion is increased by the usage of permease, a carrier molecule that is implanted within the plasma membrane. For transporting specific molecules, specific permeases are needed. The process gets saturated at a certain point as the concentration of all permease-bound transport molecules increases. Facilitated diffusion requires no energy because the movement is along the concentration gradient. If the concentration increases inside the cell, then the molecules tend to move to the outside environment. Hence, it is a reversible process. Facilitated diffusion for uptake of nutrients is more significant in eukaryotes compared to prokaryotes (Figure 7.1).

Active Transport

In active transport, the molecules move against the concentration gradient into the cell (Figure 7.2). Hence, active transport is an energy-dependent process and it requires energy

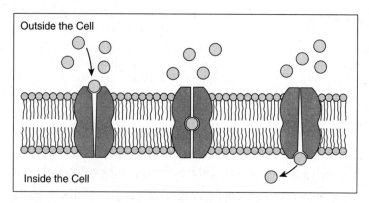

Figure 7.1 Facilitated Diffusion (See page 346 for the colour image)

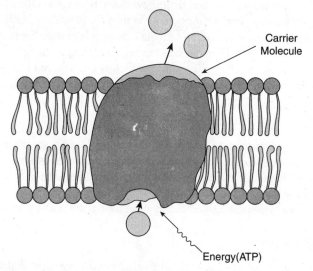

Figure 7.2 Active Transport (See page 346 for the colour image)

in the form of ATP. Active transport resembles facilitated diffusion only in having the carrier protein. At high solute concentrations, the process will get saturated.

Active transport mechanism is of various kinds, which are as follows:

1. **ATP-binding cassette (ABC) transporter:** It is a kind of active transport that is seen in both eukaryotes and prokaryotes. In this mechanism, two hydrophobic domains create a pore through the plasma membrane and two nucleotide-binding domains hydrolyse ATP for the uptake of the nutrient molecule. In Gram-negative bacteria, special binding proteins are found in the periplasmic space, which help the transport molecules to be involved with the membrane cassette proteins. To help the transport molecules cross the outer membrane, Gram-negative bacteria have certain specialized mechanisms. In Gram-positive bacteria, special binding proteins are found attached to the external side of

the plasma membrane, which help the transport molecules to interact with the membrane cassette proteins. ABC transporters pump out the drug as soon as it enters the cell. Hence, it is an important mechanism of drug resistance seen in eukaryotes.
2. **Symport and antiport transporter system:** It is also a kind of active transport that requires proton gradients to drive the uptake of nutrient molecules (Figure 7.3). Symport transporter mechanism involves the transportation of two substances in the same direction. Antiport transporter mechanism involves the transportation of two substances in opposite directions. Commonly, the nutrient that has to be transported gets bound to the co-transported molecule, which causes changes in the conformation of the protein, resulting in the transportation of the molecule to its destiny. Even in eukaryotes, the symport transporter system is used; the only difference is that it will use ATP as the energy currency alternative for protons.

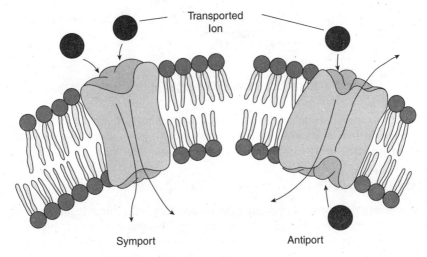

Figure 7.3 Symport and Antiport Mechanism (See page 347 for the colour image)

Group Translocation

In group translocation, the molecules that are to be transported are chemically altered in order to facilitate transportation (Figure 7.4). This is also an energy-dependent process. This mechanism is utilized by both eukaryotes and prokaryotes. The best example for group translocation is the phosphoenolpyruvate–sugar transferase system.

Iron Uptake by Siderophores

Iron uptake is very difficult when it should be from the insoluble Fe^{3+} ion, which is usually in the hydroxide form. Both eukaryotes and prokaryotes overcome this challenge by secreting specialized high-affinity iron-binding molecules called 'siderophores'. For the uptake of iron, there are two general types of siderophores, namely, hydroxamates (in ferrichrome) and phenolates-catecholates (in enterobactin).

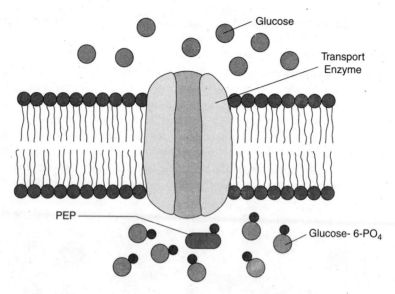

Figure 7.4 Group Translocation (See page 347 for the colour image)

Bacteria has very specialized transport systems for siderophores. Once the molecule gets transported, it releases iron in its reduced ferrous form (Fe^{2+}) into the cell. Siderophores can be recycled and used frequently to facilitate the uptake of iron.

7.3.3 Growth Curve

For understanding the growth in bacteria, the graph is plotted against logarithmic number of bacterial cells grown versus the time taken for growth. This graph is plotted only for the bacteria grown in laboratory conditions. After plotting the graph, a curve is obtained, which is called the growth curve (Figure 7.5). When favourable conditions are provided, bacterial populations

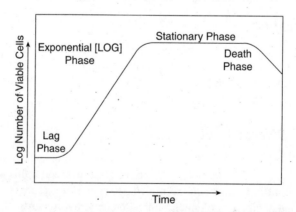

Figure 7.5 Bacterial Growth Curve

double at regular intervals in laboratories. The progression in growth is recorded geometrically, that is, $2^0, 2^1, 2^2, 2^3, 2^4, ..., 2^n$, where '$n$' stands for the number of generations. This is known as exponential growth and is not representative of the normal pattern of growth of bacteria in nature but of the growth of bacteria under laboratory conditions.

The population growth is observed over a period of time for a fresh medium inoculated with a known number of cells. A typical bacterial growth curve can be obtained by plotting the data on a graph.

Growth of bacteria in a closed batch culture system ensures the cell population follow a systematic growth curve. During the lag phase, the cells initially get adjusted to the new culture medium and then they start dividing as usual by binary fission. During the exponential phase, the cells divide continuously. Then, the cells stop dividing as their growth becomes limited; this is called the stationary phase. Due to the loss of nutrients, growth gets stopped during the death phase. The growth is expressed by plotting the graph against logarithmic number of viable cells versus time. The growth cycle of bacteria grown under laboratory conditions shows four characteristic phases, which are as follows:

Lag Phase

The population remains unchanged temporarily just after the inoculation of bacterial cells into the fresh culture medium. The cells grow in mass, synthesizing enzymes, proteins, RNA, and so on, and the metabolic activities of the cell will be increased. A wide variety of factors are responsible for determining the length of the lag phase. Firstly, the amount of inoculum inoculated into the fresh medium affects the length of the lag phase. When the inoculum is transferred to the laboratory, some physical damage may occur; the time taken by the bacteria to recover from this shock also determines the duration of the lag phase. To enter the log phase, all the essential elements required for cell division should be synthesized by the cell during the lag phase itself. Most importantly, the time required for the synthesis of essential co-enzymes (which are called division factors) and also the synthesis of inducible enzymes (which are necessary to metabolize the substrates present in the growth medium) determine the length of the lag phase.

Log Phase

During this phase, the cells that had prepared in the lag phase for division start dividing normally by binary fission. This phase provides a blueprint that shows the balanced growth of dividing cells, and the growth can be measured by geometric progression. Because of the composition of the growth medium and incubation conditions, the cells divide at a steady rate. The rate of exponential growth of a bacterial culture can be expressed by the generation time, which can be given by the following equation:

$$G = \frac{t}{n}$$

where, G is the generation time, t the time taken, and n the number of generations.

Stationary Phase

In a batch culture, continuous exponential growth cannot be expected. There are many factors that check the growth rate. Some of the factors are as follows:

1. As the cells divide repeatedly, the cell number increases; as a result, the nutrients present in the growth medium get exhausted.
2. End products released by the dividing cells also increase; these get accumulated in the growth medium and inhibit further growth.
3. As the number of cells increases, it leads to lack of biological space.

If viable cells are counted during the stationary phase, it is very difficult to know whether the population of cells in the growth culture has simply stopped growing and dividing. As in the lag phase, in the stationary phase also, active cell division cannot be seen. However, the reason behind is different in both cases: in the lag phase, due to adjustment, preparation, and soon, the cells do not start dividing, whereas in the stationary phase, the loss of nutrients or accumulation of end products stops the cells from dividing further. Bacteria release secondary metabolites, such as antibiotics, during this phase. Even the sporulation process is induced during this phase.

Death Phase

If the incubation is prolonged even after the stationary phase, all the viable cells die leading to the death phase. The number of viable cells decreases exponentially and the bacterial cell population declines; it is exactly the reverse of the growth seen during the log phase. The generation times of some bacteria are given in Table 7.2.

Table 7.2 Generation Times of Some Bacteria

Bacterium	Generation Time (Minutes)
Escherichia coli	20
Bacillus megaterium	25
Staphylococcus aureus	27–30
Lactobacillus acidophilus	66–87
Mycobacterium tuberculosis	792–932
Treponema pallidum	1980

7.3.4 Continuous Culture

Usually, batch cultures are used for the growth of bacterial cells. However, in batch cultures, the exponential growth is limited only to a few generations as the nutrients get exhausted. To overcome this, a continuous culture system is used. Here, the bacterial cells are cultured continually in a state of exponential phase over a long period of time. Continuous culture uses an apparatus called chemostat (Figure 7.6). This system can be utilized for preserving bacterial population at a constant quantity in several ways. This mimics the natural growth environment for the bacteria.

The continuous culture system contains an apparatus where the growth chamber is attached to a reservoir of sterile medium. Once the bacterial cells start to grow, fresh medium is continually added from the reservoir. The amount of fluid in the growth chamber is maintained at a constant level by using certain overflow drain. As the growth of the bacteria gets limited, fresh medium is allowed into the growth chamber to maintain the rate of growth. The rate of growth is concluded by the rate at which fresh medium is added, as fresh medium constantly contains a limiting quantity of needed nutrients. Because of this reason, the chemostat decreases the insufficiency of nutrients, accumulation of excess cells in the culture and release of toxic substances, which are the parameters responsible for the start of stationary phase in the growth cycle. Depending upon the flow rate of nutrients, the bacterial culture is grown and sustained at relatively constant conditions.

The chemostat provides favourable conditions by maintaining a constant supply of nutrients to the cells and also removing waste substances and spent cells from the culture medium.

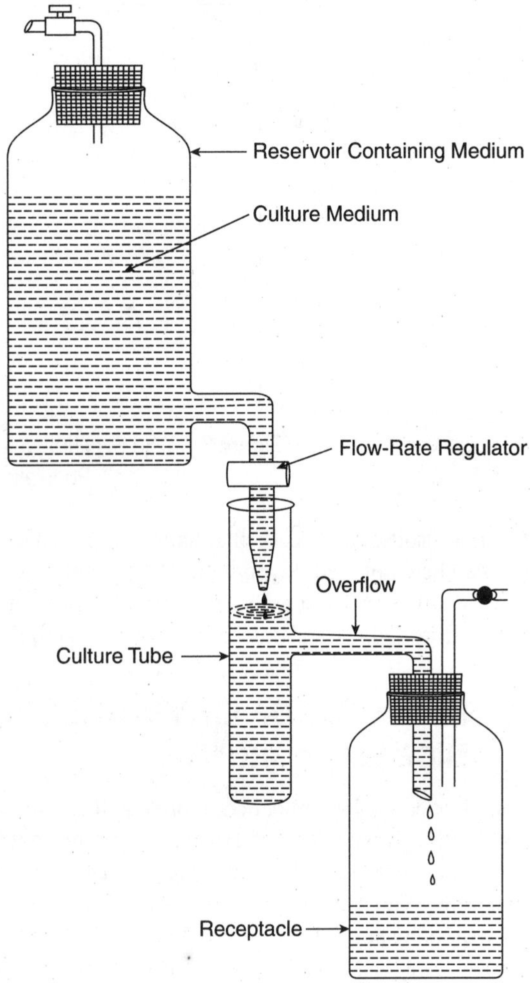

Figure 7.6 Schematic Diagram of a Chemostat—an Apparatus for Continuous Culture of Bacteria

7.3.5 Synchronous Growth

As the distribution of cell size in the bacterial population is absolutely random, the studies regarding the growth of bacterial population in either batch culture or continuous culture do not give any clear conclusion about the growth activities of a single cell. However, the growth activities of a single bacterium can still be obtained by the synchronous culture system. In this system, all the cells should be from the same stage of the bacterial cell cycle. If measurements are made on synchronized cultures, they will correspond to the measurements made on a single cell.

To obtain the bacterial populations at the same stage of the growth cycle, many methods have been employed. Environmental parameters can be managed by some methods in such a way as to stop or start the growth of bacterial population at the same point in the cell cycle.

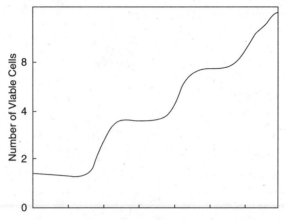

Figure 7.7 The Synchronous Growth of a Bacterial Population

For the assortment of cells that have completed cell division by binary fission, physical methods can be employed. Hypothetically, the cells that have just completed binary fission are considered as the smallest cells. In Figure 7.7, the synchronous growth of bacterial population has been illustrated. Synchrony can be maintained only for a few generations.

7.4 INFLUENCE OF ENVIRONMENTAL FACTORS ON MICROBIAL GROWTH

The physical and chemical components of the surroundings in which microbes grow affect the growth more than that by the deviations in nutrient levels, especially nutrient limitation. Hence, it is very much essential to maintain suitable environmental conditions and provide balanced and proper nutrients for the microbes to flourish. To fulfil all the criteria required for microbial growth, one should have proper knowledge about environmental influences on the growth of microorganisms. Bacteria show diverse reactions to environmental conditions and also diverse food habits. The growth and death rates of microorganisms are significantly affected by several environmental factors such as solutes and water acidity, pH, radiation, oxygen requirements, pressure, and temperature.

7.4.1 Water Acidity and Solutes

No life can survive without water; even for microorganisms, water is one of the most crucial requirements for survival. Hence, the availability of water becomes the most important factor for the growth of microbes. Water is available to microbes in the form of water from the surrounding environment or in the form of concentration of solutes.

1. The cytoplasm of a cell contains higher concentration of solutes when compared with its surrounding environment in most cases. Osmosis is the process that maintains the microbial cytoplasm; here, water always moves from the region of higher concentration to the

region of lower concentration. So when bacterial cells are kept in a hypertonic solution, the cell membrane shrinks because of the loss of water.

2. Microorganisms have the ability to adjust to the low-water activity environment. For example, *Staphylococcus aureus* can withstand a wide range of water activities and is called osmotolerant.
3. Most of the microorganisms grow flourishingly only near pure water activity. Only halophiles can survive in high salt concentrations. For example, *Halobacterium* survives in Dead Sea, which has high salt concentration.
4. The Water activity is nothing but the quantitative availability of water. For a sample solution, the water activity is given by the ratio of the vapour pressure of the sample solution to the vapour pressure of water at the same temperature.

$$aw = \frac{P_{solution}}{P_{water}}$$

where, aw is the water activity, $P_{solution}$ the vapour pressure of the sample solution, and P_{water} the vapour pressure of water.

5. Only after sealing the test sample in a closed chamber, relative humidity can be obtained at equilibrium. By this, the water activity of the solution can be determined. For example, after the sample is treated using the above method, relative humidity is 95% and water activity is 0.95.

7.4.2 Temperature

Microorganisms are poikilothermic as they change their temperature according to the surrounding temperature (Figure 7.8). Microorganisms are very much sensitive to temperature, and the rate of chemical reactions and protein structures are affected by temperature. At low

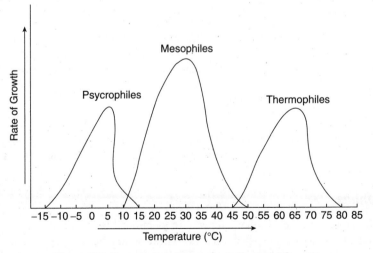

Figure 7.8 Temperature Parameter in Bacterial Growth

temperatures, enzymes do not get denatured; if the temperature is increased by say 10°C, metabolic activity and growth can be enhanced. However, at high temperatures, enzymes get denatured, leading to the death of microbes. Microorganisms show different characteristics for temperature dependence and possess their own cardinal temperatures as follows:

1. Maximum growth temperature
2. Optimal growth temperature
3. Minimal growth temperature

Fundamental values vary extensively between one bacterium and another. For example, bacteria that dwell in hot springs can withstand temperatures of 100°C and above, whereas the ones that are isolated from the snow can survive at −10°C. Depending on their susceptibility to thermal conditions, microorganisms are classified as follows (Table 7.3):

1. Thermophiles
2. Mesophiles
3. Psychrophiles

Table 7.3 Temperature Ranges Seen in Prokaryotes

Cardinal Temperature	Groups		
	Thermophiles	Mesophiles	Psychrophiles
Maximum temperature (°C)	60–90	35–47	15–20
Optimal temperature (°C)	55–75	30–45	12–15
Minimal temperature (°C)	40–45	5–15	−5 to 5

Thermophiles

Thermophiles can grow flourishingly in the higher temperature range stated in the above table. Most of the thermophiles belong to prokaryotes. Few microorganisms are hyperthermophiles, that is, their optimal temperature ranges between 80°C and 113°C (e.g., *Pyrococcus abyssi* and *Pyrodictium occultum*).

Mesophiles

The majority of microorganisms come under this category. The temperature range for mesophiles is given in the above table. Almost all pathogens are mesophiles. The most suitable temperature for mesophiles is 37°C.

Psychrophiles

Psychrophiles survive in the cardinal temperature mentioned in the above table. Few can survive even under 0°C (facultative psychrophiles). These organisms are responsible for the spoilage of refrigerated food.

7.4.3 Oxygen Requirement

Oxygen is one of the major elements of our atmosphere. About 20% of the atmospheric air is oxygen. Some microorganisms require oxygen for their metabolic activities, whereas some can grow even in the absence of oxygen. Based on their need for oxygen, microorganisms are categorized into the following groups:

1. **Aerobes:** These microorganisms are capable of living in the presence of atmospheric oxygen.
2. **Anaerobes:** These microorganisms can grow in the absence of atmospheric oxygen.
3. **Obligate aerobes:** These microorganisms are completely dependent on atmospheric oxygen, and they cannot survive in the absence of oxygen (e.g., *Pseudomonas*).
4. **Obligate anaerobes:** These microorganisms grow only in the absence of atmospheric oxygen. For these organisms, oxygen is lethal (e.g., *Bacteroides*, *Clostridium pasteurianum*, and *Fusobacterium*).
5. **Facultative anaerobes:** These microorganisms can grow flourishingly in the presence of oxygen but can also grow in the absence of atmospheric oxygen (e.g., *E. coli*).
6. **Aerotolerants:** These microorganisms are capable of growing both in the presence and in the absence of atmospheric oxygen (e.g., *Enterococcus faecalis*).
7. **Microaerophiles:** These microorganisms require oxygen at very low concentrations (2%–10%). Normal atmospheric level of oxygen is lethal for such organisms (e.g., *Campylobacter*).

Relationship between Oxygen and Growth

These differing relationships between bacteria and atmospheric oxygen are due to dissimilar factors such as protein inactivation and the end product of toxic oxygen derivatives. Sometimes, bacterial enzymes can get inactivated in the presence of atmospheric oxygen. Nitrogenase is a nitrogen-fixing enzyme that is very sensitive to oxygen. When metabolic processes are considered, flavoprotein reduces oxygen to form superoxide radical, hydrogen peroxide, and hydroxyl radical. These compounds, being powerful oxidizing agents, can damage cellular macromolecules as they are extremely toxic.

$$O_2 + e^- \rightarrow O_2^- \text{ (superoxide radical)}$$

$$O_2^- + e^- + 2H \rightarrow = H_2O_2 \text{ (hydrogen peroxide)}$$

$$H_2O_2 + e^- + H^+ \rightarrow H_2O + OH^- \text{ (hydroxyl radical)}$$

Bacteria must be able to protect themselves from these oxidizing agents so as to survive. Superoxide dismutase and catalase are the enzymes produced in all aerobes and facultative anaerobes for shielding themselves against the deadly effects of oxygen by-products. The enzyme catalase decomposes toxic hydrogen peroxide into water and oxygen. In aerotolerant bacteria such as lactic acid bacteria, peroxidase enzyme replaces catalase in order to break down the accumulated hydrogen peroxide.

$$2O_2^- + 2H^+ \xrightarrow{\text{Superoxide dismutase}} O_2 + H_2O_2$$

$$2H_2O_2 \xrightarrow{\text{Catalase}} 2H_2O + O_2$$

$$H_2O_2 + NADH + H^+ \xrightarrow{\text{Peroxidase}} 2H_2O + NAD^+$$

Obligate anaerobes do not possess these enzymes; even if they produce them, it will be in very low concentrations. Because of this reason, obligate anaerobes are susceptible to oxygen.

7.4.4 pH

pH is the hydrogen ion concentration and can be known using the pH scale (with readings 0–14). pH of an environment can fall into any of the following three categories:

1. Acidic (high hydrogen ion concentration)
2. Neutral (neither high nor low hydrogen ion concentration)
3. Basic or alkaline (low hydrogen ion concentration)

Natural environments such as volcanic soil have very low pH and hence, the volcanic soil is acidic in nature. Even our stomach has an acidic pH (1–2). The majority of soils are slightly acidic to neutral in nature. Salt lakes and soap solutions have alkaline pHs. The microorganisms that can live in acidic pHs are called acidophiles, those that can grow in neutral pHs are known as neutrophiles, and the ones that can survive in alkaline pHs are termed as alkaliphiles. Table 7.4 gives us the optimum pHs of the above-mentioned pH-sensitive organisms.

Table 7.4 pH Requirements of Microorganisms

pH-Sensitive Microorganisms	Optimum pH
Acidophilic organisms	1–5.5
Neutrophilic organisms	5.5–8.5
Alkaliphilic organisms	8.5–11.5

7.4.5 Pressure

Whether microorganisms live on the land or under the water, they are always subjected to atmospheric pressure. Almost all microorganisms can live normally at the pressure of 1 atmosphere. If the pressure exceeds 1 atmosphere, it is lethal. However, few microorganisms can be seen surviving in the extremes of hydrostatic pressure under the deep sea. Such microbes that not only survive but also grow more rapidly at high pressures are called barophiles (e.g., protobacterium, *Colwellia*, and *Shewanella*). Some Archaebacteria such as *Pyrococcus* spp. and *Methanococcus jannaschii* are thermobarophiles. These bacteria are capable of growing in about 400–500 atmospheric pressure at high temperatures.

7.4.6 Radiation

Ionizing radiations such as X-rays and gamma rays and some of the electromagnetic radiations are very toxic for the growth of microorganisms. High levels of such radiations are lethal to microbes, whereas low levels of exposure can cause mutations, sometimes indirectly leading to the death of microorganisms. Due to the exposure to ionizing radiations, ring structures get destroyed, hydrogen bonds get broken down, double bonds get oxidized, and certain molecules are polymerized. Ultraviolet rays are harmful to everyone, and they are lethal to microorganisms. The most lethal UV radiation is with the wavelength of 260 nm. UV radiation mainly affects the bacterial DNA. Foremost, it damages the thymine dimers in the DNA. As the two adjacent thymines are linked to each other covalently in the DNA and help in many metabolic activities, when these are damaged, all functions of the DNA get inhibited. The microbial photosynthetic pigments such as chlorophyll, bacteriochlorophyll, cytochromes, and flavins absorb the radiations from the Sun and become excited. These excited pigments act as photosensitizers, and due to them, singlet oxygen is produced while transferring the energy to the oxygen molecule. This singlet oxygen is a very powerful oxidizing agent and is very reactive; due to it, the cells get destroyed rapidly. Phagocytes utilize this singlet oxygen as their major weapon to demolish the engulfed bacteria.

MULTIPLE CHOICE QUESTIONS

1. Which is the most promoting feature of pili present on pathogenic Gram-negative bacteria to infect the host?
   (a) Inhibit complement activation
   (b) Facilitate the adherence of bacteria
   (c) Transport nutrients into the cell
   (d) Transfer DNA between two bacterial cells
   Ans. b

2. A person gets urinary tract infection by faecal–oral route followed by colonization of the intestine and then the perineum. Which common causative organism could be responsible for such an infection?
   (a) *Mycobacterium* spp.
   (b) *Staphylococcus aureus*
   (c) *Escherichia coli*
   (d) *Pseudomonas aeruginosa*
   Ans. c

3. Which of the following characteristics does not hold good for bacterial growth curve?
   (a) Every bacterial growth curve consists of five distinct phases.
   (b) Graph is plotted against time versus number of microbes.
   (c) Bacterial growth curve shows the growth of microbial population under laboratory conditions.
   (d) Grow is measured by geometric progression.
   Ans. a

4. Generation time of *Escherichia coli* when cultured in glucose–salt medium is _____.
   (a) 200 hours   (b) 20 minutes
   (c) 20 hours   (d) 20 days
   Ans. b

5. The microorganisms that obtain their energy from a chemical source are called as _____.
   (a) Organotrophs   (b) Phototrophs
   (c) Chemotrophs   (d) Autotrophs
   Ans. c

6. In the bacterial cell, biological structures and content of nutrients are considered as _____.
   (a) Processing factors
   (b) Implicit factors

(c) Intrinsic factors
(d) None of the above
Ans. c

7. The microorganisms whose optimum temperature is 45°C are called _____.
   (a) Thermophiles
   (b) Mesophiles
   (c) Psychrophiles
   (d) None of the above
   Ans. a

8. In the lag phase, growth is _____.
   (a) Rapid and continuous
   (b) Very slow
   (c) Not observed
   (d) None of the above
   Ans. b

9. The apparatus used to grow microorganisms in continuous culture is _____.
   (a) Petroff-Hausser chamber
   (b) Coulter counter
   (c) Thermostat
   (d) Chemostat
   Ans. d

10. The microorganisms that have the capability to withstand a wide range of water activity are called _____.
    (a) Microaerophiles
    (b) Aerotolerants
    (c) Osmotolerants
    (d) Neutrophiles
    Ans. c

11. Binary fission is _____.
    (a) Locomotion in bacteria
    (b) Reproduction in bacteria
    (c) Nutrition in bacteria
    (d) Excretion in bacteria
    Ans. b

12. The microorganisms that are capable of growing in high hydrogen ion concentration are _____.
    (a) Acidophiles
    (b) Alkaliphiles
    (c) Neutrophiles
    (d) Thermophiles
    Ans. a

13. The phase of growth curve during which cells that are dividing are equal to the cells that are dying due to depletion of nutrients is the _____.
    (a) Lag phase
    (b) Log phase
    (c) Stationary phase
    (d) Death phase
    Ans. c

14. The growth rate in exponential phase is _____.
    (a) Increasing
    (b) Decreasing
    (c) Steady
    (d) None of the above
    Ans. a

15. What is generation time?
    (a) Time interval over number of generations
    (b) Time needed for the population to get doubled
    (c) Time needed for initial adjustment
    (d) Both (b) and (c).
    Ans. d

16. When a microbe is inoculated in a test tube containing culture broth, after incubation if turbidity is seen only at the surface, the organism is _____.
    (a) Anaerobe
    (b) Aerobe
    (c) Barophile
    (d) Halophile
    Ans. b

17. Facultative anaerobes are organisms that _____.
    (a) Are killed by oxygen
    (b) Use oxygen when present and can grow even without oxygen
    (c) Requires less oxygen
    (d) None of the above
    Ans. b

18. The microorganisms that can synthesize their own food are _____.
    (a) Autotrophs (b) Decomposers
    (c) Heterotrophs (d) Parasites
    Ans. a

19. The period during which the cell gets prepared for cell division is _____.
    (a) Death phase
    (b) Stationary phase
    (c) Log phase
    (d) Lag phase
    Ans. d

20. *Clostridium perfringens* are _____.
    (a) Obligate aerobes
    (b) Facultative aerobes
    (c) Obligate anaerobes
    (d) Facultative anaerobes
    Ans. c

SHORT NOTES

1. Wound colonization
2. Mechanisms used for the uptake of nutrients
3. Continuous culture and the apparatus used for it
4. Cardinal temperature and cardinal adaptations
5. Effects of pH on microbial growth
6. Role of macroelements and microelements on growth

ESSAYS

1. Describe the types of microbes based on their nutritional types.
2. Describe colonization in bacteria.
3. Describe the environmental factors that influence the growth of microbes.

8 Culture Media

CHAPTER OBJECTIVES

8.1 Introduction
8.2 History of Culture Media
8.3 Importance of Culture Media
8.4 Common Components Used in Culture Media
8.5 Classification of Culture Media
8.6 Preparation and Storage of Culture Media

8.1 INTRODUCTION

For several reasons, microorganisms are artificially cultured on culture media in laboratories. Culturing the organism from clinical samples helps the microbiologists to identify and diagnose infectious organisms and also enable chart a treatment regime. Culturing of microorganisms is one of the preliminary steps in identifying bacteria towards defining their morphology and classifying them.

Most bacteria are heterotrophic organisms; to grow such organisms artificially, a growth medium with nutrients and a growth environment similar to their natural growth conditions should be provided. Hence, culture medium can be defined as 'a special medium containing all the essential nutrients required for the growth of many different types of microorganisms used in the microbiological laboratory'.

8.2 HISTORY OF CULTURE MEDIA

Initially, microorganisms were cultured using simple broths. Louis Pasteur was the first to use such simple broths for culturing microbes; he used urine and meat extract as broths. Robert Koch used potato pieces as solid media to culture microbes. Later, Fannie Eilshemius, wife of Walther Hesse, who was an assistant to Robert Koch, suggested using agar as the solidifying agent. Before agar was introduced, gelatin was used as the solidifying agent. However, certain bacteria were capable of digesting gelatin, and at normal incubating temperatures (i.e., 35°C–37°C), gelatin existed in the liquid form. Agar melts at 98°C and does not possess nutritive value (i.e., it is inert). Hence, agar has been used as the solidifying agent till today.

8.3 IMPORTANCE OF CULTURE MEDIA

Culture media are used for the following purposes:

1. For the clinical diagnosis and identification of bacteria
2. For isolating pure cultures by growing the specimens on culture media
3. For developing vaccines by isolating antigens from special culture media used to grow specific bacteria
4. For genetic studies and studying manipulations of cells

8.4 COMMON COMPONENTS USED IN CULTURE MEDIA

In microbiological culture media, some ingredients are purely chemical compounds and some are complex materials such as digested animal or plant tissues and even extracts of animal or plant tissues. Few such ingredients are discussed below.

8.4.1 Agar

Agar is the solidifying agent used in culture media. The name 'agar' comes from the Malay word 'agar agar', which means jelly. Agar is also known as kanten or China grass. It is an unbranched polysaccharide extracted from the cell membranes of some species of red algae coming under the following genera: *Gelidium*, *Sphaerococcus*, and *Gracilaria*. Commercially, agar is obtained from *Gelidium amansii*. Agar is made up of two long-chain polysaccharides, consisting of 70% agarose and 30% agaropectin. Usually, 1%–3% agar is employed in solid agar media. Agar is preferable for culture media because of the following reasons:

1. It has no nutritive property.
2. It does not produce any growth-retarding or growth-promoting substances.
3. It is not hydrolysed by most bacteria.
4. It melts at 98°C and sets at 42°C.
5. It is sometimes a source of organic ions and calcium to the microbes.

8.4.2 Peptones

Peptones are proteins obtained by the digestion of plant or animal tissues. These are often present in heart muscle, soya bean meal, fibrin, and casein, and with the help of proteolytic enzymes such as pepsin, papain, or trypsin, the by-products including peptones, amino acids, and proteoses can be obtained. Along with these by-products, some inorganic salts such as magnesium, phosphates, and potassium are also produced.

8.4.3 Water

Even tap water can be used for preparing culture media, but it should not contain high amount of minerals. If the mineral quantity of tap water is high, demineralized water or distilled water should be used.

8.4.4 Extracts

Extracts from eukaryotic tissues such as brain, heart, liver, and beef muscle are used. For obtaining extracts from these tissues, the tissues have to be boiled till they are concentrated to form a paste or dried to make a powdered form. These extracts serve as the source of vitamins, coenzymes, and amino acids. Fastidious organisms can be grown if these extracts are provided in culture media.

Casein Extract

On hydrolysing milk using trypsin or HCl, casein hydrolysate can be obtained.

Meat Extract

Hot water extraction of lean beef is done, and this extract is subjected to evaporation to get concentrates of meat extract. Meat extract contains purines, creatinine, amino acids, proteases, peptones, albumoses, gelatin, and accessory growth factors.

Yeast Extract

Yeast extract can be obtained by washing the cells of baker's yeast. It contains various amino acids, inorganic salts, and growth factors.

Malt Extract

By extracting the soluble material from sprouted barley in water at about 55°C and then by evaporation, concentrates of malt extract are obtained. Malt extract contains maltose, glucose, dextrin, starch, a small amount of protein and protein by-products, and some growth factors.

8.4.5 Body Fluids

Defibrinated or complete serum, plasma, blood, or other body fluids are used to prepare culture media. The media with such ingredients are especially used for the isolation and cultivation of various pathogens. These body fluids are responsible for contributing certain growth factors and substances to culture media, which sometimes detoxify certain inhibitors.

Depending on the need of the bacteria that have to be cultured, selective agents, reducing agents, pH indicators, and buffer are also added to culture media in addition to a few of the above ingredients.

8.5 CLASSIFICATION OF CULTURE MEDIA

To grow different types of microbes in laboratory conditions, culture media are used. They should contain all the essential nutrients needed by the bacteria for their growth. As there are numerous varieties of microorganisms and each microbe has some unique property and requires specific nutrients, culture media are also of different types. Culture media can be classified based on their consistency, nutritional components, and functional use.

8.5.1 Classification Based on Consistency

Based on their consistency, culture media are classified as follows:

Liquid Media

In liquid media, microbes are visibly seen, generally in the form of turbidity, which is due to the uniform bacterial growth. Liquid media do not require solidifying agents. For culturing in liquid media, test tubes, bottles, or flasks are suitable. Most commonly, liquid media are used to culture large amount of bacteria. They are also helpful in diluting any inhibitors of bacterial growth. Liquid media are referred as 'broths'. Few examples of the microbes that grow in liquid media are stated below:

1. *Vibrio* and *Bacillus*, which possess fimbriae, grow on the surface of undisturbed broth as surface pellicle.
2. *Bacillus anthracis* grows on ghee-containing broth and produces stalactite.
3. In *Pneumococci*, initial turbidity gets clarified because of autolysis.
4. When chains of streptococci are cultured in a liquid medium, granular deposition can be seen in the clear medium.

Drawbacks of Liquid Media: The demerits of liquid media are as follows:

1. Properties of the bacteria cannot be known when cultured in liquid media.
2. As one or more types of bacteria might be present, isolation of a particular species cannot be done.
3. Only when the inoculum is low, liquid culture media are suitable.

Semi-solid Media

Semi-solid media contain 0.2%–0.5% agar. For demonstrating bacterial motility, semi-solid media are used. In the case of semi-solid media, consistency will be fairly soft. They are used for separating motile bacteria from non-motile bacteria. Some of the semi-solid media used for culturing are transport media such as Stuart and Amies medium, mannitol motility medium, and Hugh and Leifson medium, used as an oxidation–fermentation test medium.

Solid Media

Solid media contain 1%–3% agar by weight. For preparing a solid medium, 2% agar is added to a nutrient broth. This broth is then autoclaved at 50°C. The hot broth is then poured onto sterile Petri plates and allowed to solidify. To prepare solid media in a test tube, broth containing 2% agar is poured into the test tube and heated along with continuous stirring. The test tubes are then autoclaved and allowed to set.

Biphasic Media

Biphasic media refer to the culture system that comprises both the solid and the liquid medium in the same bottle. The inoculum is inoculated into the liquid medium. Whenever subcultures have to be obtained, the bottle with biphasic media is tilted so that the liquid flows over the solid medium. However, this makes it inconvenient to frequently open the bottle for subculturing.

Other Solidifying Agents

Apart from agar, serum and egg yolk are also used to solidify culture media. When the serum and egg yolk are subjected to heat, the media get solidified due to coagulation. As the egg yolk and serum are normally in the liquid phase, heating is necessary. Lowenstein–Jensen medium and Dorset egg medium are examples of media that contain egg, and Loeffler's serum slope is an example of serum-containing medium.

8.5.2 Classification Based on Nutritional Components

Based on their nutritional components, culture media are classified into the following types.

Simple Media

Simple media are those media that provide minimal requirements for the growth of non-fastidious bacteria. Non-fastidious bacteria are those that can grow with minimal requirements in the media (e.g., nutrient agar and peptone water).

Complex Media

Complex media are those whose exact chemical composition is not known. These media are made up of very complex substances (e.g., peptone, body fluids, tissue extracts, and infusions).

Synthetic Media

Synthetic media are those whose chemical composition is known both qualitatively and quantitatively. Hence, they are also known as 'chemically defined media'. These media are used to grow fastidious organisms. Fastidious bacteria grow only when extra nutrients are supplied through the media. Synthetic media are specially prepared for research purposes as they have great value in studying the nutritional requirements and also the various metabolic activities of microbes (e.g., Davis and Mingioli medium).

8.5.3 Classification Based on Functional Use

Based on their functional use, culture media are classified into the following types.

Basal Media

Basal media contain minimal requirements and support the growth of non-fastidious bacteria. This medium is also called minimal medium. Microbiologists and geneticists use basal media to grow 'wild type' organisms. These media are also used to select recombinants or exconjugants (e.g., peptone water, nutrient agar, and nutrient broth).

Enriched Media

The addition of extra nutrients to basal media gives us enriched media. The extra nutrients include blood, serum, and egg yolk. Enriched media support the growth of fastidious bacteria. Few examples of such media are stated in the next page.

Blood Agar: Blood agar is prepared by the addition of blood to the blood agar base. Blood cannot be sterilized; hence, it should be collected aseptically in a sterile container. Blood can be collected from rabbit, ox, horse, or sheep; among these animals, sheep is the most preferable. Blood should be added to the autoclaved blood agar base just above the solidifying point of the autoclaved medium. This mixture should be poured onto Petri plates and allowed to set. Blood agar media are helpful in demonstrating the haemolytic property of some organisms. Beta, alpha, and gamma haemolysis can be seen on blood agar. Complete lysis of RBC with clearing zone around the colony can be seen in the case of beta haemolysis. Partial lysis of RBC resulting in greenish discolouration around the colonies can be observed in alpha haemolysis. Non-haemolytic colonies indicate gamma haemolysis.

Chocolate Agar: Chocolate agar is also called lysed blood agar or heated blood agar. Here also, the blood agar base is autoclaved and then blood is poured into it while it is still hot. As the blood agar base is hot, the blood cells get lysed and their contents will be released into the medium. Due to this lysis, the colour of the medium changes to brown. Therefore, it is called chocolate agar. *Haemophilus* and *Neisseria* are grown on this medium.

All-Purpose Media

All-purpose media contain most of the nutrients including various growth factors and are therefore suitable for the growth of a wide variety of microorganisms. (e.g., penassay agar, brain–heart infusion agar, heart infusion agar, All Purpose medium with Tween (APT) agar, nutrient agar, and plate count agar).

Selective Media

Selective media are also called enrichment media. Selective media support the growth of organisms of interest. Selective media are designed in such a way that they inhibit the growth of other organisms that are unwanted commensals or contaminants, thereby helping the growth of the desired organism. This can be achieved by the addition of certain selective agents such as antibiotics, the alteration of pH, and the addition of some chemicals and dyes, which will inhibit the growth of unwanted bacteria. Using selective media, one can isolate the desired organism from a mixed culture.

Few examples of selective media are stated below:

1. **Thayer–Martin Agar:** It contains vancomycin, colistin, and nystatin and helps to recover *Neisseria gonorrhoeae*.
2. **Mannitol salt agar and salt milk agar:** These media contain 10% NaCl and help to recover *Staphylococcus aureus*.
3. **Potassium tellurite medium:** It contains 0.04% potassium tellurite and is used to recover *Corynebaterium diphtheriae*.
4. **MacConkey agar:** It contains bile salt and is used for recovering the members of Enterobacteriaceae.
5. **Cetrimide agar or pseudosel agar:** It contains cetrimide and helps to recover *Pseudomonas aeruginosa*.
6. **Crystal violet blood agar:** It contains 0.0002% crystal violet and is used to recover *Streptococcus pyogenes*.

7. **Lowenstein–Jensen medium:** It contains malachite green and is used to recover *Mycobacterium tuberculosis*.
8. **Wilson and Blair agar:** It contains brilliant green dye and is used to recover *Salmonella typhi*.
9. **Thiosulfate Citrate Bile Salts-sucrose (TCBS) agar and Monsur's tellurite taurocholate gelatin agar:** They help in isolating *Vibrio cholerae*, due to the elevated pH (8.5–5.6).

Differential Media

Differential media are used to distinguish between two or more colonies, which are recognized on the basis of their colony colour. This can be done by the incorporation of metabolic substrates and dyes, which are utilized by different bacteria resulting in a variety of end products forming different coloured colonies. Differential media are also known as indicator media.

Few examples of differential media are stated below:

1. When a mixture of bacteria is inoculated into the medium that has been incorporated with a particular carbohydrate, acid is produced by the organism that is capable of fermenting that carbohydrate. When a pH indicator is added to this medium, the bacterial colony that had fermented the carbohydrate appears in a different colour. MacConkey agar, Cystine Lactose Electrolyte Deficient (CLED) agar, TCBS agar, and Xylose-Lysine Deoxycholate (XLD) agar are the media used for this approach.
2. MacConkey agar contains lactose sugar, pH indicators such as neutral red, peptone, selective agents (especially bile salts), agar, and water. The members of Enterobacteriaceae can be identified by using this medium. When cultured on this medium, lactose-fermenting bacteria produce pink-coloured colonies and non-lactose-fermenting bacteria produce colourless colonies.
3. TCBS medium contains sucrose sugar. When cultured on this medium, *V. cholerae* ferments sucrose and produces yellow-coloured colonies.
4. Potassium Tellurite agar contains potassium tellurite. When *C. diphtheriae* is grown on this medium, it reduces potassium tellurite to metallic tellurium resulting in black-coloured colonies.
5. When *S. typhi* is grown on Wilson and Blair medium, it produces H_2S resulting in black-coloured colonies.
6. Bile esculin agar contains esculin. When *Enterococcus faecalis* is grown on this medium, it reduces esculin to esculetin resulting in black-coloured colonies.

Transport Media

Transport media are employed for the transportation of clinical specimens to the laboratory immediately after collection from the patient. This medium should be able to inhibit the overgrowth of unwanted bacteria. This medium should also prevent the desiccation of specimens. Some of the common transport media used in laboratories are given in the next page.

1. **Stuart and Amies transport medium:** It is semi-solid in consistency. Inhibitory factors can be neutralized by the addition of charcoal.

2. **Cary–Blair medium and Venkatraman Ramakrishnan medium:** These media are used for the transportation of faeces from suspected cholera patients.
3. **Sach's buffered glycerol saline:** It is used for the transportation of faeces from bacillary dysentery patients.
4. **Pike's medium:** It is used for the transportation of streptococci from throat samples.

Anaerobic Media

Anaerobic media is a special media used to culture anaerobes. They should possess low oxygen content, extra nutrients, and reduced oxidation–reduction potential. Nutrients such as haemin and vitamin K are added to anaerobic media. Physical reduction and chemical reduction are necessary for anaerobic media. The addition of 0.05% cysteine, 0.1% ascorbic acid, 0.1% thioglycollate, 1% glucose, or red hot iron filings will help in reducing the medium.

Few examples of anaerobic media are given below:

1. Robertson's cooked meat medium contains 2.5-cm column of bullock heart meat and 15 ml of nutrient broth. This medium is commonly used to culture *Clostridium* spp. Before using this medium, it must be boiled in the water bath in order to remove dissolved oxygen; after that, it should be sealed immediately with sterile liquid paraffin.
2. Thioglycollate broth contains glucose, cystine, yeast extract, casein hydrolysate, and sodium thioglycollate. Resazurin or methylene blue, which is an oxidation–reduction potential indicator, is added to the medium. Methylene blue becomes colourless under reduced condition.

Resuscitation Culture Media

Resuscitation culture medium is a special type of medium used for growing bacteria that are damaged or have lost their ability to reproduce because of harsh environmental conditions. This medium provides all the essential nutrients necessary for the growth of the organism and helps it to regain its metabolic activity.

Few examples of resuscitation culture media are stated below:

1. When a kind of bacteria that need histamine for their growth are subjected to histamine-deprived conditions, their growth gets inhibited. If the same bacteria are grown in the medium containing histamine, they grow again normally. Here, the histamine-containing medium acts as the resuscitation medium.
2. Most commonly used resuscitation medium is tryptic soya agar.

8.6 PREPARATION AND STORAGE OF CULTURE MEDIA

During the preparation and storage of culture media, one should take care of the following points:

1. Before autoclaving, proper care should be taken to adjust the pH of the medium.
2. Reconstitution as per manufacturer's recommendation must be done for commercially available dehydrated media.

3. Culture media must be sterilized using the autoclave.
4. Heat-labile components such as antibiotics, urea, blood, and serum should not be autoclaved.
5. Media that are not autoclaved must be filter sterilized.
6. Highly selective media should not be autoclaved (e.g., Wilson and Blair medium and TCBS agar).
7. The prepared media can be stored at 4°C–5°C in the refrigerator for 1–2 weeks.
8. The liquid media that are screw-capped or cotton-plugged can be stored at room temperature for about few weeks.

MULTIPLE CHOICE QUESTIONS

1. Chocolate agar is used for culturing _____.
   (a) Streptococci
   (b) Enterococci
   (c) *Haemophilus*
   (d) *Salmonella*

   Ans. c

2. Lowenstein–Jensen medium is used for growing which of the following organisms?
   (a) *Salmonella*
   (b) Enterobacteria
   (c) *Mycobacterium tuberculosis*
   (d) *Neisseria*

   Ans. c

3. Which one of the following is a basal medium?
   (a) Blood agar
   (b) Chocolate agar
   (c) Nutrient broth
   (d) MacConkey agar

   Ans. c

4. Sucrose fermenters produce which coloured colonies on TCBS agar?
   (a) Pink coloured
   (b) Brown coloured
   (c) Yellow coloured
   (d) Black coloured

   Ans. c

5. Which medium is used to culture *Salmonella*?
   (a) Selenite broth
   (b) Penassay agar
   (c) Pike's medium
   (d) Thioglycollate broth

   Ans. a

6. Which of the following is not a selective medium?
   (a) Cary–Blair medium
   (b) Crystal violet blood agar
   (c) Salt milk agar
   (d) Cetrimide agar

   Ans. a

7. Which medium is suitable for the transportation of faeces from suspected cholera patients?
   (a) Wilson and Blair medium
   (b) TCBS agar
   (c) Venkatraman Ramakrishnan medium
   (d) Potassium tellurite medium

   Ans. c

8. Resuscitation culture media are employed for _____.
   (a) The transportation of faeces from bacillary dysentery patients
   (b) The microorganisms that are damaged due to harsh environmental conditions
   (c) The growth of a wide variety of microorganisms
   (d) The microorganisms that are capable of fermenting carbohydrates

   Ans. b

9. Who suggested using agar as a solidifying agent?
   (a) Walther Hesse
   (b) Fannie Eilshemius
   (c) Louis Pasteur
   (d) Robert Koch

   Ans. b

10. The chemical composition of which of the following media is known both qualitatively and quantitatively?
    (a) Simple media
    (b) Complex media
    (c) Synthetic media
    (d) None of the above

    Ans. c

SHORT NOTES

1. Basal media
2. Biphasic media
3. Drawbacks of liquid media
4. Common components used in culture media
5. Anaerobic media
6. Suitability of agar in comparison with gelatin

ESSAYS

1. Explain the history and importance of culture media.
2. Explain the classification of culture media with examples.

9 Laboratory Methods for the Identification of Microorganisms

CHAPTER OBJECTIVES

9.1 Good Laboratory Practices
9.2 Five I's in a Microbiology Laboratory
9.3 Molecular Techniques and Typing
9.4 Biochemical Identification Techniques

9.1 GOOD LABORATORY PRACTICES (GLP)

The use of good laboratory practices is very essential in safeguarding the health and safety of the laboratory personnel. The bacteria that are cultured in microbiological laboratories are capable of causing diseases. Therefore, microbiologists working in laboratories are exposed to high risk. Adequate precautions have to be taken to overcome such risks. This can be achieved by following certain good practices in microbiological laboratories, which are as follows:

1. Sterilizing all media and equipment
2. Avoiding recontamination
3. Cleaning of working surface frequently using a proper disinfectant
4. Frequent hand washing using a disinfectant
5. Preventing the entry of any insects or flies inside the laboratory, to prevent vector-borne infections
6. Avoiding mouth pipetting
7. Avoiding eating, drinking, or smoking in the laboratory
8. Wearing protective clothing (at all times)
9. Disposing microbiological waste properly

9.2 FIVE I'S IN A MICROBIOLOGY LABORATORY

Five I's are techniques employed for characterizing, observing, growing, and situating microorganisms. Five I's are used in microbiological laboratories.

The five I's stand for the following:

1. Inoculation
2. Incubation
3. Isolation
4. Inspection
5. Identification

Specimen collection can be done through body fluids, foods, water, or soil. Once the specimen is collected, five I's are performed on it.

9.2.1 Inoculation

The sample collected for examination should be kept in a sterile container. From this container, a small amount is transferred aseptically to a suitable nutrient medium where the organism will grow. This process of aseptic transfer is known as inoculation.

9.2.2 Incubation

Incubation is providing suitable growth conditions to the inoculated sample. It could be maintained in the laboratory using an incubator. The incubator is an instrument employed to regulate appropriate growth conditions. As almost all bacteria grow well in the temperatures between 20°C and 40°C, the incubator provides the suitable temperature for the sample that is subjected to incubation. After incubation, colonies can be seen on solid media and cloudiness can be seen in liquid media. Both colonies and cloudiness indicate the growth of microorganisms.

9.2.3 Isolation

Isolation can be defined as the separation of one species from another and giving it suitable media and conditions for growth. A small amount has to be picked from only one colony on the incubated agar plate and inoculated into another sterile agar plate to obtain the colony formed by just one species of bacteria. Thus, the organism of interest can be obtained. Several isolating techniques are employed for the isolation of microorganisms, which are discussed below.

Streak Plate Method

A small amount of sample or a small droplet from the broth is spread over the Petri plate containing suitable culture media (Figure 9.1). A sterile loop is used to spread the sample all over.

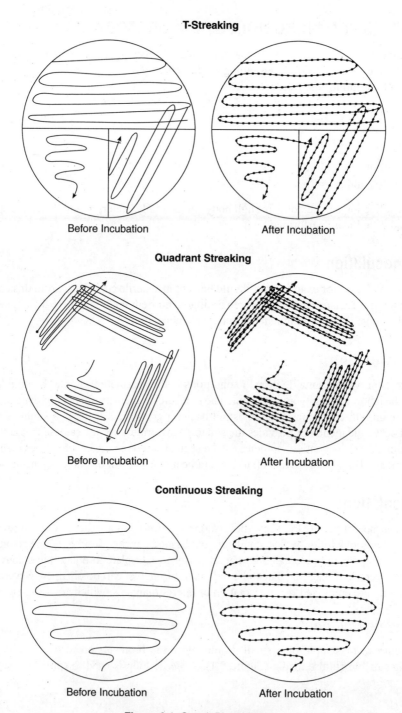

Figure 9.1 Streak Plate Method

Laboratory Methods for the Identification of Microorganisms | 85

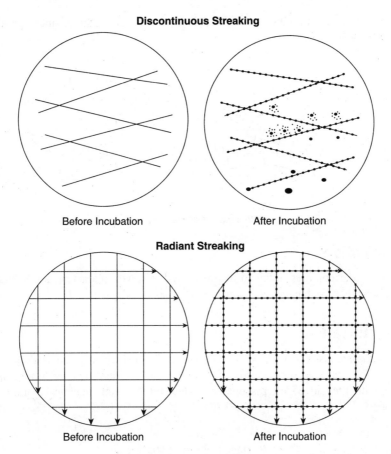

Figure 9.1 (*Continued*)

Spread Plate Method

This method is used for liquid medium. The sample should be diluted using distilled water. This diluted sample should be pipetted out onto the surface of an agar plate. And by using a sterile spreading tool, the medium should be constantly spread all around (Figure 9.2).

Pour Plate Method

A loop full of inoculum is taken from the sample and inoculated into a liquid agar tube. Then, from this tube, the sample is inoculated into another liquid agar tube; the sample is inoculated sequentially into a series of liquid agar tubes using a sterile loop in order to dilute the number of cells in each successive tube. Hence, the technique is also called loop

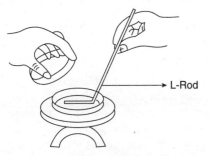

Figure 9.2 Spread Plate Method

dilution method. Then, the tubes are poured onto sterile Petri plates, which are allowed to solidify (Figure 9.3).

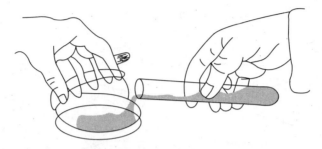

Figure 9.3 Pour Plate Method

9.2.4 Inspection

After isolation, the colonies obtained are inspected. Growth characteristics, which help to analyse the specimen contents, are viewed.

9.2.5 Identification

Microorganisms are usually verified to the species level. Identification is done using certain criteria; for example, metabolism of bacteria can be identified by biochemical tests, appearance with the aid of microscope, and the culture using genetic or immunological analysis.

9.3 MOLECULAR TECHNIQUES AND TYPING

The methods used for expressing microbiological characteristics are very important and improvement and refinement of the techniques are much necessary. Molecular biology is a boon as it provides a set of powerful tools that facilitate the microbiologists to identify even the smallest difference among species and also within individual strains. This technology can provide satisfactory results to the microbiologists by employing advanced techniques.

9.3.1 Species Identification

Species-level identification gives just a partial characterization of the isolate. However, species identification is valuable as it allows the laboratory to access the existing knowledge about the species. It also aids in finding the pathogen responsible for contamination. There are more than 8000 bacterial species, as the number of identified species of microorganisms has steadily increased. Microbiology laboratories identify many genera to their species level by some of the best key tests done in very less time as their routine work. In order to identify isolates to the species level, laboratories have developed and improved many simple test procedures, such as colony characteristics and cell morphology, Gram staining and several other staining techniques, physical and nutritional needs for growth, pathogenicity factors, and metabolic characteristics.

9.3.2 Typing

In a research laboratory, typing of the organism is important. Typing is differentiating between unlike strains of the same species. It is important to characterize a bacterial isolate ahead of species level in order to get its sub-species, strain, or even sub-strain because of the following reasons:

1. It helps to relate an individual case to an epidemic of infectious disease.
2. An association between an outbreak of food poisoning and a specific food carrier can be set up with the aid of typing.
3. Variations in pathogenicity, antibiotic resistance, and virulence of individual strains within a species can be studied.
4. Within the developed process, we can identify the source of contamination.
5. The microbial ecology of complex communities such as biofilms can be studied.
6. Microorganisms can be characterized, and the useful strains can be employed for industrial applications.

Typing has been performed for decades to type few isolates on the basis of serological testing. For example, Kauffmann–White scheme has been used for *Salmonella*. There are 2500 serovars for only two species of *Salmonella*. Typing can be done using the reaction between specific antigens and antibodies on the cell wall and flagella. To type some of the bacterial pathogens, phage typing has been used for several years. Some of the phenotypic typing techniques that mainly depend on expressed characteristics are as follows:

1. Bacteriocin typing
2. Protein typing
3. Biotyping (which is based on detailed biochemical characteristics)

These traditional typing systems have been elaborated into highly sophisticated systems with a wide range of applications. Some of them are stated below:

1. Matrix-assisted laser desorption–ionization time-of-flight (MALDI-TOF) mass spectrometer (which is a mass spectrometry technique used to carry out typing of extracted cell proteins)
2. Fatty acid methyl esters (FAME, which is the analysis of cellular fatty acid methyl esters by gas chromatography and the application of the results can be used to type bacteria)
3. Rapid and semi-automated *Salmonella* serotyping systems
4. Some systems consisting of profile databases and software for exact typing (which are developed into commercial typing)

9.3.3 Genotyping

Genotyping is a technique developed for the direct study of the microbial genome. Over the last two decades, the advent of PCR based arrays has led to a reliable & consistent methods for typing microorganisms. To make available definitive typing, one must perfectly sequence the complete genome of an isolate. Some of the more extensively used techniques are mentioned as follows:

1. Multilocus sequence typing (MLST)
   (a) MLST can sequence about 400–500 base pair fragments of DNA.
   (b) Sequencing is done at seven different conserved genes.
   (c) It permits only small deviations within a species to be identified.
   (d) It is a costly and time-consuming process.
   (e) At times, it can be extremely discriminatory if the genes are correctly chosen.
2. Pulsed-field gel electrophoresis (PFGE)
   (a) PFGE is a technique that aids the electrophoretic separation of less numbers of large DNA fragments.
   (b) These DNA fragments are generated using restriction enzymes leading to the availability of highly discriminatory genetic fingerprints.
   (c) This technique is widely used in typing the bacteria that can cause diseases in humans and even helps in the investigation of disease outbreaks.
   (d) At least three days are needed to get the results, and it is a pretty costly process.
   (e) The degree of discrimination depends on the restriction enzymes.
3. Ribotyping
   (a) Ribotyping is a technique that depends on the relative stability of 16S rRNA and 23S rRNA genes, which code for the ribosomal RNA.
   (b) DNA fragments, which are the resultant of restriction enzyme digestion, are separated by electrophoresis. This fingerprint is noticed by using fluorescent probes.
   (c) Ribotyping has been commercially developed into an automated system with dedicated databases.
   (d) This technique is quick and reproducible.
   (e) Ribotyping can be performed for a wide range of bacterial species.
   (f) Ribotyping is relatively costly as the equipment used are expensive.
4. Repetitive sequence-based PCR (Rep-PCR)
   (a) Rep-PCR technique depends on amplicons of varying lengths produced by the amplification of repetitive sequences, which can be separated by electrophoresis.
   (b) After the fragments get bound to an intercalating dye, they can be visualized under different intensities of fluorescence.
   (c) It also has been developed into a commercial typing system, and it gives rapid results by using dedicated software that help typing.
   (d) It is used for typing bacterial and fungal genomes, which include several repetitive DNA sequences, non-coding sequences, separating longer sequences, and single-copy sequences, with their arrangement differing among strains.
   (e) It is broadly utilized for typing human pathogens.

9.4 BIOCHEMICAL IDENTIFICATION TECHNIQUES

Standardized techniques are necessary to identify the exact bacteria, but initially, utilization of Gram staining or other staining processes will partially characterize the organism. Through partial characterization, one can recognize the genus of a particular bacterium or even the species, but further tests are very necessary to recognize the exact organism.

Note: Refer Chapter 10 to learn more on the identification of microorganisms. In addition, refer Chapter 5 to study the various physical methods of identification of microorganisms.

As metabolic activities involve many chemical reactions within the cell and as one cannot understand the underlying chemistry of tests without the knowledge of chemistry, some knowledge of chemistry is essential. Biochemical identification techniques involve biochemical testing that is based on metabolic activities of bacteria. These biochemical tests are arranged into groups, and the results are computerized. Testing formats are available to facilitate biochemical identification tests. All Purpose medium with Tween (APT) and Microbact are examples of standardized testing formats, with all tests supplied in a plastic strip or tray and reagents in handy dropper bottles.

9.4.1 Biochemical Tests

Enteric bacteria are the main contributory agents of intestinal infections. These organisms play a significant role in the contamination of food. Enteric bacteria are Gram-negative organisms. For the differentiation of the members of Enterobacteriaceae family, typically four tests are used. Enterobacteriaceae family comprises *Shigella, Proteus, Klebsiella, Salmonella, Escherichia, Enterobacter,* and so on.

Indole Test

Indole test is used to detect the production of tryptophanase enzyme in bacteria (Figure 9.4). If the bacteria produce tryptophanase, it will digest the amino acid tryptophan present in the media in which the bacteria are grown. After the digestion of the amino acid, indole, ammonia, and pyruvic acid are produced. Indole reacts with Kovac's reagent and produces a cherry red complex, indicating a positive result for indole test. The absence of red colour indicates a negative result as the bacteria do not produce the enzyme tryptophanase.

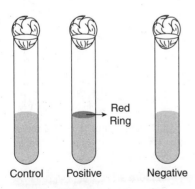

Figure 9.4 Indole Test (See page 348 for the colour image)

Methyl Red Test

Methyl red test is used to determine the capability of microorganisms to ferment glucose (Figure 9.5). From glucose metabolism, enteric bacteria produce pyruvic acid. Utilizing the mixed acid pathway, pyruvic acid is metabolized into other acidic products such as formic acid, lactic acid, and acetic acid. As a result, the pH of the media get decreased. Methyl red, a pH indicator, changes its colour with the change in pH. When the pH of the media is acidic, that is, pH value is below 4.4, the colour of the media changes to red due to the indicator. And when the pH of the media is above 7, that is, alkaline, the colour of the media turns yellow. The positive result of methyl red test is indicated by the formation of red colour (as pyruvic acid is metabolized into acidic end products making the media acidic).

Voges–Proskauer Test

Voges–Proskauer test also determines the capability of microorganisms to ferment glucose (Figure 9.6). However, in this test, the end product is different. Glucose metabolism leads to pyruvic acid formation; this further gets metabolized into neutral end products such as acetoin and 2,3-butanediol. The metabolism of pyruvic acid is carried out by the butylene glycol pathway. To detect the neutral end products, 40% KOH solution (Barritt's reagent A) and 5% alpha naphthol solution (Barritt's reagent B) are added to the media. In the presence of the catalyst alpha naphthol, oxygen, and Barritt's reagent A, acetoin gets oxidized into diacetyl. Diacetyl then reacts with peptone, and cherry red colour is imparted to the media.

Citrate Utilization Test

Citrate utilization test is employed to determine the capability of microorganisms to utilize citrate (Figure 9.7). Few bacteria have the ability to convert the salts of organic acids. In Krebs cycle, the chief metabolite is sodium citrate. For utilizing citrate, special membrane transporter and citrate lyase activity are needed. For few bacteria, citrate is the sole source of carbon. Citrate is converted into oxaloacetate by the enzyme citrate lyase. Oxaloacetate is further converted into pyruvate by the enzymatic action of oxaloacetate decarboxylase. Carbon dioxide gets released during this reaction, and it reacts with sodium and water to form sodium carbonate.

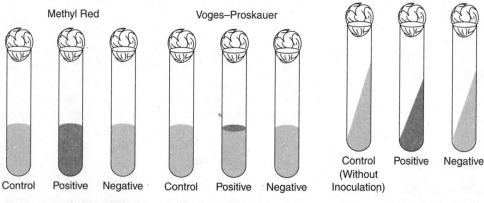

Figure 9.5 Methyl Red Test (See page 348 for the colour image)

Figure 9.6 Voges–Proskauer Test (See page 348 for the colour image)

Figure 9.7 Citrate Utilization Test (See page 349 for the colour image)

Triple Sugar Iron Agar Test

Triple sugar iron (TSI) agar test is used to distinguish different genera of Enterobacteriaceae family (Figure 9.8). This family includes all the Gram-negative bacilli. TSI agar test is done to distinguish the bacteria that ferment glucose from the bacteria that reside in intestines. This discrimination is done based on the capability of different bacteria to ferment carbohydrates. Carbohydrate fermentation pattern and hydrogen sulphide production help to detect various groups of intestinal organisms. Carbohydrate fermentation is identified by the release of gas, and when the indicator phenol red is added, the colour changes from red to yellow.

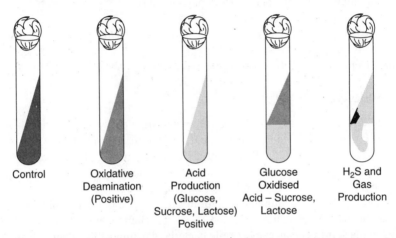

Figure 9.8 TSI Agar Test (See page 349 for the colour image)

TSI agar comprises three sugar (fermentors), that is, lactose in 1% concentrations, sucrose also in 1% concentrations, and glucose in 0.1% concentration. pH usually decreases during fermentation, so the acid–base indicator phenol red is added for detecting carbohydrate fermentation, which is indicated by the colour change in the medium from orange red to yellow. For the oxidative decarboxylation of peptone, alkaline products are produced. The pH of the medium rises due to this, which is indicated by the colour change from orange red to deep red. The production of hydrogen sulphide is indicated by the blackening in the butt of the test tube. The presence of hydrogen sulphide in the medium can be indicated by sodium thiosulphate and ferrous ammonium sulphate.

Urease Test

Urease test is used to detect the production of the enzyme urease in bacteria (Figure 9.9). Urease is a hydrolytic enzyme. Urea is produced during the decarboxylation process of amino acid arginine in the urea cycle. Urea is a nitrogen-containing compound highly soluble in water. Urea supplies nitrogen to the human body. If excess urea gets accumulated, it is removed from the body in urine. In order to break down urea into ammonia and carbon dioxide, certain bacteria produce urease during their metabolic activities. *Proteus*, *Providencia*, and *Morganella* are the few genera that can be identified using urease test.

Although many organisms produce urease enzyme, *Proteus* sp. tends to show more rapid action on substrate urea. One can rapidly distinguish the members of this genus from other non-lactose-fermenting enteric organisms using this test. Urease hits the nitrogen and carbon bond present in the amide compounds such as urea and forms the alkaline end product ammonia. When the pH indicator phenol red is added to the urea broth, the presence of an alkaline environment causes the solution to turn into deep pink colour. This indicates a positive result for the presence of urease; if deep pink colour does not develop, it indicates a negative result.

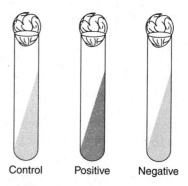

Figure 9.9 Urease Test (See page 349 for the colour image)

Sulphide Indole Motility Test

Sulphide indole motility (SIM) test is used to detect the motility in bacteria. In eukaryotes, locomotion is facilitated by flagella, pseudopods, or cilia. In prokaryotes, most bacteria show motility by means of propeller-like flagella or gliding movement. Some bacteria have the ability to move along certain chemical gradients, that is, chemotaxis. Almost all spiral and half of the rods are motile. However, almost all cocci are non-motile.

Sulphide indole motility medium is used for this test, which is a combination of differential media. Using this medium, sulphur reduction, indole production, and motility can be detected. As this medium is soft in consistency, motile bacteria, when inoculated, tend to swim all over creating cloudiness in the medium. And non-motile bacteria tend to grow only along the inoculated region as they lack the locomotory organ. Cloudiness indicates the positive result for this test.

Sulphide indole motility medium is also used to detect the production of hydrogen sulphide. The medium contains peptones and sodium thiosulphate as substrates and hydrogen sulphide as the indicator, that is, ferrous ammonium sulphate. When hydrogen sulphide is produced, it combines with ferrous ammonium sulphate and insoluble black precipitate is formed along the line of stab inoculation in the semi-solid medium. In addition, if the organism is motile, then the entire tube turns black.

Gelatin Hydrolysis Test

Gelatin serves as the solidifying agent in food; it is derived from the animal protein collagen. Earlier, Robert Koch used gelatin as the solidifying agent but some microorganisms hydrolysed gelatin and so it was replaced by agar. Some microorganisms are capable of producing the enzyme gelatinase, which hydrolyses gelatin into amino acids. After incubation at room temperature for one week, the media should be checked for gelatinase activity. The tube that is incubated should be placed on ice for a few minutes. If the medium does not solidify, it is the positive result for this test.

Nitrate Reduction Broth

Depending on the capability of bacteria to reduce the nitrate provided in the nutrient medium into nitrite or nitrogenous gases, bacteria might be grouped diversely. In some bacteria, anaerobic respiration is coupled with the reduction of nitrate. Nitrate is present in the broth; when organisms are inoculated into it, they reduce nitrate into nitrite and this nitrite in turn gets reduced into nitrous oxide, nitric oxide, or nitrogen. When the nitrate reacts with sulphanilic acid (reagent A), a red-coloured nitrite–sulphanilic acid complex is formed (Figure 9.10). When this reacts with α-naphthylamine (reagent B), a red precipitate of Prontosil is obtained. The reduction of nitrate into nitrite is favoured by a catalyst, that is, zinc powder.

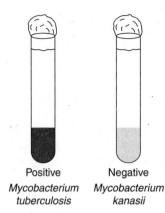

Positive *Mycobacterium tuberculosis* Negative *Mycobacterium kanasii*

Figure 9.10 Nitrate Reduction Broth (See page 350 for the colour image)

Catalase Test

Anaerobes are incapable of producing the enzymes such as catalase, peroxidase, and superoxide dismutase. Due to this incapability, they cannot survive in the presence of oxygen. These enzymes are responsible for degrading the lethal hydrogen peroxide. As these enzymes are not produced in anaerobes, oxygen is also lethal to them. Hydrogen peroxide is usually made up of a tetramer surrounded by four polypeptide chains (about 500 amino acids long) and an iron group. Catalase enzyme degrades it very rapidly. In aerobes and facultative anaerobes, catalase is produced in most cytochromes. Only one molecule of catalase can degrade hydrogen peroxide into water and oxygen in a second. Catalase production and activity can be visualized by using a tryptic soy agar slant culture. To this slant, hydrogen peroxide should be added and incubated for about 18–24 hours. The slant will contain water and air bubbles if the hydrolysis of hydrogen peroxide has occurred, indicating a positive result. If there is no change in the slant, it indicates a negative result.

Coagulase Test

Coagulase test is done to detect the production of exoenzyme in microorganisms. These enzymes are responsible for the clotting of blood plasma. *Staphylococcus aureus* is the only significant human pathogen that produces coagulase enzyme. When *S. aureus* is inoculated into rabbit plasma and incubated for a specific period of time, formation of clots can be seen, indicating the positive result of this test. This also indicates the virulence factor of *S. aureus* strain. If no clots are seen in the plasma, then it indicates the negative result for coagulase test.

Oxidase Test

Oxidase test is an important test performed on all Gram-negative bacteria for quick identification. Certain bacteria are capable of producing indophenol blue from the oxidation of dimethyl-p-phenylenediamine and α-naphthol; oxidase test depends upon this oxidizing capability of microorganisms. All staphylococci are oxidase negative. *Pseudomonas aeruginosa* is an oxidase-positive organism. In the presence of the enzyme cytochrome oxidase, N,N-dimethyl-p-phenylenediamine oxalate and α-naphthol react with indophenol blue.

Starch Hydrolysis Test

Amylase enzymes are capable of digesting the glycosidic linkages in starch. These enzymes are present in almost all living organisms. The enzymes differ in requirement, activity, and specificity from species to species and even from tissue to tissue in similar organisms. α-Amylase acts upon larger polymers of starch at internal bonds and digests them into small glucose polymers. α-Amylase catalyses the hydrolysis of internal bonds of polysaccharides, resulting in the formation of a mixture of maltose and glucose. The enzyme amyloglucosidase digests the shorter polymers in order to obtain single glucose sugar. Bacterial α-amylase is used in industries and thus is economically important.

Starch agar is a simple nutritional medium containing starch. The organism of interest should be inoculated into the medium in the Petri plate and incubated. Then, iodine solution must be added to the incubated plate. If no colour change is observed, it indicates a positive result. As bacterial growth proceeds, the organism starts to hydrolyse the starch, clearing around. If the colour changes to blue, black, or purple, it indicates a negative result. The colour depends on the concentration of iodine used by the organisms.

Lipid Hydrolysis Test

Lipid hydrolysis test is employed for the identification and enumeration of lipolytic microorganisms in food and other materials. The medium used for testing such bacteria is tributyrin agar.

The growth of microorganisms on different nutrient media is given below:

1. Selective media allows the growth of only few types of organisms, and they do not permit other organisms to grow on them (e.g., mannitol salt agar, Hektoen enteric agar, and phenylethyl alcohol agar).
2. Differential media are used for differentiating closely related groups as well as organisms. The organisms tend to adapt some special characteristic changes or growth patterns depending upon the presence of suitable dyes or certain chemicals in the growth media. This can be utilized for further steps of differentiation or identification [e.g., eosin methylene blue (EMB) agar and MacConkey (MCK) agar].
3. Enriched media are the media that are prepared by the addition of highly nutritious materials such as serum or yeast extract and blood to grow certain fastidious organisms (e.g., chocolate agar and blood agar).
4. Mannitol salt agar medium serves as both a differential and a selective medium. It is used for the isolation of pathogenic staphylococci from mixed cultures.
5. EMB agar also serves as both a selective and a differential medium. However, it is employed for detecting and isolating Gram-negative pathogens that reside inside the intestines.
6. MCK agar is used to grow Gram-negative bacteria. It again acts as both a selective and a differential medium. It helps to differentiate the bacteria that are capable of lactose fermentation.

MULTIPLE CHOICE QUESTIONS

1. Which one of the following sense organs can help microbiologists?
   (a) Eyes (b) Nose
   (c) Tongue (d) Skin
   Ans. a

2. Which is the most common problem faced by microbiologists while studying microorganisms?
   (a) Culture is contaminated.
   (b) Microbes are invisible.
   (c) Microbes are widely distributed.
   (d) Microbes must be grown artificially.
   Ans. a

3. Among the five I's, which 'I' promotes the growth of microorganisms?
   (a) Inoculation (b) Incubation
   (c) Isolation (d) Identification
   Ans. c

4. The most commonly used isolating technique is _____.
   (a) Streak plate method
   (b) Loop dilution method
   (c) Pour plate method
   (d) Spread plate method
   Ans. a

5. Semi-solid media are used for _____.
   (a) Isolating discrete colonies
   (b) Subculturing microbes
   (c) Determining motility
   (d) Obtaining growth all over the tube
   Ans. c

6. For the preparation of solid media, which component is used as the solidifying agent?
   (a) Methylene blue (b) Gelatin
   (c) Agar (d) Indole
   Ans. c

7. Which media is used for culturing fastidious microorganisms?
   (a) Anaerobic growth media
   (b) Mannitol salt agar
   (c) MacConkey agar
   (d) Enriched media
   Ans. d

8. When the culture media contains any unwanted unknown organism, the culture is called _____.
   (a) Mixed (b) Pure
   (c) Contaminated (d) Auxenic
   Ans. c

9. APT system used for the identification of microorganisms is a _____.
   (a) Microbiological test
   (b) Immunological test
   (c) Biochemical test
   (d) Molecular test
   Ans. c

10. The growth of Gram-negative bacteria is inhibited in _____.
    (a) MacConkey agar
    (b) Hektoen enteric agar
    (c) Eosin methylene blue agar
    (d) None of the above
    Ans. a

SHORT NOTES

1. Five I's
2. Genotyping
3. Tests included in included in Indole, Methyl Red, Voge–Proskauer, Citrate (IMViC) series of reactions

ESSAYS

1. Describe the molecular techniques used for microbial identification with appropriate examples.
2. Describe the biochemical techniques used for the identification of microorganisms with suitable examples.

10 Staining Techniques

CHAPTER OBJECTIVES

10.1 Definition and Protocol
10.2 Simple Staining
10.3 Differential Staining

10.1 DEFINITION AND PROTOCOL

Cellular staining techniques are employed for improving the visibility of cell components and enhancing the contrast in observation under the microscope. Staining can be done using special dyes to observe clear details of cell components, such as the cell wall or the nucleus. The majority of stains are employed on fixed cells or non-living cells. Only few stains can be used on living cells. Some stains can stain both living and non-living cells. Staining techniques help identifying metabolic processes and also determining the viability of cells (to distinguish dead from live cells).

The steps involved in the staining procedure are as follows (Figure 10.1):

1. **First step:** Cells are treated with a mild surfactant to allow larger dye molecules to enter inside the cell by dissolving the cell membrane.
2. **Second step:** Cells are fixed to preserve cell or tissue morphology. This process involves a chemical fixative, which creates chemical bonds between proteins to increase their rigidity, and is known as fixation. The commonly used fixatives are formaldehyde, ethanol, methanol, and picric acid.
3. **Third step:** This process is called mounting. Here, the fixed sample is mounted on a glass slide for observation and analysed under the

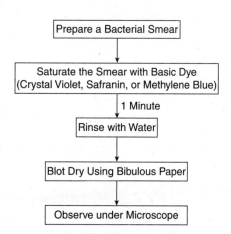

Figure 10.1 General Flowchart for Cell Staining

microscope. Using sterile techniques, cells may be directly grown on a slide, or free cells can be placed on a sterile slide. If tissues from multicellular organisms, such as plants or animals, are to be mounted, then thin sections of materials have to be taken.

4. **Fourth step:** It involves the application of the stain to the mounted specimen, for colouring cells, tissues, cell components, or metabolic processes. The specimen can be immersed in the dye solution before or after fixation or mounting based on the analysis techniques used for different organisms. Then, the specimen is rinsed and observed under the microscope. For some dyes, mordants are used because the dyes are easily washed off from the specimen. Mordant is a chemical compound that reacts with the stain to form an insoluble coloured precipitate.

10.1.1 Uses of Stains

The uses of stains are mentioned below:

1. Stains make the examination (i.e., viewing under the microscope) of cells very easy.
2. Differentiation of organisms can be achieved by staining.
3. Viable staining techniques enable one to distinguish between live and dead bacterial cells in a given sample.
4. Staining can provide details on structure, cell composition, and metabolic processes.
5. Staining also provides data on the morphology of the cell (such as shape, size, and identification), but only for certain species.

10.1.2 Some Commonly Used Stains

Some of the commonly used stains and their applications are mentioned below:

1. Bismarck Brown
   (a) It colours acid mucins, which are a kind of protein.
   (b) It stains the specimen in yellow colour.
   (c) It is used for live cells.
2. Carmine
   (a) It is used to stain glycogen or animal starch.
   (b) It stains the specimen in red colour.
3. Coomassie blue
   (a) It stains proteins in a brilliant blue colour.
   (b) It is more often used in gel electrophoresis.
4. Crystal violet
   (a) It is used in Gram staining.
5. 4',6 - Diamidino-Phenylindole, Dihydrochloride (DAPI)
   (a) It is a fluorescent nuclear stain that is excited by UV light, showing blue fluorescence when bound to DNA.
   (b) It can be used on fixed cells.

6. Eosin
   (a) It is a counterstain to haematoxylin.
   (b) It is used to stain red blood cells, cell membrane, extracellular structures, and cytoplasmic material.
   (c) It stains the specimen in red or pink colour.
7. Ethidium bromide
   (a) It is used to identify cells in the final stages of apoptosis.
   (b) It is a DNA-intercalating agent.
   (c) It stains cells in fluorescent red to orange colour.
8. Fuchsin
   (a) It is used to stain collagen, smooth muscles, and mitochondria.
9. Haematoxylin
   (a) It is a nuclear stain.
   (b) When used along with a mordant, it stains nuclei in blue-violet or brown colour.
10. Hoechst stains
    (a) It is a combination of Hoechst 33258 and Hoechst 33342 fluorescent stains.
    (b) It is used to stain DNA in living cells.
11. Iodine
    (a) It is most commonly used as an indicator for starch. The presence of starch is indicated by the solution turning into dark blue colour.
12. Malachite green
    (a) It is a counterstain to safranin.
    (b) It is used to stain spores.
13. Methylene blue
    (a) It is used to stain animal cells.
    (b) It stains nuclei in blue, making them more visible.
14. Neutral red or toluylene red
    (a) It is used to stain living cells.
    (b) It stains nuclei in red colour.
15. Nile blue
    (a) It stains nuclei in red colour.
16. Nile red–Nile blue oxazone
    (a) It is prepared by boiling Nile blue along with sulphuric acid.
    (b) It is a mixture of Nile red and Nile blue.
    (c) It is used on living cells.
    (d) It stains intracellular lipid globules in red colour.
17. Osmium tetroxide
    (a) It stains lipids in black colour.
    (b) It is used in optical microscopy.
18. Rhodamine
    (a) It is a fluorescent dye.
    (b) It is used as a protein-specific dye.

(c) After staining with this dye, the specimen can be observed with the help of fluorescence microscope.
19. Safranin
   (a) It is a nuclear stain.
   (b) It is used as a counterstain.
   (c) It stains collagen in yellow colour.

10.1.3 Staining Bacteria

To stain bacteria, fundamentally, basic stains are used. Basic stains are positively charged when they are administered to negatively charged molecules such as polysaccharides, proteins, and nucleic acids. Negatively charged dyes get bound to positively charged molecules. Some of the commonly used basic stains are crystal violet, safranin, and methylene blue. Basic stains can be used alone or in combination depending upon the organism.

10.2 SIMPLE STAINING

10.2.1 Staining of Bacteria from the Colony

The steps involved in the staining of bacteria from the colony are stated below:

1. A clean, grease free glass slide is taken and the slide is marked off centre.
2. Add one small drop of water to the centre of the slide (It should not overlap the pencil mark.) using a sterile loop.
3. Air-dry the drop of water.
4. Sterilize the loop and touch only one colony; then, transfer the bacteria into the water droplet.
5. Mix well and air-dry it.
6. Heat-fix the slide containing the bacterial smear using a flame (Bunsen burner flame).
7. Stain the slide with the dye of choice and remove the excess stain; blot dry and observe under the microscope.

10.2.2 Staining of Bacteria from the Broth

The steps involved in the staining of bacteria from the broth are stated below:

1. Take a clean glass slide and mark the slide off centre.
2. Mix the broth culture well to ensure uniform distribution.
3. Add a drop of broth culture on the marked area and air-dry.
4. Now, hold the slide by one edge and pass it gradually through Bunsen burner flame (Heat-fixing is done to just fix the bacteria to the slide.).
5. Stain the slide with the dye of choice and remove the excess stain; blot dry and observe under the microscope.

10.3 DIFFERENTIAL STAINING

Differential staining differentiates between the types of bacteria. Depending on the dissimilarities in the physical and chemical characters of microorganisms, this technique is helpful in characterizing microorganisms. This technique facilitates the discrimination of acid-fast and non-acid-fast cells. It is also helpful in providing an image of the intracellular constituents of microbial cells such as capsules, flagella, endospores, and metachromatic granules.

The basic principle lying behind differential staining is that the staining reagents will react differently, according to the different chemical and physical properties of the cell. More than one stain is required for this technique. In some cases, stains are used in combination and in some, stains are applied separately. Most significant differential staining techniques are acid-fast staining technique and Gram staining technique.

10.3.1 Gram Staining

This technique was discovered by the Danish scientist and physician Hans Christian Gram in 1884. Gram staining technique is one of the most frequently used staining procedures in microbiology.

Principle of Gram Staining

Gram staining technique is employed to improve the clarity of the microscopic image. It is done to highlight the structure of cells, tissues, biological specimens, and so on.

Importance of Gram Staining

Gram staining is important for the following reasons:

1. It is the preliminary step in the initial characterization and classification of bacteria.
2. It is a very significant key procedure in the recognition of bacteria using staining characteristics and allowing it to be observed under the microscope.
3. Most important, it is a vital technique used in the screening of infectious agents in clinical specimens.

Mechanism for Gram Staining

It is extensively the first step in the identification of an unidentified prokaryotic organism. So it has got great significance in taxonomy. This technique chiefly divides bacteria into two groups, which are as follows:

1. **Gram-positive bacteria:** Gram-positive bacteria are those bacteria that retain the primary dye crystal violet and appear deep violet in colour.
2. **Gram-negative bacteria:** Gram-negative bacteria are those bacteria that retain the counterstain safranin and appear red in colour.

Gram-positive bacteria have a thick multilayered cell wall, and about 50%–90% of it is made up of peptidoglycan. Whereas Gram-negative cell wall is thinner, and only 10% is made up of peptidoglycan. Gram-negative bacteria contain lipids in the additional outer membrane.

In aqueous solutions, crystal violet (CV) dissociates into CV⁺ and Cl⁻ ions. These ions tend to migrate all the way through the cell wall and cell membrane of both Gram-positive and Gram-negative bacteria. The negatively charged peptidoglycan of bacterial cells interacts with the CV⁺ ion of crystal violet. When the mordant iodine is applied, iodine (I^- or I_3^-) interacts with CV⁺ ion to form bigger complexes of crystal violet and iodine (CV–I) within the cell wall. Iodine actually traps the CV–I complex and prevents its removal. Hence, when decolourized with alcohol or acetone, this complex remains intact in the Gram-positive cell wall making the cell appear purple in colour.

In the case of Gram-negative bacteria, the CV–I complex gets washed off along with the outer membrane. Only the lipopolysaccharide layer will be left exposed after it has been washed off with alcohol or acetone solution. When these cells are subjected to a counterstain, typically positively charged safranin, Gram-negative bacteria appear pink in colour.

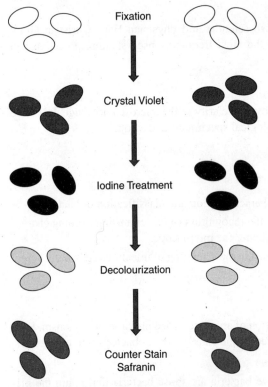

Figure 10.2a Flowchart Representation of Gram Staining (See page 350 for the colour image)

Stain Reaction

The four basic steps of Gram staining are as follows (Figure 10.2a and b):

Step 1 (Application of the Primary Stain): Crystal violet stain is applied to heat-fixed bacterial smear. This dye will react with the cell wall of both Gram-positive and Gram-negative bacteria.

Step 2 (Addition of Gram's Iodine Solution): Iodine is a mordant in this technique, and it acts as a trapping agent. A mordant is a substance that increases the affinity of the cell wall for a stain by binding the primary dye, resulting in the formation of an insoluble complex (CV–I), which is responsible for the purple colour of the cell.

Step 3 (Decolourization): Usually, 95% ethyl alcohol is used for decolourizing. The cells with the CV–I complex on the slide are subjected to alcohol. In Gram-positive bacteria, the CV–I complex formed within the cell wall does not get washed away, and the cells remain purple coloured. In Gram-negative bacteria, the alcohol makes the wall porous; so the lipopolysaccharide gets dissolved resulting in the removal of the outer thin peptidoglycan layer. Hence, the CV–I complex that lies within the peptidoglycan layer also gets washed away retaining the Gram-negative bacteria as colourless so that they could take the colour of the counterstain.

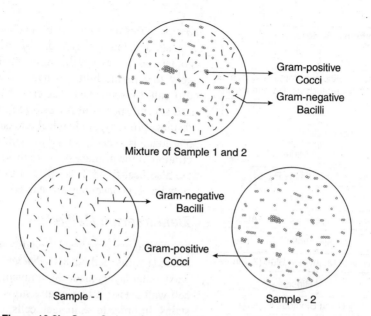

Figure 10.2b Gram Stained Organisms (See page 351 for the colour image)

Step 4 (Counterstaining): Gram-negative cells will be colourless after decolourization. When the negatively charged counterstain safranin is applied, Gram-negative bacteria take the counterstain and appear pink in colour. The crystal violet colour of Gram-positive bacteria masks the counterstain safranin, retaining their purple colour.

Limitations of Gram Staining Technique

The limitations of Gram staining technique are as follows:

1. Some bacteria cannot take up the dyes used in the Gram staining process (e.g., *Mycoplasma*, *Rickettsia*, and *Chlamydia*), and some get weakly stained (e.g., mycobacteria).
2. Gram staining techniques are restricted to be observed with the light microscope only.

10.3.2 Acid-Fast Staining

In microbiology, acid-fast technique is another most extensively used differential staining technique (Figure 10.3). Paul Ehrlich developed this technique in 1882. He was working on the aetiology of tuberculosis when he developed this procedure. Acid and alcohol are very strong decolourizing agents. Some bacteria resist such strong decolourizers, and hence, they are termed as acid-fast organisms. The technique used to stain such organisms is called acid-fast staining technique. This technique divides organisms into two groups, which are as follows:

1. Acid-fast
2. Non-acid-fast

By using this technique, one can diagnose tuberculosis and leprosy. Because of the high lipid content of the cell wall, these organisms have very low permeability to dyes, and it is very difficult to stain such bacteria. Mycobacteria and several species of *Nocardia* have such acid-fastness property. A chemical intensifier, that is, 5% aqueous phenol, is used to facilitate the penetration of the primary dye into these cells. Cells are also heat-fixed. It is difficult to decolourize such bacteria once they are stained.

Ziehl Neelsen Method

In clinical specimens, *Mycobacterium* spp. is stained by this method. *Mycobacterium* has thick waxy outer layers made up of mycolic acid. This cell wall acts as a barrier and stops the entry of stains. In order to stain such cells, concentrated staining solutions are required. The primary dye used in this technique is carbol fuchsin. This dye is used on the clinical specimen, which is then subjected to continuous heat. The heat will soften the waxy covering allowing the stain to enter into the cell. The cell cytoplasm gets stained in this method. These cells are then subjected to decolourization followed by the application of

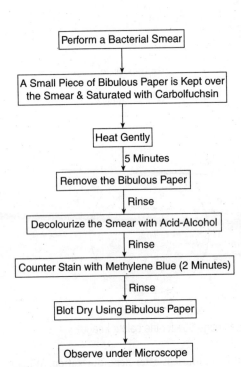

- → Acid-Fast Organisms Stain Fuchsia
- → Non-Acid-Fast Organisms Stain Blue

Figure 10.3 Flowchart Representation Acid-Fast

the counterstain methylene blue. Cells that are pink coloured are acid-fast bacteria, whereas those that are stained blue are non-acid-fast bacteria.

10.3.3 Endospore Staining

Some bacteria have the ability to produce endospores, which gives significant evidence in bacterial identification. To demonstrate the spore structure in bacteria and also free spores, endospore staining is done (Figure 10.4). Endospore-producing bacterial genera show positive results for this technique (e.g., *Bacillus* spp. and *Clostridium* spp.).

Schaeffer–Fulton Method

Due to unfavourable environmental conditions, some bacteria tend to produce endospores. These endospores are highly resistant, inactive, and remain in such an inactive state for years together. When conditions turn favourable, these endospores germinate into vegetative cells.

To determine these highly resistant spores of certain microorganisms within their vegetative cells, endospore stains are used. Schaeffer–Fulton method uses the primary dye malachite green and the counterstain safranin. In this method, the primary dye is applied to the bacterial smear and the specimen is heated for few minutes using the steam bath. The coat covering of endospores is very thick and resists the staining of cells. This hard coat softens during heating, making the primary dye to stick to the endospore. The slide is taken out of the steam bath and cooled. During cooling, the coat covering hardens again. By now, both the vegetative cell and the endospore inside would have taken the primary dye and will be green in colour. Decolourization is done using water; water washes away the primary dye colour from the vegetative cells but fails to remove it from the spores. When the specimen is subjected to counterstain safranin, all the vegetative cells become pink, and endospores retain their green colour.

Dorner Method

This is one more method used to stain endospores. In this staining technique, equal volumes of basic fuchsin and the broth culture of bacteria are taken. This mixture is heated to about 100°C for 10 minutes on a water bath. Because of the heating, the coat covering of endospores soften and allow the basic fuchsin dye inside. After this, cooling of the microbial culture is done to harden the coat covering. In order to differentiate between vegetative cells and endospores, a thin film of a loopful of boiled mixture is taken on a glass slide and nigrosin is added. After some time, nigrosin takes up all the stain from the vegetative cells, and they appear colourless. Endospores can be terminal or sub-terminal; they appear red coloured, and the background appears dark as it takes up all the dye from the vegetative cells.

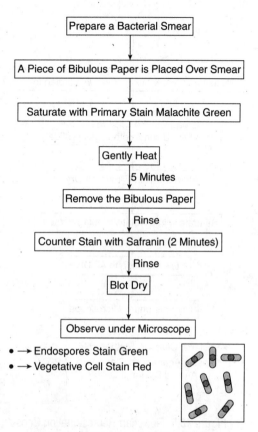

Figure 10.4 Flowchart Representation Endospore

10.3.4 Capsule Staining

Capsules are the gelatinous outer layers of bacterial cells, and these structures cannot retain the colour of the staining agents (Figure 10.5). Capsules can be visualized by means of two methods, which are discussed below.

Positive Capsule Staining

As capsules are water soluble in nature, it is very complicated to stain capsules with usual staining methods. The positive capsule staining method uses two reagents to stain the capsular material. The primary stain crystal violet is applied over a non-heat-fixed bacterial smear so that both the bacterial cells and capsular material take the colour of the primary stain. The ionic nature of the bacterial cell binds the crystal violet stain more strongly, whereas the non-ionic nature of the capsule makes the crystal violet stain just hold on. When the decolourizing agent copper sulphate is added over the bacterial smear, the loosely adhered crystal violet stain is washed off from the capsular material without removing the tightly bound crystal violet from the cell wall. The capsular material absorbs the light blue colour of copper sulphate in contrast to the purple colour of the bacterial cell.

Negative Capsule Staining

Bacterial capsules can be stained by another simple method using the negative staining technique. During staining, the bacterial smear (non-heat fixed) is subjected to an acidic stain, such as nigrosin, which will not penetrate into the cell. The stain gets deposited around the bacterial cells and creates a dark background. The bacteria remain unstained with a clear area around them, which is the capsule.

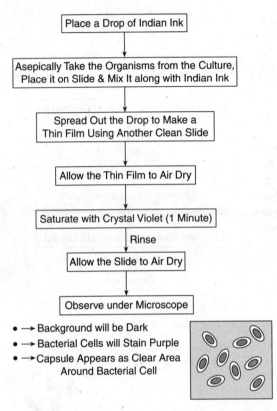

Figure 10.5 Flowchart Representation Capsule

10.3.5 Metachromatic Granule Staining (Albert's Staining)

The presence or absence of metachromatic granules can be determined by Albert's staining. Metachromatic granules are a characteristic feature of *Corynebacterium diphtheriae*. If the cells are V-shaped or rod-shaped with a slight curve and if they stain purple during Gram staining, then they are suspected to be *Corynebacterium diphtheriae*. In order to confirm the finding, this method is used. This technique uses two stains, namely, Albert stain A and Albert stain B. Albert stain A is the combination of toluidine blue, malachite green, glacial acetic acid, alcohol, and distilled water. Albert stain B is the combination of iodine, potassium iodide, and distilled water. After staining, the vegetative cells will be green in colour and the granules will appear bluish black in colour.

10.3.6 Flagella Staining

Bacteria are motile by means of flagella. Flagella are too thin to be seen using ordinary stains; hence, special stains and techniques are required to picture flagella. Specialized stains are generally found in microbiology laboratories to identify the presence or absence of flagella and indicate the nature of the bacterium (motile or non-motile).

Leifson Stain

This technique is used to stain flagella. When the bacterial culture is subjected to Leifson stain, the tannic acid component of the stain forms a colloidal precipitate. This colloidal precipitate is easily taken up by the flagella, and they are colourized. Although the alcohol concentration in this stain maintains the solubility of components, the concentration of tannic acid and dye succeeds in staining the flagella. When observed under the microscope, flagellar arrangements can be visualized effortlessly. The cell and flagella will be stained in red. However, to use this staining technique effectively, the age of the bacterial culture, the concentration and conditions of the staining solution and so on should be appropriate.

This staining method is used to validate the presence of bacteria in blood cultures when Gram staining results are complex to interpret or when the presence of bacteria is highly suspected but none are detected using light microscopy. Acridine orange binds to nucleic acid and stains them. It is also used for the detection of *Mycoplasma* (a cell wall-deficient bacterium).

Auramine–Rhodamine Technique

This technique is used to enhance the detection of mycobacteria. Auramine–rhodamine technique is nothing but the fluorochrome staining method. This method is used to detect the organism directly from the patient specimen. Initial characteristics of the organism can be obtained by growing the cells in the culture.

10.3.7 Calcofluor White Staining

It is commonly used to directly detect fungal elements and observe the subtle characteristics of fungi grown in the culture. The stain calcofluor white binds to the cell wall of fungi. This greatly enhances the visibility of fungal elements in tissue or other specimens.

MULTIPLE CHOICE QUESTIONS

1. Metachromatic granules can be stained with _____.
   (a) Safranin
   (b) Methylene blue
   (c) Crystal violet
   (d) Picric acid

   Ans. b

2. Bacterial spores are _____.
   (a) Weakly acid-fast
   (b) Strongly acid-fast
   (c) Alcohol-fast
   (d) Non-acid-fast

   Ans. a

3. Endospores can be stained with _____.
   (a) Safranin
   (b) Crystal violet
   (c) Methylene blue
   (d) Malachite green

   Ans. d

4. The order of stains in Gram staining procedure is _____.
   (a) Crystal violet, iodine solution, alcohol, safranin
   (b) Iodine solution, crystal violet, safranin, alcohol

(c) Alcohol, crystal violet, iodine solution, safranin
(d) All the above

Ans. a

5. The percentage of alcohol used in Gram staining is _____.
   (a) 75%
   (b) 90%
   (c) 60%
   (d) 25%

Ans. b

6. Gram-positive bacteria appear as _____.
   (a) Pink
   (b) Violet
   (c) Both (a) and (b)
   (d) None of the above

Ans. b

7. Gram-negative bacteria appear as _____.
   (a) Pink
   (b) Violet
   (c) Both (a) and (b)
   (d) None of the above

Ans. a

8. The action of alcohol during Gram staining is _____.
   (a) It allows colour
   (b) It adds colour
   (c) It decolourizes the cells
   (d) None of the above

Ans. c

9. If only one stain is used for staining the specimen, it is _____.
   (a) Simple staining
   (b) Negative staining
   (c) Differential staining
   (d) None of the above

Ans. a

10. If the portion other than the sample (specimen) is stained, then it is called _____.
    (a) Simple staining
    (b) Negative staining
    (c) Differential staining
    (d) None of the above

Ans. b

11. When more than one stain is used, such staining is called _____.
    (a) Simple staining
    (b) Negative staining
    (c) Differential staining
    (d) None of the above

Ans. c

12. Mycobacteria are stained by _____.
    (a) Gram staining
    (b) Simple staining
    (c) Both (a) and (b)
    (d) Ziehl Neelsen staining

Ans. d

13. Ziehl Neelsen stain is a _____.
    (a) Simple stain
    (b) Counterstain
    (c) Differential stain
    (d) None of the above

Ans. c

14. *Rickettsiae* are stained with _____.
    (a) Giemsa and Castaneda's stains
    (b) Macchiavello and Gimenez stains
    (c) Both (a) and (b)
    (d) Malachite green

Ans. c

15. All of the following are acid-fast bacteria except _____.
    (a) *Cryptosporidium*
    (b) *Mycoplasma*
    (c) *Mycobacterium*
    (d) *Nocardia*

Ans. a

16. Which one of the following staining methods is an example of negative staining?
    (a) Gram staining
    (b) Fontana's staining

(c) India ink preparation
(d) Ziehl Neelsen staining
          Ans. c

17. In negative staining, _____.
    (a) The structure to be demonstrated is stained.
    (b) The structure to be demonstrated is not stained.
    (c) The background is not stained.
    (d) The background and the structure are stained.
              Ans. b

18. The dye used in fluorescent microscopy is _____.
    (a) Congo red
    (b) Brilliant blue
    (c) Eosin
    (d) Auramine
              Ans. d

19. Dark ground microscopy is used to see _____.
    (a) Refractile organisms
    (b) Flagella
    (c) Capsules
    (d) Fimbriae
              Ans. c

20. Bacteria can be seen using _____.
    (a) Microscopy
    (b) Stained preparations
    (c) Both (a) and (b)
    (d) None of the above
              Ans. c

SHORT NOTES

1. Steps involved in simple staining
2. Commonly used counterstain
3. Stain used for bacterial flagella
4. Limitations of Gram staining
5. Difference between simple staining and differential staining

ESSAYS

1. Explain in detail with examples the various staining techniques adopted in microbiology.
2. Explain in detail the mechanism of Gram staining with a neat labelled diagram.
3. Give a brief account on acid-fast stain and endospore stain.

Unit 3

INFECTION CONTROL

Chapter 11	Sources, Portals, and Transmission of Infections	113
Chapter 12	Asepsis, Disinfection, and Sterilization—Types and Methods	119
Chapter 13	Chemotherapy and Antibiotics	137
Chapter 14	Standard Safety Measures and Biomedical Waste Management	152
Chapter 15	Hospital Acquired Infection and Hospital Infection Control Programme	163

11 Sources, Portals, and Transmission of Infections

CHAPTER OBJECTIVES

11.1 Introduction
11.2 Classification of Infections
11.3 Sources of Infections
11.4 Portals of Entry and Exit
11.5 Modes of Transmission of Infections

11.1 INTRODUCTION

Infection can be stated as the interaction that occurs between the host and the infecting microorganism. Microorganisms can be classified as saprophytes and parasites depending on their relationship with their hosts. The term 'saprophyte' is derived from the Greek words *sapros*, meaning decayed, and *phyton*, meaning plant. Saprophytes are free-living microorganisms that exist on dead and decaying organic matter, soil, and water. They play an important role in the degradation of organic materials but are incapable of multiplying in living tissues. Unlike saprophytes, parasites are microbes that can multiply inside hosts. Parasites can be of two types: pathogens and commensals. Pathogens are disease-causing microbes, that is, they are capable of causing diseases in the host. Commensals are microbes that live completely within the host but do not cause any harm to it. Commensals are largely found in the body as normal bacterial flora.

11.2 CLASSIFICATION OF INFECTIONS

Infections can be classified in several ways and one such classification is given below:

1. **Primary infection:** When a microorganism causes an initial infection in a host, it is called as 'primary infection'. Subsequent or recurrent infections caused by the same organism in the host are known as 'reinfections'.
2. **Secondary infection:** A secondary infection occurs when a new microbe, say a parasite, causes an infection in a host whose immunity (resistance to infection) is lowered by a pre-existing infectious disease.

3. **Focal infection:** A condition where infection at localized sites (such as tonsils and appendix) produces generalized effects is known as focal infection. It is also known as focal sepsis.
4. **Cross infection:** A cross infection occurs when a new infection is set up in a patient already suffering a disease from another host. Cross infections occurring in hospitals are called nosocomial infections.
5. **Iatrogenic infection:** Physician-induced infections or infections resulting from investigative or therapeutic procedures are known as iatrogenic infections.

Based on whether the source of infection is from the host's own body or is external, infections are classified into two types, namely, endogenous and exogenous.

Based on the clinical effects, infections are classified as follows:

1. **Inapparent infections:** These are infections in which clinical effects are not apparent. They are also known as subclinical infections.
2. **Atypical infections:** These are infections in which the characteristic clinical manifestations of a particular infectious disease are not present.
3. **Latent infections:** These are infections that occur when some microbes, following an infection, remain in the tissues in a hidden form proliferating and producing clinical disease when the host resistance is lowered.

11.3 SOURCES OF INFECTIONS

11.3.1 Humans

Humans are the most common source of infections for themselves. The parasite may arise from a patient or a carrier. The term 'carrier' refers to a person harbouring the pathogenic microbe without suffering from any disease due to it. Different types of carriers are listed below.

1. **Healthy carriers:** Healthy carriers are those who harbour the pathogen but are not affected by it.
2. **Convalescent carriers:** Convalescent carriers are those who have recovered from the disease caused by the pathogen but continue to harbour the pathogen in the body.
3. **Contact carriers:** Contact carriers are those who have acquired the pathogen from a patient.
4. **Paradoxical carriers:** Paradoxical carriers are those who have acquired the pathogen from another carrier.

Based on the duration of carriage, carriers can be categorized into temporary and chronic carriers. Temporary carriers last for less than 6 months, whereas chronic carriers last for several years (in some cases even for the rest of the life).

11.3.2 Animals

Animals also serve as sources of human infections. Infections are caused in humans not only by diseased animals but also by animals that are asymptomatic. Such animals serve as reservoirs of human infection by maintaining the parasite in nature. They are called as reservoir hosts. Infections transmitted to humans from animals are commonly known as zoonoses. Zoonotic diseases are of several types, which are as follows:

1. Bacterial (e.g., plague from rats)
2. Viral (e.g., rabies from dogs)
3. Protozoal (e.g., toxoplasmosis from cats)
4. Helminthic (e.g., hydatid disease from dogs)
5. Fungal (e.g., zoophilic dermatophytes from cats and dogs)

11.3.3 Insects

Insects, mainly blood-sucking insects, also transmit pathogens to humans. Diseases caused by such insects are known as arthropod-borne diseases. Insects that transmit infections, such as mosquitoes, ticks, mites, lice, flies, and fleas, are called as vectors. Vectors are of two types, namely, mechanical vectors and biological vectors.

Mechanical vectors are those in which the transmission is mechanical, for instance, transmission of dysentery (typhoid bacilli) by domestic fly. Biological vectors are those in which the pathogen multiplies and undergoes a part of its developmental cycle in the body of the vector (e.g., during the transmission of malaria by *Anopheles* mosquito). Only after the pathogen has multiplied in the vector sufficiently and has undergone a part of its developmental cycle, these biological vectors transmit the infection. The duration between the time of entry of the pathogen into the vector and the time of the vector becoming infective is called as 'extrinsic incubation period'. Insects can also serve as 'reservoir hosts' like animals (e.g., ticks in relapsing fever).

11.3.4 Soil and Water

Soil and water are also among the predominant sources of infection. Few pathogens can survive in soil for even decades, such as spores of tetanus bacilli, fungi like *Histoplasma* and *Nocardia*, and some parasites like roundworm and hookworm.

Contamination of water with pathogens such as *Vibrio cholerae* and infective hepatitis virus or by the presence of aquatic vectors makes water a source of infections.

11.3.5 Food

Infections are transmitted to humans when they consume contaminated food or meat and other animal products with pre-existing infections (e.g., food poisoning by *Staphylococcus* sp.).

11.4 PORTALS OF ENTRY AND EXIT

The portal of entry to cause an infection differs with pathogen. In some cases, the portals of entry and exit will be the same, whereas in others they are different. For example, staphylococcal bacteria may escape from a person's respiratory tract and infect another person's skin lesion or the bacteria may escape from a person's skin lesion and contaminate the food, which when consumed by another person can cause food poisoning.

Some of the most common portals of entry are as follows:

1. **Respiratory tract:** If the pathogen enters through the respiratory tract, the modes of transmission of infection may be airborne droplets and fomites. Diseases caused include measles and common cold.
2. **Alimentary tract:** The modes of transmission of infection include water, food, flies, and fomites. Diseases caused include typhoid, shigellosis, and polio.
3. **Skin and genital membranes:** Transmission occurs through direct contact, fomites, and sexual intercourse. Diseases caused include syphilis and gonorrhoea.
4. **Ocular mucous:** The modes of transmission are flies and fomites. The disease commonly caused is trachoma.
5. **Injured skin or skin lesions:** Infection is transmitted through blood-sucking arthropod vectors. Diseases caused include yellow fever and malaria.

The exit or escape of a pathogen from a reservoir (human or animal) may be through one or many portals. Below mentioned are some of the most common portals of exit.

1. **Respiratory tract:** It is the most important portal of exit and is also very difficult to control. Some of the diseases that use this portal include common cold, influenza, tuberculosis, measles, mumps, rubella, pertussis, and pneumococcal disease.
2. **Genitourinary tract:** It is involved in transmitting syphilis, gonorrhoea, Chlamydia, and HIV. Parasitic diseases such as schistosomiasis and bacterial infections like leptospirosis also use this route.
3. **Alimentary tract:** Enteric diseases generally use this portal. Some of the diseases that use this portal include hepatitis (mainly hepatitis type A), salmonellosis, shigellosis, and cholera.
4. **Skin:** It acts as a portal of exit through superficial lesions and percutaneous penetration (e.g., smallpox, chicken pox, syphilis, chancroid, and impetigo).
5. **Transplacental tract:** Exit from mother to foetus occurs in the transmission of rubella, HIV, syphilis, and cytomegalovirus.

11.5 MODES OF TRANSMISSION OF INFECTIONS

The mode of transmission is important to bridge the gap between the portal of exit of the pathogen from a reservoir and the portal of entry into the host. Following are the modes of transmission of infections.

1. **Contact:** Infections may be transmitted through direct or indirect contact. Diseases transmitted through direct contact include sexually transmitted diseases such as syphilis and gonorrhoea. Such diseases are said to be contagious diseases. Indirect contact occurs through inanimate objects such as clothing and toys, which are contaminated by the pathogen from one person and thus transmit the infection to another person. Diseases such as diphtheria and trachoma can be caused through indirect contact.

2. **Inhalation:** Respiratory secretions through coughing and sneezing shed pathogens that spread to the environment and cause respiratory infections through inhalation. Some infections transmitted by inhalation of pathogens are influenza and tuberculosis. Pathogens resistant to drying remain viable in the dust and serve as potential sources of infection. Large droplets of secretions fall on the ground and dry there, whereas small droplets evaporate quickly and form minute particles or droplet nuclei, which remain suspended in the air for long periods.
3. **Ingestion:** Ingestion of contaminated food and drinks causes intestinal infections. Infections transmitted by ingestion may be waterborne, food borne, and hand borne. Some infections transmitted by ingestion include cholera, food poisoning, and dysentery.
4. **Inoculation:** Microbes can also be directly inoculated into the tissues of the host (e.g., tetanus spores implanted in deep wounds and rabies virus deposited subcutaneously by dog bites). The use of unsterile syringes and surgical equipment may also cause infections via inoculation. Some examples of such infections are hepatitis B and HIV.
5. **Insects:** They act as vectors for the transmission of infectious diseases.
6. **Congenital:** Some microbes are capable of crossing the placental barrier and infect the foetus in the uterus. This type of transmission is known as vertical transmission and may lead to abortion, miscarriage, and stillbirth. In some cases, the foetus survives but may develop serious manifestations such as congenital syphilis. Some infections such as congenital rubella may interfere with the organogenesis (organ development) of the foetus and lead to congenital malformation. Such infections are considered as teratogenic infections.
7. **Iatrogenic and laboratory transmission:** Infections may also be transmitted by the administration of injections, lumbar puncture, and catheterization due to the lack of care in asepsis. Other modern methods of treatment like dialysis, organ transplantation, and transfusion also have increased the possibilities of iatrogenic infections.

MULTIPLE CHOICE QUESTIONS

1. Infections at localized sites are referred to as _____.
   (a) Focal infections
   (b) Primary infections
   (c) Secondary infections
   (d) Tertiary infections

   Ans. a

2. Hospital-acquired infections are referred to as _____.
   (a) Focal infections
   (b) Primary infections
   (c) Secondary infections
   (d) Nosocomial infections

   Ans. d

3. Which of the following is a mechanical vector?
   (a) Housefly   (b) Mosquito
   (c) Dog   (d) Cat

   Ans. a

4. Which of the following organisms is transmitted through soil?
   (a) Tetanus bacilli
   (b) *Histoplasma*
   (c) *Nocardia*
   (d) All the above

   Ans. d

5. Which of the following is a waterborne infection?

6. Respiratory tract infections are caused by _____.
   (a) Inhalation
   (b) Ingestion
   (c) Congenital transmission
   (d) Insects

   Ans. a

7. Which of the following is an example of transplacental transmission?
   (a) Rabies   (b) Cholera
   (c) Polio   (d) HIV

   Ans. d

8. Which of the following is an example of genital tract infection?
   (a) Cholera   (b) Malaria
   (c) Filaria   (d) Rabies

   Ans. a

   (a) HIV   (b) Syphilis
   (c) Gonorrhoea (d) All the above

   Ans. d

9. Which of the following is an example of food poisoning?
   (a) Salmonellosis   (b) Cholera
   (c) Tetanus   (d) Tuberculosis

   Ans. a

10. Diseases that are transmitted through animals are referred to as _____.
    (a) Zoonoses   (b) Protozoal
    (c) Fungal   (d) All the above

    Ans. a

SHORT NOTES

1. Classification of infections
2. Sources of infections
3. Modes of transmission of infections
4. Portals of entry of infections
5. Portals of exit of infections

ESSAYS

1. Explain in detail the various sources and modes of transmission of infections. Add a note on the classification of infections.
2. Describe the various portals of entry and exit of infections.

12 Asepsis, Disinfection, and Sterilization—Types and Methods

CHAPTER OBJECTIVES

12.1 Introduction
12.2 Asepsis
12.3 Sterilization and Disinfection
12.4 Types of Sterilization
12.5 Types of Disinfection

12.1 INTRODUCTION

Good laboratory practices (GLP) include certain rules and regulations that have to be followed in the microbiology laboratory to ensure the safety of technicians and the judicial use of laboratory equipments by maintaining proper record of the work planned, executed, and reported. All these details are obtained to maintain quality in the laboratory. Personal care and hygiene should be strictly maintained while handling cultures, and utmost care must be taken when working with pathogens. Pathogens are infectious agents that are responsible for causing infections and play a major role in contamination; hence, they have to be strictly eradicated after being handled. There are guidelines formulated for eradicating pathogenic microorganisms by following methods like asepsis, sterilization, and disinfection.

12.2 ASEPSIS

Asepsis is the process of complete removal of microorganisms to form a germ-free environment. This aseptic technique evolved in the 19th century. Earlier, antisepsis was followed for the eradication of pathogens as it was believed that infections were transmitted from a surface via contacts, and hence, they used chemical disinfectants on the surface for inhibiting the growth of microorganisms. But these techniques were not effective when compared with the aseptic technique, which actually minimizes the load of microorganisms.

12.2.1 Practice of Aseptic Techniques

Aseptic techniques should be practised before and after laboratory procedures to maintain a germ-free environment. The methods adopted are as follows:

1. Washing hands with proper antiseptics and using sterile gloves before beginning any procedure.
2. Using sterile aprons or scrubs and surgical masks for external protection.
3. Sterilizing all equipments by either physical or chemical treatment before any procedure to avoid contamination.
4. Sterilizing the work area before and after any procedure.
5. Following strict aseptic techniques while performing surgical procedures to avoid infections.
6. Documenting each and every procedure and maintaining the laboratory according to established guidelines to ensure good laboratory practices.

12.3 STERILIZATION AND DISINFECTION

Disinfection is the process of removal of pathogenic microorganisms by chemical methods. The chemicals used for disinfecting are called as disinfectants. These disinfectants have both narrow and broad spectrum of activity towards microorganisms. When the number of microorganisms decreases and they lose their viability but are not killed, such an effect is called as bacteriostasis; most of the broad-spectrum disinfectants belong to this category. On the other hand, the narrow-spectrum disinfectants kill pathogenic microorganisms and hence are bactericidal in nature.

Sterilization is a technique that eradicates microorganisms like bacteria, fungi, viruses, and parasites including spores that are the most resistant of all. Sterilization and disinfection (Figure 12.1), may be classified as:

1. Physical sterilization is the destruction of pathogens using heat, radiation or filtration. It is used for the sterilization of instruments and materials used for various procedures. Various media and reagents are also sterilized using this method.
2. Chemical disinfection is the removal of microorganisms using antimicrobial agents and is used for surface sterilization. The disinfection is carried out by using gaseous and liquid disinfectants.

12.4 TYPES OF STERILIZATION

The various types of sterilization are given in Figure 12.1.

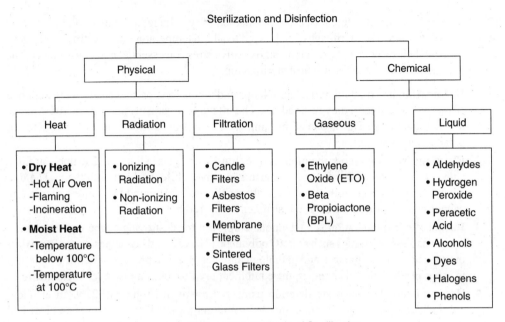

Figure 12.1 Methods of Sterilization

12.4.1 Physical Sterilization

Heat Sterilization

Heat sterilization is the most effective method of sterilization as pathogens are destroyed at high temperatures. A natural source of sterilization is sunlight, which actually eliminates pathogens by ultraviolet rays.

Factors influencing heat sterilization

1. Time of exposure (where the temperature and time play an important role; higher the temperature, lesser the time).
2. Type of heat applied (high temperature or low temperature).
3. Load of microbial contamination.
4. Type of microorganism (Spores are very resistant and hence, they require high temperature and more time for destruction.).
5. Type of material to be sterilized (e.g., surgical instruments and laboratory wares require higher temperatures, whereas media and reagents require lower temperatures).

The two types of heat sterilization employed in laboratories depending on the tolerance of the material used are as follows:

1. Dry heat sterilization
2. Moist heat sterilization

Dry heat sterilization: It is the most effective method of sterilizing instruments and laboratory wares. The materials to be sterilized are subjected to a high temperature that destroys pathogens by denaturing their protein and producing an oxidative stress on essential cell constituents. The following methods are used for dry heat sterilization.

1. **Red heat:** In this method, inoculation loops and wires, forceps, and spatulas are kept in a flame and heated until they turn red hot. The infective organisms in these instruments are killed in the flame. This is one of the simplest methods of surface heat sterilization to kill pathogenic organisms.
2. **Flaming:** In this method, the mouths of culture vials, glass slides, and scalpels are just showed in a flame to render the organisms inactive while working with them. The disadvantage of this method is that vegetative cells are inactivated but spores remain and cannot be inactivated unless the instruments turn red hot.
3. **Incineration:** Incineration is a technique employed for destroying contaminated clothes, animal carcasses, plastics such as PVC polythene (They should be incinerated because they produce poisonous smoke.), and pathological material. It is a process of burning materials to ashes inside a closed chamber; the equipment used for burning is called incinerator.
4. **Hot air oven:** Dry heat sterilization using hot air oven (Figure 12.2) is an efficient method adopted for sterilizing equipments and glassware. Hot air oven is an electrical device that operates in the temperature range of about 160°C–180°C, thereby destroying pathogens by an oxidative stress on cellular constituents.

Construction and operation of hot air oven

Construction: Hot air oven consists of an insulated chamber surrounded by an outer case of electrical heaters that produce heat, which is circulated uniformly inside the insulated chambers with the help of a fan. The temperature is maintained inside by a double-walled metallic chamber and is stabilized by a thermostat. The rise in temperature is periodically checked by a temperature sensor. Adjustable metallic trays, which are made of aluminium, are arranged inside and they serve as a compartment for the arrangement of glassware. Finally, it has a door locking control that has a grip and does not allow the air to escape out.

Operation: The materials to be sterilized are placed on the metallic tray inside the oven. The materials to be sterilized should be free of moisture. The door is then tightly closed due to the presence of door locking control, the device is switched on, and the thermostat knob is adjusted to set the temperature at 160°C. The oven is heated to the set temperature and switched off after an hour. The materials are then allowed to cool inside the oven before being removed for use.

Precautions

1. Care should be taken to switch off the mains as soon as the heating period is over.
2. Proper maintenance should be done before and after use.
3. The door should not be opened as soon as the power is switched off as sudden exposure to cooling may cause cracks in the glassware. Hence, the oven has to be left for cooling before removing the materials kept for sterilization from the oven.

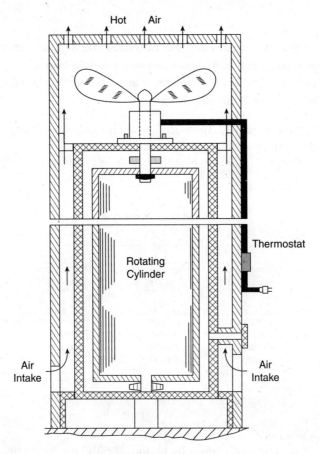

Figure 12.2 Hot Air Oven

4. Sterility check has to be done for each and every lot to enhance the quality of sterilization. For this purpose, the biological indicator organism *Geobacillus stearothermophilus* is used. This organism is commercially available in the name of Raven ProSpore. These commercially available spores are packed in ampoules with growth media along with the indicator. These ampoules are placed along with the articles to be sterilized and autoclaved. This is then subjected to incubation and checked for the colour change due to acid formation. This kind of sterility check should be done periodically to maintain the quality of the machine.

Advantages

1. It is used for sterilizing glassware and metal items.
2. As the method does not require water, the metal items that are sterilized do not corrode.
3. It is used for sterilizing chemicals, powders, and oil.
4. The sterilization is very effective.
5. High temperature can be attained when compared with other means of sterilization.

Disadvantages

1. Plastics cannot be sterilized as they will melt due to the high temperature involved.
2. Certain resistant organisms might not be killed when the heat does not penetrate within the stipulated time; but if the temperature and time are increased, the paper used for covering the materials might get charred.

Moist heat sterilization **(Figure 12.3):** Moist heat sterilization is a technique wherein steam is used to sterilize materials by heating or boiling; this method kills microorganisms by coagulating or denaturing their proteins. Moist heat sterilization may be done at different temperatures as follows:

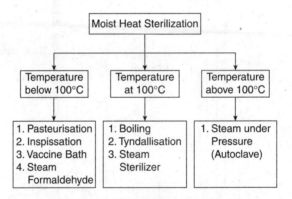

Figure 12.3 Moist Heat Sterilization below 100

Temperatures below 100°C

Pasteurization: Pasteurization is done by heating at 60°C for 30 minutes (holder method) and then cooling. This method kills vegetative cells but does not kill spores. It eliminates pathogenic bacteria like *Mycobacterium*, *Brucella*, *Salmonella*, and *Staphylococcus* from milk and milk products. This method might not be that effective for certain heat-resistant bacteria like *Coxiella burnetii*, but they might be killed if the temperature is slightly increased to 72°C for 20 minutes and then decreased quickly (flash method) or using ultra high temperature, which involves the application of a temperature of 140°C for 15 seconds. A quality control check is carried out by performing methylene blue test and phosphatase test for checking the quality of the product.

Inspissation: This method is used for sterilizing media. As very high temperature might coagulate the media owing to their high protein content, sterilization is done by applying a temperature of 80°C–85°C for half an hour for three consecutive days. Vegetative cells are killed on the first day, and spores are killed on the second and third day; though the protein gets coagulated due to the heating during this process, the chemical nature of the media is not altered. This process of repeated sterilization is termed as inspissation, and it is done using an instrument called inspissator. Löwenstein–Jensen medium and Loeffeler's serum slope, which is used for the isolation of *Mycobacterium tuberculosis*, are heat sterilized using this method.

Vaccine bath: A water bath heated to 60°C for an hour is used for eradicating vegetative cells in the vaccines or the serum to be used.

Steam formaldehyde sterilization: This technique is used for sterilizing the media that cannot withstand a temperature above 100°C. Temperature of 70°C–75°C is used to heat formaldehyde, and the vapours formed under pressure are used for sterilization.

Temperature at 100°C

Boiling: This is an age-old method of sterilization where water is boiled to 100°C. It is considered as disinfection and not sterilization because this method kills vegetative cells but resistant microorganisms and spore-forming organisms are not eliminated. This method can be used for disinfecting needles and stoppers but is not recommended for equipments and laboratory wares.

Steam sterilizer: It is also termed as Koch–Arnold steamer and looks much like an autoclave but works with steam at 100°C for 90 minutes. The technique is just the same as Tyndallisation but involves a higher exposure time. The steamer is made of a perforated metal cabinet with a heater at the bottom and a discharge tap at the top for expulsion of steam. This was the traditional steamer that was used for sterilization before the progression of Tyndallisation.

Tyndallisation: This is a steam sterilization process where the temperature is maintained at 100°C and the materials are subjected to steam for 20 minutes, cooled, and incubated for 24 hours. This process is carried out for two consecutive days. Vegetative bacteria, viruses, and even resistant spores are killed by this process. John Tyndall was the scientist who discovered this technique, and therefore it is named after him. This process is also called fractional sterilization or intermediate sterilization. Certain media that have sugar and gelatin as ingredients, on exposure to high temperature, may get decayed; hence, to avoid the decay, such media can be subjected to Tyndallisation.

Temperature above 100°C When a liquid, say water, is boiled in a closed container, the temperature inside increases, the water gets converted into steam and this, in turn, starts building pressure inside the closed vessel; due to the increase in pressure, the temperature also increases. This principle of steam under pressure has been functional in the autoclave.

Autoclave: Autoclave (Figure 12.4) is a device formulated for moist heat sterilization. It works on the principle of the sterilizing action of steam under pressure (121°C for 15 minutes at 15 lbs/inch^2). This type of sterilization is very successful because the rise in temperature above 100°C kills not only vegetative cells but also resistant spores by coagulating their protein and thereby leading to the death of pathogenic organisms.

CONSTRUCTION: An autoclave is made up of a double-walled stainless steel vessel with a pressure gauge, a steam valve (exhaust valve), a safety valve, and clamps on all sides of the vessel to close the container while autoclaving. A rubber gasket is placed inside the lid, which makes it airtight and does not allow the steam to escape while autoclaving. Before autoclaving, the heating coil present at the bottom of the vessel has to be immersed in water. When the device is switched on, the coil gets heated and in turn heats the water to produce steam. The materials to be autoclaved should be placed inside the stainless steel basket and then inside the vessel. The whole unit can be closed tightly using the clamps.

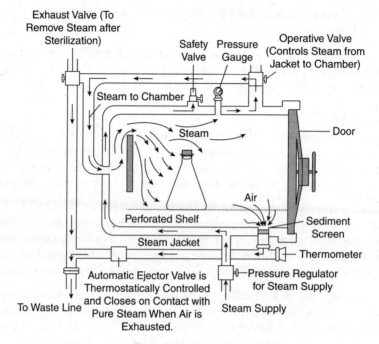

Figure 12.4 Autoclave

OPERATION

1. Water should be filled to a level just below the bottom of the basket in such a way that the coil is totally immersed.
2. Materials to be sterilized should be neatly wrapped, kept inside the stainless steel basket, and then placed inside the vessel.
3. Then the rubber gasket should be placed and the lid closed tightly using the clamps.
4. The safety valve must be opened and the power, switched on.
5. As soon as the steam is generated and the pressure starts building inside the autoclave, the steam tends to release through the open valve; hence, the valve should be closed.
6. The pressure is checked till it reaches 15 lbs/inch^2, then the time is noted and after 15 minutes, the power is switched off.
7. The steam should then be slowly released from the autoclave through the safety valve.
8. When the steam is completely released, the lid should be opened and the materials taken out with care.

PRECAUTIONS

1. The water level should always be maintained to a point where the coil is immersed, or else the coil will burn.
2. The materials kept for sterilization have to be tightly packed.

3. The materials to be sterilized must be packed with thick paper.
4. Before autoclaving, the air should be released completely or else the pressure gauge will not display the pressure properly.
5. Correct pressure should be maintained for the prescribed time, or else the sterilization will not be complete.
6. Sterility check has to be done for each and every lot to enhance the quality of sterilization that has been done. Hence, the indicator organism *Geobacillus stearothermophilus*, which is a common contaminant in the laboratory but a more-resistant bacterium with an ability to produce spores, is used as a test organism for quality check. These organisms are inoculated on a strip, which is dried and packed carefully and placed along with the things to be sterilized. This is then subjected to the media and incubation and checked for growth. Such kind of sterility check should be done periodically.

ADVANTAGES

1. This device is used for sterilizing media, rubber items like gloves, stoppers, tips, lids, and so on.
2. Uniform penetration of heat to all the materials kept inside the device is possible, and this is mainly achieved due to the humidity inside.
3. Spores are also killed as they do not withstand the temperature inside the autoclave.

DISADVANTAGES

1. As the items to be sterilized are packed with thick sheets, they get drenched because of the humidity, and water might also enter the items if they are not packed properly.
2. The sterilized items take longer time to cool down.
3. Stainless steel materials might corrode when subjected to repeated sterilization due to the moisture.

Radiation

Radiation is one of the effective methods of sterilization as energy waves take only seconds to penetrate microorganisms and kill them. There are two types of radiation that are used widely, and they are discussed below.

Ionising radiation: Ionising radiation has high penetrating power and is of two types by nature, namely, particulate and electromagnetic. Some of the ionising rays used are X-rays, gamma rays, and cosmic rays. These rays are very toxic and may cause genetic mutations in human beings; so proper precautions should be taken while using these rays. Gamma rays, which are electromagnetic in nature and obtained from a source such as cobalt-60, are commercially used for sterilization in laboratories. Exposure to these rays causes damage to the DNA of the pathogenic organisms, thereby causing their death. The use of this radiation to sterilize used or contaminated laboratory wares without the use of heat is termed as cold sterilization.

Particulate rays such as electron beams are also used for sterilizing materials that are packed, but the disadvantage of this method is that they do not penetrate well and hence sterilization is partial.

Advantages

1. There is no use of heat.
2. Less time is required.
3. High penetration is observed.
4. Effective sterilization is done.

Disadvantage

1. As they are very harmful to humans, precautionary methods have to be followed while using these rays.

Non-ionising radiation: Non-ionising radiation are low-energy rays, and their penetrating power is also very low. These rays are used for surface sterilization as they have no effect on spores. Some of the ionising rays used in laboratories are UV rays and infrared rays. UV rays, with wavelengths ranging from 200 to 280 nm, emitted from a high-pressure mercury lamp have a germicidal effect (especially on the DNA-forming thymine dimers) on microorganisms like bacteria, viruses, and fungi. These rays are used for sterilizing biosafety cabinets, operation theatres, and closed spaces before and after work. Infrared radiation is considered as a rapid type of sterilization and hence, it is used for sterilizing syringes and catheters.

Advantages

1. Very good bactericidal activity is seen.
2. The method is cost effective.

Disadvantage

1. Spores are not killed because of the low penetrating power.

Filtration

This method is used for separating contaminants from liquids without the use of heat when the materials to be sterilized are heat sensitive. Substances like sera, antibiotics, vaccines, toxins, and media are sterilized by this method. Filter discs with a pore size of 0.2 μm are used and hence are very effective in the separation of contaminants from filtrates. The separated contaminants can then be autoclaved for complete destruction. These filters are also used for the purification of water. There are different types of filters, which are as follows:

1. Candle filters (Figure 12.5) are used for the separation of impurities from water on a large scale, mainly in industries. These filters are of two types, namely, diatomaceous earth filters and porcelain filters. Diatomaceous earth filters are made of either porcelain or ceramic (e.g., Berkefeld filters). Porcelain filters are also called as Chamberland filters. These filters are made with different porosities and are constructed like hollow candles.

The liquid that has to be filtered is passed through the chamber that contains the filters with different porosities.

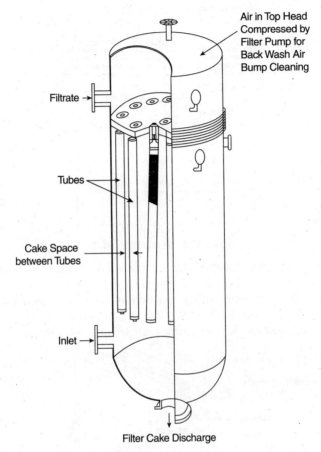

Figure 12.5 Candle Filter

2. Asbestos disc filters contain a special type of disc that can be used only once and should not be reused, and hence, each time a fresh disc has to be used for filtering. They are made up of asbestos containing magnesium silicate. The disc is placed on a metal holder that is attached to a flask and the flask, in turn, is connected to the vacuum pump that provides either positive or negative pressure; the liquid is filtered in this set up. Seitz filters are the commonly used asbestos filters for filtering microbial contaminants from pharmaceutical products.

3. Membrane filters (Figure 12.6) are a combination of polymeric compounds such as cellulose esters, polycarbonates, and polyesters. Laboratories use these membrane filters mostly for purifying water, complicated media, and contaminated sera. The membrane used is made of nitrocellulose, with a pore size of 0.22 μm. These filters effectively filter microorganisms as the pore size is much smaller when compared with the size of microorganisms.

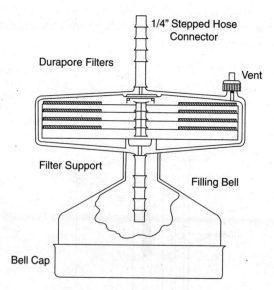

Figure 12.6 Membrane Filters

4. Air filters are used in laminar hoods for sucking and recycling the air to provide a sterile environment for working. Figure 12.7 represents the use of laminar air flow in a microbiology laboratory in ensuring safety and sterility. The filters used are high-efficiency particulate air filters. They are 99.99% effective in removing microorganisms during air circulation. They are capable of removing particles of size beginning from 0.3 μm.

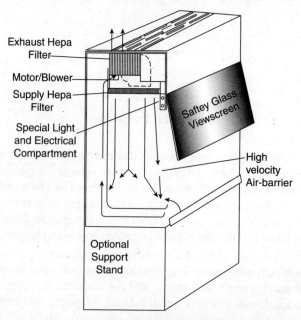

Figure 12.7 Laminar Air Flow

12.5 TYPES OF DISINFECTION

This is a method that employs chemicals or disinfectants to kill pathogenic microorganisms from inert surfaces of living and non-living things. Chemicals that are safe to use on the skin are termed as antiseptics. These chemosterilants must have a broad spectrum and hence should be effective against all microorganisms. A good antiseptic or disinfectant should eradicate both vegetative and spore forms. There are two types of chemical sterilization, which are as follows:

1. Gaseous disinfectants
2. Liquid disinfectants

Gaseous Disinfectants

Gaseous vapours are used for sterilizing closed areas like microbiology laboratories or operation theatres before and after work. These gases should have certain characteristics to be good disinfectants, such as their concentration and time of action should be good, they should have high penetrating power, and they should be able to destroy microorganisms within a practical period of time.

Ethylene oxide (ETO) is a highly effective sterilant that rapidly kills spores on exposure. It is a colourless liquid that, on heating, produces fumes that react with the DNA of pathogens and destroy them. It is used to sterilize plastics, gloves, dental equipments, and clothing. ETO is carcinogenic; it also has an explosive effect and hence, it is not used for fumigating laboratories. The materials that are resistant to heat are sterilized by this method. This method employs a humid chamber with a heater where ETO is combined with inert gases such as carbon dioxide and heated above 30°C. Fumes are formed inside the chamber, and the materials to be sterilized are kept in it for at least 2–3 hours for proper sterilization. The chamber used is called as chemiclave or gas chamber. Large packed materials can be sterilized using this method.

Formaldehyde is a widely used gas sterilant employed for fumigating laboratories, operation theatres, and wards. The gas is formed when potassium permanganate is added to formalin; the quantities of both these chemicals are based on the volume of the room to be fumigated. Entry should be restricted for at least 48 hours because of the toxic nature of the gas. This effect can be nullified by using ammonium vapours.

Betapropiolactone is a condensation product of ketone and formaldehyde. It is boiled at a very high temperature, and the fumes produced are very effective when compared with that of formaldehyde; 0.2% of the gas is effective and hence has a sporicidal effect. The main disadvantage is that it is carcinogenic and has low penetrating power. It is used for inactivating pharmaceutical products such as vaccines.

Liquid Disinfectants

Here, liquid chemical substances are used as antiseptics or disinfectants. They have three different levels of activity and hence are of three types, namely, high-level disinfectant, intermediate-level disinfectant, and low-level disinfectant. The level depends on the potency of the chemical substance used. Ideally, 2% of the chemical is used for sterilization.

Aldehydes: Glutaraldehyde is a chemosterilant or disinfectant used for the eradication of resistant bacteria, viruses, and fungi including spores. Formaldehyde is another common liquid disinfectant used for sterilization.

Advantages

1. They are high-level disinfectants.
2. They are very effective.
3. Glutaraldehyde is less irritating when compared with formaldehyde.
4. They have a longer shelf life when compared with other sterilants.
5. They have good penetration capacity.
6. They are used for the sterilization of equipments like endoscopes, cystoscopes, bronchoscopes, and retinoscopes.
7. Aldehydes are also used for the sterilization of plastics as they do not cause any damage to them.

Alcohols: Alcohols such as ethyl alcohol, methyl alcohol, and isopropyl alcohol are used as skin antiseptics. The mode of action of alcohols towards microorganisms is very effective as they denature bacterial protein and disrupt their membrane, killing vegetative cells; but they have no effect on spores.

Advantages

1. They are used as antiseptics.
2. As they have broad spectrum of action when compared with absolute alcohol, 70% of ethyl, methyl, and isopropyl alcohol are used instead of absolute alcohol.
3. They are non-corrosive.
4. As they are highly volatile, no residue is left.

Disadvantages

1. The vapours are toxic.
2. They are inflammable.
3. They do not have effect on spores.

Oxidizing agents: Hydrogen peroxide is widely used in hospitals because it is efficient in damaging the protein and DNA of pathogenic microorganisms using free hydroxyl radicals that link with the DNA and protein and kill them. It eradicates not only vegetative cells but also spores. It is used as a disinfectant for instruments and also as an antiseptic on skin, but the concentration levels vary for both (6% for disinfecting instruments and 3% when used as an antiseptic solution).

Advantages

1. It is used for the sterilization of plastic materials used in operational procedures, lenses, and dental materials.
2. It is sporicidal in nature.
3. It is used for advanced sterilization.

Disadvantage

1. It is light sensitive and loses its ability on exposure.

Peracetic acid: Peracetic acid is a good complex oxidizing agent that is effective against all microorganisms including spores. It is the combination of acetic acid and hydrogen peroxide and is used as an antimicrobial agent.

Advantages

1. It is a potent germicidal agent and is also sporicidal in nature.
2. It is non-toxic.
3. It is more effective than hydrogen peroxide.

Dyes: Dyes are being used as antimicrobial agents due to their interference with DNA replication or synthesis of cell wall of pathogenic microorganisms. causing a germicidal and bacteriostatic effect on the organisms. Dyes such as aniline and acridine are used as antiseptics against skin infections.

Advantages

1. They act against Gram-positive and Gram-negative bacterial skin infections.
2. They are non-toxic and painless when applied on skin.
3. They are even used as selective ingredients for the growth of organisms such as *Mycobacterium tuberculosis*.

Halogens: Halogens are bactericidal in nature. Two commonly used halogens are chlorine and iodine. Chlorine is the most common disinfectant used for water purification. Chlorine, when added to water, reacts to form hypochlorite which is very toxic to vegetative cells as well as spores. They are bactericidal, fungicidal, and sporicidal in nature. Sodium hypochlorite and chloramines are also used as antiseptics. Iodine is used as a skin disinfectant, and it is also called as tincture when 2% of it is used with 70% of alcohol. Iodophor is a compound of iodine combined with a surface active agent such as polyvinylpyrrolidone. It is more effective than the combination of iodine and alcohol. Betadine is one of the commercially available iodophors that is used as an antiseptic.

Advantages

1. They are very effective antiseptics.
2. They are sporicidal in nature.

Disadvantages

1. They are corrosive.
2. They are inactivated in the presence of organic matter.

Phenols: Phenols were introduced by Joseph Lister in 1867 and were the first disinfectants used for surface sterilization during operations. They are produced by the distillation of coal tar at a temperature of 170°C or even more. They act on the cell membrane of pathogenic microorganisms, causing destruction of cells. Commercially used phenol derivatives are cresols, hexachlorophene, chlorohexidine, and chloroxylenol.

Advantages

1. They are used for the sterilization of glassware, as antiseptics for wounds, as disinfectants against contaminants, and so on.
2. They are less irritating.
3. They are active against Gram-positive and Gram-negative bacteria.

Surfactants: Surface active agents or surfactants are antibacterial agents that act on the surface of the cell membrane and denature the protein of the microorganisms causing damage to their cells. The most widely used surfactants are quaternary ammonium compounds (cationic detergents), which are most effective against Gram-positive organisms; but these compounds are not sporicidal.

Heavy metal salts: Heavy metal salts are fine disinfectants because of their antimicrobial nature. Salts such as those of copper, mercury, and silver are commonly used for the eradication of microorganisms. The microbial protein gets coagulated, and there is also loss of intracellular components of the microbial cell. Mercuric chloride was used as a disinfectant but is being replaced by copper salts (fungicides) as it is highly toxic, while thimerosal and phenylmercuric nitrate are used as mild antiseptics owing to their low toxicity.

CONCLUSION

Sterilization and disinfection are the processes employed to maintain sterility by removing or eradicating the vegetative or spore forms of microorganisms from materials or surfaces. These aseptic techniques are very useful while performing surgical or diagnostic procedures, for conducting experiments in a germ-free environment especially in laboratories, and also in the food and pharmaceutical industries to ascertain that the products are free of contamination by pathogenic microorganisms. Quality control is done for all the above-mentioned sterilization techniques and recommended concentrations of the disinfectants are used. Newer methods are also being identified in the recent fields of research to develop good aseptic techniques.

MULTIPLE CHOICE QUESTIONS

1. Dry heat sterilization is performed by _____.
   - (a) autoclave
   - (b) hot air oven
   - (c) ethylene oxide
   - (d) formaldehyde

   Ans. b

2. Standard temperature and pressure used for autoclave are _____.
   - (a) 121°C at 15 lbs/inch^2 for 15 minutes
   - (b) 110°C at 14.5 lbs/inch^2 for 20 minutes
   - (c) 115°C at 16 lbs/inch^2 for 20 minutes
   - (d) 116°C at 16 lbs/inch^2 for 20 minutes

   Ans. a

3. The process of complete removal of microorganisms to form a germ-free environment is _____.

(a) sterilization
(b) disinfection
(c) asepsis
(d) decontamination
Ans. c

4. The destruction of pathogens by heat, radiation, or filtration is termed as _____.
   (a) chemical sterilization
   (b) physical sterilization
   (c) filtration
   (d) disinfection
   Ans. b

5. The equipment in which materials are burnt to ashes inside a closed chamber is called as _____.
   (a) autoclave   (b) hot air oven
   (c) incinerator   (d) incubator
   Ans. c

6. The process of heating milk to 60°C for 30 minutes and then cooling to 13°C for sterilizing it is termed as _____.
   (a) intermittent sterilization
   (b) pasteurization
   (c) boiling
   (d) autoclaving
   Ans. b

7. An intermittent process of sterilization is termed as _____.
   (a) tyndallisation
   (b) pasteurization
   (c) autoclaving
   (d) steam sterilization
   Ans. a

8. The method used for separating contaminants from liquids without the application of heat is termed as _____.
   (a) filtration   (b) sedimentation
   (c) distillation   (d) crystallization
   Ans. a

9. Sterilization using gamma rays is called as _____.
   (a) heat sterilization
   (b) cold sterilization
   (c) pasteurization
   (d) tyndallisation
   Ans. b

10. UV rays are _____.
    (a) ionising radiation
    (b) non-ionising radiation
    (c) chemoradiation
    (d) cosmic radiation
    Ans. b

11. Ceramic candle filters are _____.
    (a) chamberland and Doulton filters
    (b) berkefeld filters
    (c) asbestos disc filters
    (d) membrane filters
    Ans. a

12. *Bacillus subtilis* spores are used for sterility check in _____.
    (a) dry heat sterilization
    (b) moist heat sterilization
    (c) chemical sterilization
    (d) filtration
    Ans. a

13. A mixture of iodine and surface active agent is _____.
    (a) iodophor
    (b) iodine
    (c) surface active agent
    (d) chloramines
    Ans. a

14. *Bacillus stearothermophilus* spores are used for sterility check in _____.
    (a) dry heat sterilization
    (b) moist heat sterilization
    (c) chemical sterilization
    (d) filtration
    Ans. b

15. Air filters used in laminar air flow are _____.
    (a) high-efficiency particulate filters
    (b) highly effective positive filters
    (c) high-efficiency particulate flow
    (d) high-efficacy particulate flow

    Ans. a

SHORT NOTES

1. Autoclave
2. Hot air oven
3. Types of sterilization
4. Gaseous sterilization
5. Radiation
6. Inspissation

ESSAYS

1. Brief about the types of physical sterilization and their applications.
2. Brief about the types of chemical sterilization and their applications.

13 Chemotherapy and Antibiotics

CHAPTER OBJECTIVES

13.1 Introduction

13.2 History of Antibiotics

13.3 General Properties of an Ideal Antibiotic

13.4 Classification of Antibiotics

13.5 Antifungal Drugs

13.6 Antiviral Drugs

13.7 Antiprotozoal Drugs

13.8 Antibiotic Resistance

13.1 INTRODUCTION

Infectious diseases and causative pathogens are one of man's most dreaded nightmares; infectious diseases have been associated with diverse modes of infection and severity ranging from an asymptomatic disorder to a highly debilitating disease leading to mortality. According to WHO reports, infectious diseases account to almost 25% of the global deaths annually. Although pathogenic infectious diseases have been a constant worry, their spread and associated mortality rates have been reduced due to the systematic usage of specific chemical agents produced naturally, by microorganisms and plant sources, and synthetically.

The chemical agents that are synthesized chemically are referred to as chemotherapeutic agents. The procedure of treatment is referred to as chemotherapy; Paul Ehrlich, a German scientist, coined the term chemotherapy and defined it as follows: 'Chemotherapy refers to the use of chemical substances to kill pathogenic organisms without injuring the host'. The modern-day definition of a chemotherapeutic agent is 'any chemical substance used in medical treatment'; they are also generally referred to as drugs.

Antibiotics refer to the chemical molecules obtained from natural sources such as microorganisms possessing the capacity to inhibit infectious diseases. (They are not synthesized chemically.) The word antibiotics originated from the term antibiosis with 'anti' meaning 'against' and 'biosis' meaning 'life'. The term antibiotic was coined by Selman Waksman in 1945; he defined it as 'any chemical substance produced by a microorganism possessing the capacity to inhibit the growth of other microorganisms or even destroy them in minute concentrations or dilute solutions'.

13.2 HISTORY OF ANTIBIOTICS

The history of antimicrobial chemotherapy dates back to Egyptians, who have documented the use of mouldy bread to treat wounds. American Indians have been treating malaria using the bark of cinchona tree. The first organized attempt to treat microbial infections was carried out by Paul Ehrlich to treat syphilis; Ehrlich, in 1910, tried an arsenic compound named Salvarsan, which effectively cured syphilis with no severe damage to the host, thus was born the science of chemotherapy to treat infections. Further progress in chemotherapy was observed with various scientists trying out different chemical agents to treat infections. An important breakthrough came from Gerhard Domagk, a German scientist, in 1935, who discovered the antimicrobial property of sulphur molecules. Prontosil, a red dye molecule synthesized by Domagk, was specifically effective against Gram-positive bacteria, and later research proved the action to be associated with sulphonamides, which led to the discovery of thousands of molecules all around the globe.

In between the search for chemotherapeutic agents came the most famous and serendipitous discovery of penicillin by Sir Alexander Fleming in 1928. Fleming, as a young researcher working on *Staphylococcus aureus*, noticed in an agar plate the inhibitory effect, marked by a clear zone of inhibition, of a mould on the growth of *S. aureus* colonies. The mould was identified as *Penicillium notatum*, and the inhibitory metabolite produced by the mould was named 'penicillin'. The discovery of penicillin, though was a very significant contribution, did not gain much significance due to the hardships in the cultivation, isolation, and extraction of the active principle from the mould. It was in 1940 that penicillin discovery regained importance as a result of the work of Ernst Chain and Howard Florey, who successfully carried out the mass production of *P. notatum* and the isolation and characterization of the active principle penicillin. This production of penicillin became a prominent discovery and saved thousands of lives during the Second World War.

This was followed by the extensive screening procedures for microbial and plant sources of antibiotic molecules, the most prominent being the discovery of streptomycin from *Streptomyces griseus* by Selman Waksman. The discovery of streptomycin from *Streptomyces* led to the extensive research on soil-borne actinomycetes, especially on the group *Streptomyces*, which produce more than half of the antibiotic molecules used in the present day. Statistical models have suggested that *Streptomyces* can produce antibiotic molecules in the order of lakhs and that we have just touched the tip of the ice berg.

13.3 GENERAL PROPERTIES OF AN IDEAL ANTIBIOTIC

To be used in the control of infectious agents, antibiotics should have the below-listed attributes.

1. **Selective toxicity:** The antibiotic molecule of interest should be selectively toxic, that is, it should harm only the pathogen and not cause any damage to the host.
2. **Chemotherapeutic index:** It is an important property of an antibiotic molecule; the efficacy of the molecule on pathogens should always prevail over the toxic reactions or

adverse effects arising in the host. The chemotherapeutic index is calculated using the following formula:

$$\text{Chemotherapeutic index} = \frac{\text{Toxic dose (highest dose)}}{\text{Therapeutic dose (lowest dose)}}$$

Toxic dose refers to the dose at which adverse or toxic effects is observed in the host. Therapeutic dose refers to the dose at which the pathogens are eliminated with no adverse or toxic effects in the host.

3. **Solubility in body fluids:** The molecule should be readily soluble in body fluids such as blood, plasma, serum, and so on to bring in the desired effect, effective distribution, and metabolism of the molecule inside the host tissues or organs.
4. **No alterations in toxicity:** The molecule should be stable, and its toxicity should not be altered during its absorption, digestion, metabolism, and excretion in the host.
5. **No allergenicity:** The molecule should be non-allergenic to the host even in the lowest concentration. No toxic or adverse effects such as rashes, irritation, redness, fever, and so on should be observed in the host.
6. **Stability:** The molecule should be stable in body fluids with a constant therapeutic index and desired half life for the effective treatment of the infection. Molecules with very short and very long half lives are not desirable as the former lack therapeutic index and the latter require longer time for elimination.
7. **Should not develop resistance:** The molecule should have the capacity to avoid the resistance that readily develops in the pathogens.
8. **Should not harm or eliminate normal microflora:** The molecule should eliminate only the desired pathogens and not the normal microflora of the host. The elimination of normal microflora from the host may lead to an imbalance in the host metabolism and in some cases to super or secondary infections by the pathogens as the primary barrier formed against the colonization of pathogenic organisms by the normal microflora inside the host tissues is affected.
9. **Should have long shelf life:** The molecule should have a long shelf life outside the host system and not be readily degraded by exposure to changes in light, temperature, humidity, and other extrinsic or environmental factors.
10. **Reasonable cost and easy availability:** The molecule should be cost-effective and easily available.

13.4 CLASSIFICATION OF ANTIBIOTICS

Antibiotics are classified based on various characteristics such as activity, origin, mode of action, and so on. The commonly used classification criteria of antibiotics have been discussed below.

13.4.1 Based on Target Organism

Based on the target organism against which it is intended to be used, antibiotics are classified as antibacterial, antifungal, antiviral, antiprotozoal, antihelminthic, and so on.

13.4.2 Based on Spectrum of Action

Based on the variety of organisms they can inhibit or kill, antibiotics are classified as broad-spectrum antibiotics and narrow-spectrum antibiotics.

Broad-spectrum Antibiotics

These are molecules that are effective against a diverse array of organisms such as Gram-positive bacteria, Gram-negative bacteria, *Rickettsiae, Chlamydiae,* and so on. Broad-spectrum antibiotics are the most used antibiotics in treating infections of unknown aetiology. Cephalosporin, chloramphenicol, and tetracyclines are a few examples.

Narrow-spectrum Antibiotics

These are molecules that are effective against a specific group of bacteria or maybe even a single species. These molecules are highly precise in their function and are helpful in treating infections with known aetiology. Examples include penicillin G, streptomycin, and polymyxin B.

13.4.3 Based on Cidal and Static Activity

Based on their ability to kill or stop the growth of pathogens, antibiotics are classified as microbicidal antibiotics and microbiostatic antibiotics.

Microbicidal Antibiotics

These are molecules that have the ability to kill pathogens, thereby completely stopping their growth and spread; the effect of microbicidal antibiotics is permanent. Depending on the target organism, they are referred to as bactericidal, fungicidal, or virucidal. Cephalosporin, penicillin, and aminoglycosides are a few examples.

Microbiostatic Antibiotics

These are molecules that have the ability to stop the multiplication or growth of the organism for a temporary period of time. The microorganisms resume their normal life cycle once the antibiotic is removed from the site of action. Depending on the target organism, they are referred to as bacteriostatic, fungistatic, or virustatic sulphonamides, tetracyclines, and chloramphenicol are a few examples.

13.4.4 Based on Origin

Based on the source from which the antibiotics originated, they are classified as natural antibiotics, semi-synthetic antibiotics, and synthetic antibiotics.

Natural Antibiotics

These are molecules that have been directly obtained from natural sources such as plants and microbes, and they are used in their native form without any chemical or synthetic modifications.

These are usually referred to as first-generation antibiotics. Penicillin G and penicillin V are a few examples.

Semi-synthetic Antibiotics

These molecules are also obtained from natural sources but they are modified chemically by the addition of chemical groups to increase their efficacy and avoid the development of resistance against them. These are the most common type of antibiotics used. Ampicillin and methicillin are a few examples.

Synthetic Antibiotics

These are molecules that are produced synthetically, exclusive of any microbial metabolite. Chloramphenicol, isoniazid, and sulphonamides are a few examples.

13.4.5 Based on Mode of Action

Antibiotics are also classified based on their mode of action on pathogenic organisms. This classification forms the most important basis for the classification of antibacterial drugs. Listed below are a few established modes of action of antibacterial antibiotics.

1. Inhibition of cell wall synthesis
2. Damage to cell membrane function
3. Inhibition of protein synthesis
4. Inhibition of nucleic acid synthesis
5. Action as antimetabolites

Inhibition of Cell Wall Synthesis

Bacterial cells maintain their intact shape due to the presence of a rigid cell wall composed of peptidoglycan. Damage to this cell wall deforms the cell structure leading to cell disruption. Antibiotics such as penicillin (Figure 13.1) (penicillin G, penicillin V, methicillin, ampicillin, and carbenicillin) and cephalosporins (Figure 13.2) (cephalothin, cefoxitin, and ceftriaxone), referred to as β-lactam antibiotics, posses a chemical structure called β-lactam ring that specifically acts upon the enzymes such as transglycosylase, transpeptidase, and DD-carboxypeptidase, thereby interfering with the cross-linking of tetrapeptides and thus inhibiting the synthesis of peptidoglycan. They also activate the cell wall lytic enzymes leading to cell disruption.

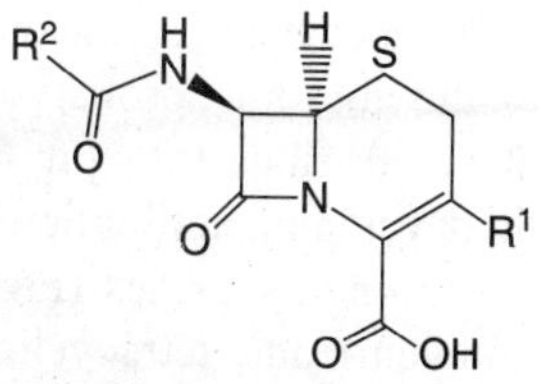

Vancomycin and cycloserine have a different mode of action in comparison with the β-lactam antibiotics. They specifically inhibit the binding of D-alanine–D-alanine subunit to the peptide cross

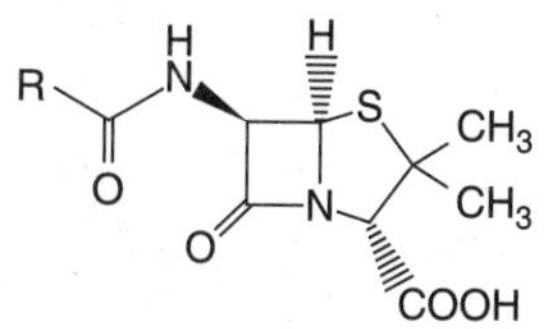

Figure 13.1 Penicillin

Figure 13.2 Cephalosporin

bridges, thus preventing the transpeptidation reaction of peptidoglycan synthesis and damaging the cell wall formation.

They are bactericidal and are specifically effective against Gram-positive bacteria as their cell wall is rich in peptidoglycan in comparison with Gram-negative bacteria but are ineffective against Archaea and fungi, which lack cell wall peptidoglycan. The probable adverse effects include allergic responses such as diarrhoea, anaemia, nausea, and renal toxicity.

Damage to Cell Membrane Function

All cells are bound by a cell membrane. Polypeptide antibiotics such as polymyxin B (Figure 13.3) act as detergent molecules and disrupt the cell membrane by binding to the phospholipids in the membrane leading to membrane distortion and loss of cellular constituents. They are specifically effective against Gram-negative bacteria, which have a cell membrane rich in phospholipids. They are bactericidal in activity. Antifungal antibiotics such as polyenes (amphotericin B) act on the fungal cell membrane, which is rich in ergosterol, and cause membrane disruption. Fortunately, bacterial, fungal, and animal cell membranes have different cell membrane composition. Animal cells have a cell membrane rich in cholesterol against which these antibiotics lack activity, thereby reducing the toxicity to the host. The possible adverse effects commonly observed are kidney damage, drowsiness, and dizziness.

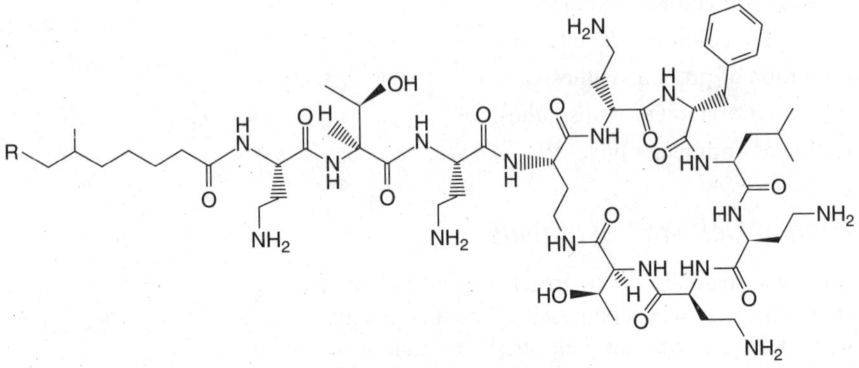

Figure 13.3 Structure of Polymyxin B

Inhibition of Protein Synthesis

In all living forms, DNA, RNA, and ribosomes form the prerequisite to protein synthesis. The difference between the eukaryotic and prokaryotic protein synthesis machinery is the variation in the ribosome: prokaryotes have 70s ribosome with 50s and 30s subunits, whereas eukaryotes have 80s ribosome with 60s and 40s subunits. This difference in ribosome has been exploited in developing antibiotic chemotherapy against bacteria.

Aminoglycosides (Figures 13.4 and 13.5) (neomycin, kanamycin, gentamicin, and streptomycin) and tetracyclines (Figure 13.6) (chlortetracycline, oxytetracycline, doxycycline, and minocycline) act on the 30s ribosomal subunit and interfere with the translation process (reading of mRNA), leading to the incorporation of wrong amino acids in the protein chain. Macrolides (erythromycin and clindamycin) and chloramphenicol act on the 23S rRNA of the 50s ribosomal subunit and inhibit the elongation of the polypeptide chain.

Aminoglycosides are bactericidal in activity, whereas tetracyclines, macrolides, and chloramphenicol are known to be bacteriostatic. The common side effects observed are allergic responses, deafness, gastrointestinal disturbances and renal and hepatic injury. Chloramphenicol (Figure 13.7) has serious side effects of bone marrow suppression and hence has very restricted usage. These groups of antibiotics have a broad spectrum of action and are effective against Gram-positive bacteria, Gram-negative bacteria, *Rickettsiae*, and *Chlamydiae*. In eukaryotes, mitochondria have 70s ribosome and hence are affected by the usage of these antibiotics.

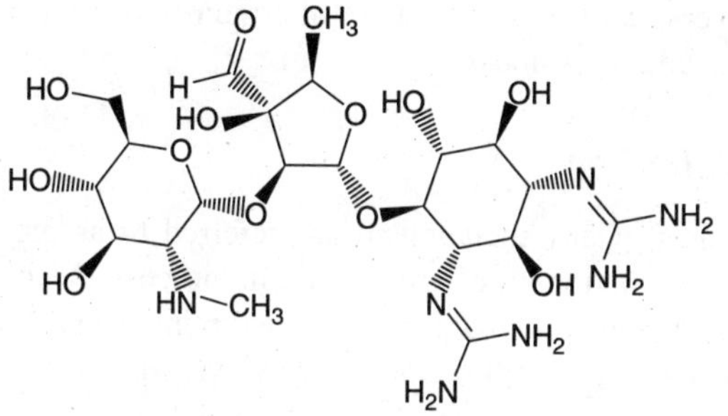

Figure 13.4 Streptomycin

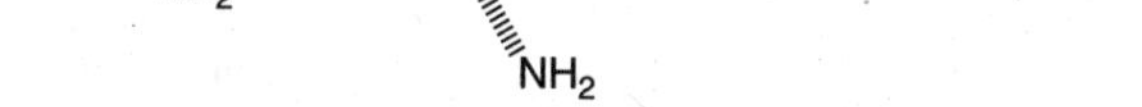

Figure 13.5 Neomycin

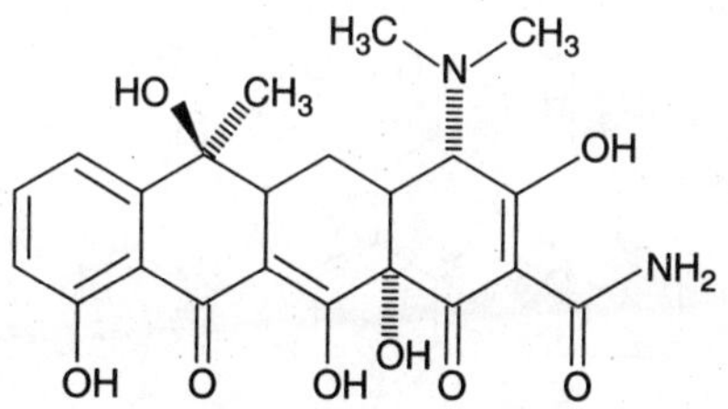

Figure 13.6 Tetracycline

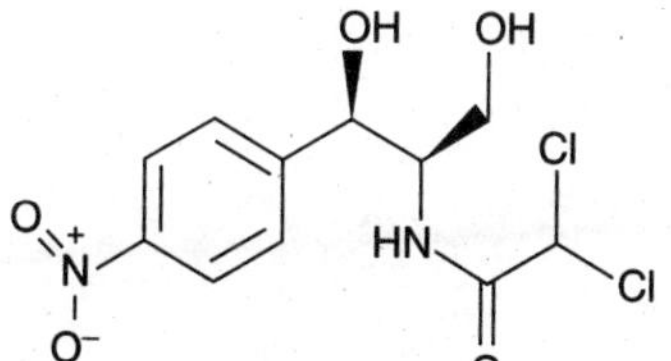

Figure 13.7 Chloramphenicol

Inhibition of Nucleic Acid Synthesis

Inhibition of nucleic acid synthesis is brought about by acting on the enzymatic system involved in the replication process. Antibiotics such as quinolones (norfloxacin, ciprofloxacin, and nalidixic acid (Figure 13.8) and fluoroquinolones (levofloxacin (Figure 13.9) act on the enzymes DNA gyrase and topoisomerase and block DNA replication and RNA synthesis (transcription), thus leading to cell death. These antibiotics have a low chemotherapeutic index due to the lack of selective toxicity. Rifampin (rifamycin (Figure 13.10)) acts on the bacterial DNA-dependent RNA polymerase and inhibits RNA synthesis and hence has selective toxicity. These antibiotics are bactericidal in activity. The most common adverse effects observed are allergic reactions, convulsions, nausea, head ache, fatigue, diarrhoea, and drowsiness.

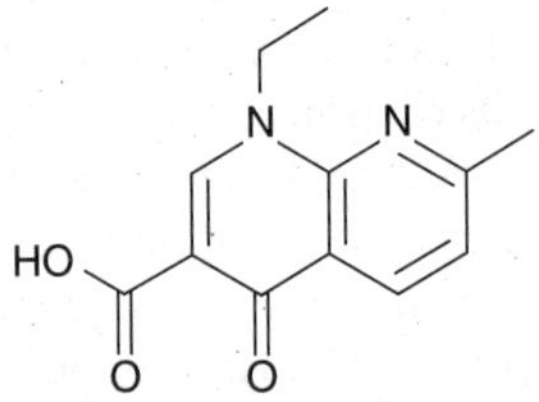

Figure 13.8 Nalidixic Acid

Action as Antimetabolites

The intermediate products of any metabolism are referred to as metabolites, and these are required for the effective functioning of any metabolic process. Antimetabolites are the molecules that affect the utilization of metabolites, thus disrupting the necessary metabolic process. Antimetabolites have been known to function in two ways, which are discussed here.

By Competitively Inhibiting Enzymes: Sulphonamides (sulphanilamide (Figure 13.11) and sulphisoxazole) inhibit folic acid synthesis. Folic acid, a precursor molecule for purines and pyrimidines, is required for nucleic acid synthesis in bacteria. Normal folic acid synthesis involves the enzymatic activity using pteridine, glutamic acid, and para-aminobenzoic acid (PABA). PABA being a co-enzyme is important in a prokaryotic system for folic acid synthesis. Sulphonamide molecules structurally mimic PABA and thus compete with the folic acid synthesis machinery in bacteria, thereby causing a decline in the folic acid concentration. The decrease in folic acid concentration in turn leads to the inhibition of purine and pyrimidine synthesis, thus affecting the replication process leading to cell death. Sulphonamides are selectively toxic and thus have a good

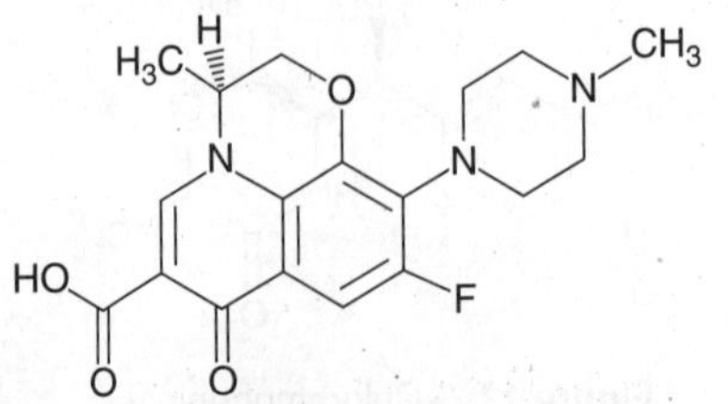

Figure 13.9 Levofloxacin

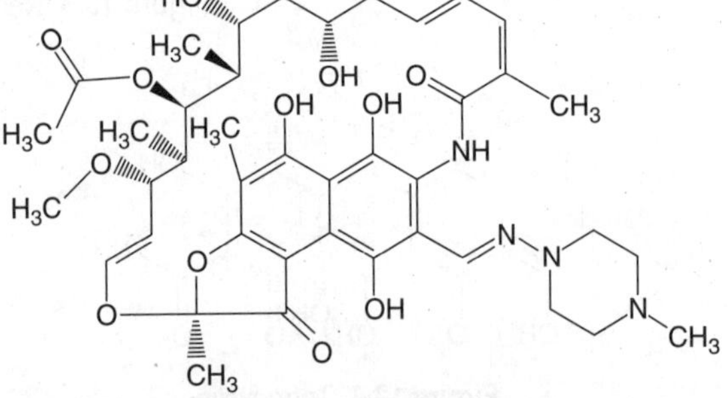

Figure 13.10 Rifampicin

chemotherapeutic index. As humans do not synthesize folic acid, it has to be obtained through the diet. The most common side effects observed are nausea, vomiting, diarrhoea, and rashes.

By Erroneous Incorporation into Nucleic Acids: Vidarabine and idoxuridine (Figure 13.12) are analogues of purines and pyrimidines, respectively. During the replication process, these molecules get incorporated into the nucleic acid and deform the information it codes leading to the mismatch of base pairs and inhibition of transcription, which causes cell death. These molecules have low chemotherapeutic indices due to low selective toxicity as both prokaryotes and eukaryotes employ the same nucleotides for the replication process.

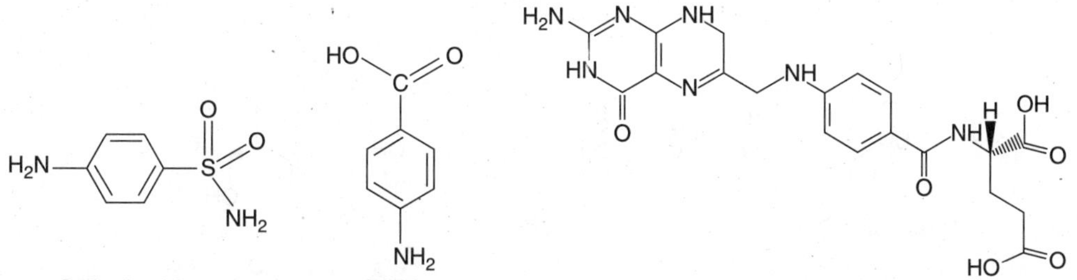

Figure 13.11 Structures of Sulfanilamide, PABA, and Dihydrofolate

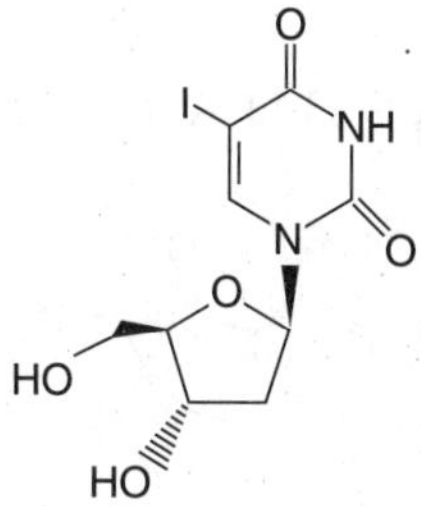

Figure 13.12 Idoxuridine

13.5 ANTIFUNGAL DRUGS

Fungal infections, unlike bacterial infections, are difficult to treat due to their close resemblance with their eukaryotic hosts. These pathogens are similar to their hosts in respect to their metabolism and cell organelle structure with minor differences, posing the problem of selective toxicity. Fungal infections, being superficial or systemic, are now commonly observed as opportunistic infections in immunocompromized patients, especially those suffering with HIV.

Antifungal antibiotics are usually prescribed and used as topical applications, to treat cutaneous infections, or oral drugs, to treat systemic infections. Polyenes (nystatin and amphotericin B (Figure 13.13)) produced by *Streptomyces* sp. and imidazoles (clotrimazole (Figure 13.14), ketoconazole, and miconazole) are a few examples of antifungal antibiotics. Synthetic antibiotics are one of the most common choices of antibiotics to treat *Candida albicans* infections, systemic fungal infections, and superficial fungal infections. The mode of action involves the disruption of cell membrane function leading to the loss of cell constituents by acting on the sterols present in the fungal plasma membrane. The toxic nature of polyenes has restricted their usage; in comparison with polyenes, imidazoles are less toxic.

Griseofulvin (Figure 13.15) is an antifungal antibiotic produced by *Penicillium griseofulvum*, and it has been found effective against superficial fungal infections of the hair and nail involving keratin. The mode of action involves action on microtubules during cell division and interference with the reproduction process. Flucytosine (Figure 13.16) is another synthetic antibiotic effective against *Candida* sp. and *Cryptococcus* sp. Flucytosine acts as an antimetabolite by mimicking cytosine and thus inhibiting nucleic acid synthesis.

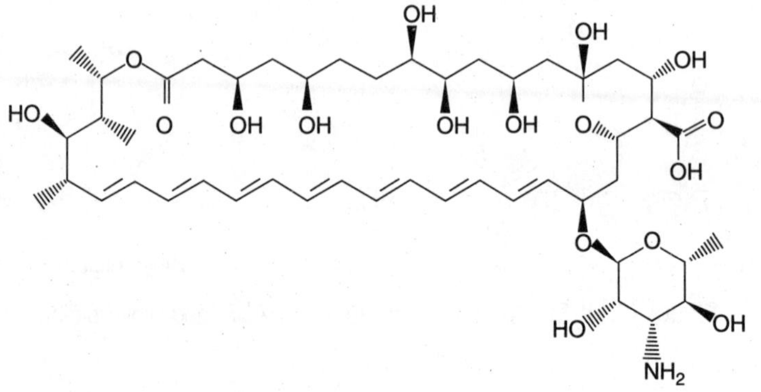

Figure 13.13 Amphotericin B

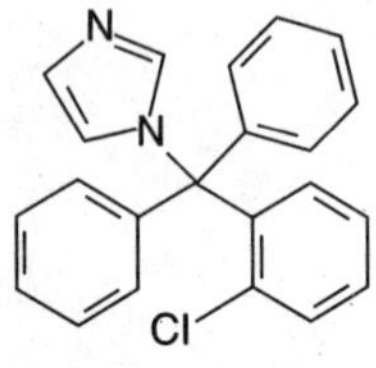

Figure 13.14 Clotrimazole

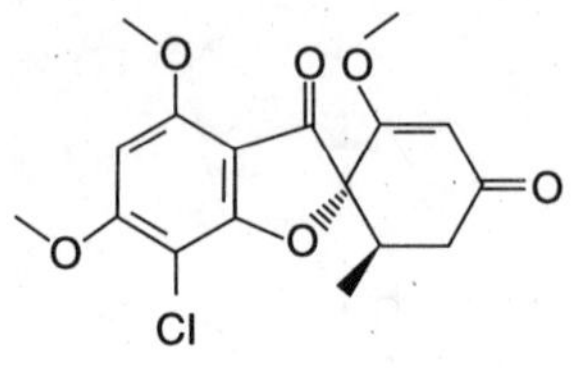

Figure 13.15 Griseofulvin

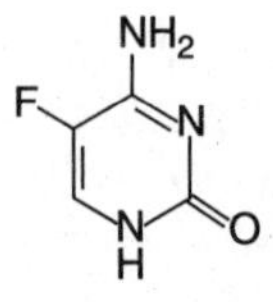

Figure 13.16 Flucytosine

13.6 ANTIVIRAL DRUGS

Viral infections, being the most common form of microbial infections, were once thought untreatable due to their complex mode of infection and the usage of the host system to propagate. The advancements in antimicrobial chemotherapy have made possible treating viral infections, which were considered toxic to the host system.

A large number of antiviral drugs are now available with different modes of action. Antiviral drugs such as amantadine (Figure 13.17), rimantadine (Figure 13.18), and pleconaril work by blocking the penetration of viral particles or uncoating

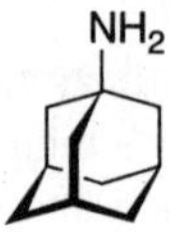

Figure 13.17 Amantadine

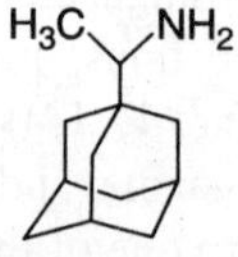

Figure 13.18 Rimantadine

them inside the host. The other most common mode of action is the inhibition of viral replication by acting as nucleoside analogues. Drugs such as acyclovir (Figure 13.19), ganciclovir, zidovudine (Figure 13.20) and ribavirin, most commonly used to treat Herpes infection, act as guanine analogues, thus inhibiting viral replication. Other nucleotide analogues are as follows: azidothymidine mimics adenine, idoxuridine mimics thymidine, dideoxycytidine mimics cytosine, and dideoxyinosine mimics adenine. These molecules are used in the treatment of a broad range of viruses including hepatitis B, hepatitis C, HIV, and other retroviruses.

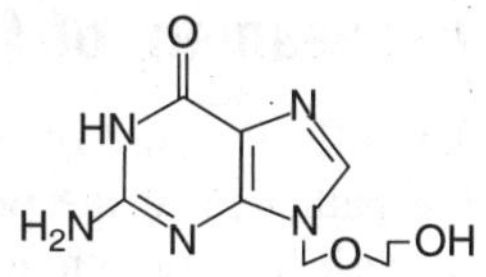

Figure 13.19 Acyclovir

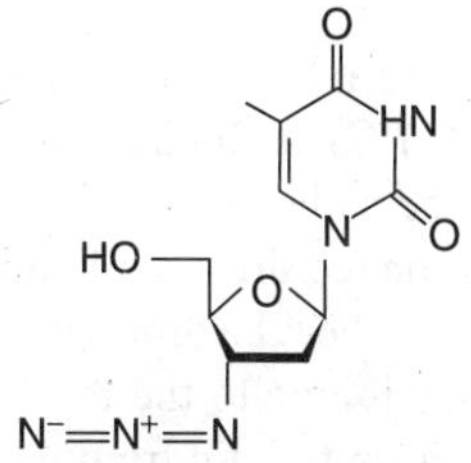

Inhibition of virion maturation and release is another mode of action observed in drugs such as indinavir, ritonavir, and saquinavir. These drugs are effectively used to treat HIV infections as they inhibit HIV protease thereby damaging the synthesis of viral polypeptides required for viral replication. Inhibition of viral particle release is observed in drugs such as oseltamivir (Tamiflu), a neuraminidase inhibitor effective against influenza viruses. Tamiflu is one of the most acclaimed antiviral drugs that have been extensively employed in treating avian influenza or 'bird flu'.

Figure 13.20 Zidovudine

13.7 ANTIPROTOZOAL DRUGS

Antiprotozoal drugs are one of the least-studied drugs with their complete mode of action still to be elucidated. Aminoquinolines comprising quinine, chloroquine (Figure 13.21), mefloquine, and primaquine are synthetic molecules used against the malarial parasite *Plasmodium* sp. The reported mode of action is the inhibition of the enzyme heme polymerase, which converts toxic heme to non-toxic heme, leading to the accumulation of toxic heme and thereby causing cell death. Artemisinin (Figure 13.22) produced by *Artemisia annua* is also effective against *Plasmodium* sp., wherein the free radicals generated lead to cellular damage within the parasite. Sulphonamides and nitroimidazoles are synthetic drugs effective against *Toxoplasma* sp. and *Trichomonas* sp., respectively. They act by inhibiting folic acid metabolism and DNA synthesis.

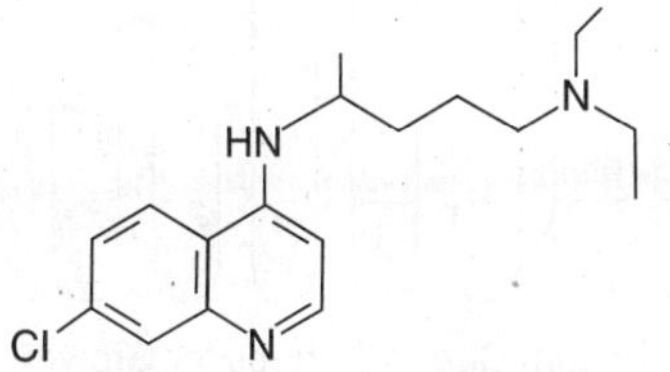

Figure 13.21 Chloroquine

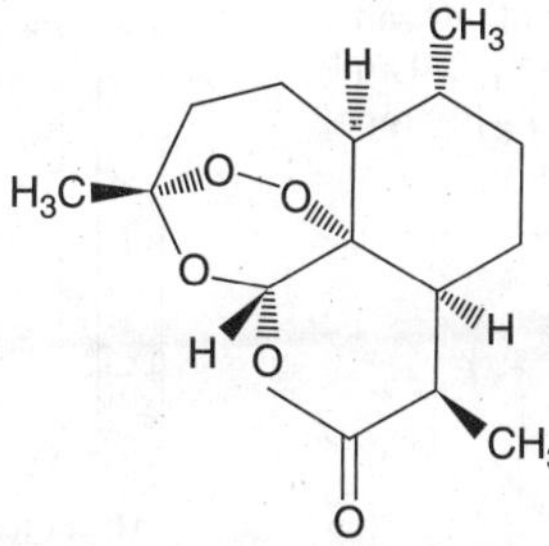

Figure 13.22 Artemisinin

Assessment of Antimicrobial Activity

The effectiveness of any antimicrobial chemotherapy is dependent on the susceptibility of the pathogen to the prescribed antibiotic. The suitability and concentration required of the prescribed antibiotic against the pathogen can be assessed by tube dilution method and disc diffusion method.

Tube Dilution Method (Figure 13.23)

The tube dilution method is carried out to determine the minimum inhibitory concentration (MIC), defined as the minimum concentration of the antibiotic required to prevent the growth of the pathogen, and the minimum lethal concentration (MLC), defined as the minimum concentration required to kill the pathogen completely. Both MIC and MLC are usually expressed in µg/ml.

The tube dilution method is performed by the addition of increasing concentrations of the test antibiotic to the broth medium inoculated with the pathogen of interest or by serially diluting the antibiotic from the highest concentration to the lowest possible concentration. The quantity of the broth, antibiotic, and inoculum should remain the same for all the tubes. The assessment of antimicrobial activity is done by observing the tubes for the growth of the organism. The tube showing no turbidity or growth of pathogen is considered to assess the least concentration of the antibiotic, that is, MIC, is obtained. From the tube showing no growth, the culture should be inoculated to antibiotic-free agar medium. The concentration in which no growth of

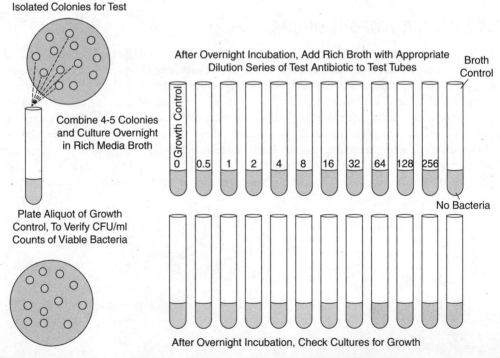

Figure 13.23 Tube Dilution Method to Determine Minimum Inhibitory Concentration (See page 351 for the colour image)

the organism is observed is referred to as MLC. The MLC values will usually be higher than the MIC values. The MIC and MLC values are less for microbicidal antibiotics in comparison with microbiostatic antibiotics.

Disc Diffusion Method (Figure 13.24)

The disc diffusion method is carried out on an agar surface. It is often performed to assess the susceptibility or resistance of the pathogen to a number of test antibiotics. In this method, antibiotic-impregnated paper discs are placed on agar plates seeded with the test organism and are incubated for 24 hours. After incubation, the plates are observed for the presence of clear zones around the discs. The disc showing the maximum zone of inhibition is considered the most potential antibiotic for the pathogen. Resistant pathogens grow till the surface of antibiotic disc with no zone of inhibition around the disc.

The most common method of disc diffusion assay followed is the Kirby–Bauer method. In this method, fresh broth culture of the test pathogen is swabbed onto Muller–Hinton Agar surface; further, the discs of the antibiotics of interest are placed on the swabbed agar surface by using either a sterile forceps or a disc applicator. The plates are incubated at 35°C for 16–18 hours. After incubation, they are observed for the presence of clear zones around the discs, and the zone is measured using an antibiotic zone measurement scale to the closest millimetres (Regular scales can also be used for measurement.). The readings are then tabulated, and a graph of MIC (µg/ml) versus zone of inhibition (mm) is plotted. The more the sensitivity of the pathogen to the test pathogen, the lower the MIC values and bigger the zone of inhibition. If the pathogen is resistant, the MIC value will be higher and the zone of inhibition smaller.

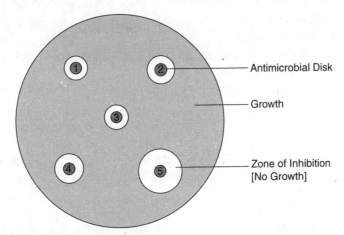

Figure 13.24 Disc Diffusion Method (See page 352 for the colour image)

13.8 ANTIBIOTIC RESISTANCE (FIGURE 13.25)

The increase in the number of antibiotics and their low cost and easy availability led to the problem of unscientific usage; for example, trivial infections were also prescribed with antibiotics. In addition to this, over-the-counter usage of antibiotics, lack of proper knowledge among the common mass regarding the dose and dosage completion, and also the use of these molecules

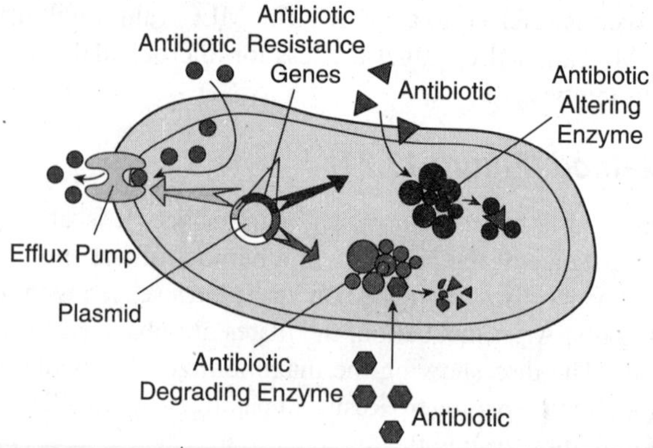

Figure 13.25 Mechanism of Antibiotic Resistance (See page 352 for the colour image)

in poultry and dairy industry to increase the animal body weight led to the gradual development of resistance among the once-susceptible group of pathogens. Numerous pathogens have been reported to have gained resistance to more than one antibiotic, thereby making their treatment difficult and costly. These multi-drug-resistant organisms have also been a cause of high morbidity and mortality and are commonly referred to as 'superbugs'.

Pathogens acquire resistance by various molecular mechanisms and mutations. Few of the known mechanisms are as listed below.

1. Alteration of the target site by modifying the surface protein to which the antibiotic binds.
2. Synthesis and upregulation of antibiotic-degrading enzymes, thereby leading to the degradation of the antibiotic (e.g., β-lactamases).
3. Alteration or downregulation of channel proteins, thereby blocking the portal of entry inside the cell for the antibiotics.
4. Expulsion of antibiotics through efflux pumps before the desired activity has been achieved.

All these mechanisms of resistance are categorized as 'vertical evolution' and are due to the presence of resistant plasmids acquired through chromosomal mutations. Bacteria can also transfer their resistant genes to other bacteria of the same or other genera by conjugation, transformation, or transduction. This process of transfer of resistance from one organism to another is termed as 'horizontal evolution'.

MULTIPLE CHOICE QUESTIONS

1. Penicillin was discovered by _____.
   (a) Selman Waksman
   (b) Gerhard Domagk
   (c) Alexander Fleming
   (d) Paul Ehrlich

   Ans. c

2. _____ is a β-lactam antibiotic.
   (a) Tetracycline   (b) Griseofulvin
   (c) Sulphanilamide (d) Penicillin

   Ans. d

3. _____ acts as an antimetabolite.
   (a) Sulphonamides (b) Cephalosporin
   (c) Amphotericin B (d) Polymyxin B

   Ans. a

4. _____ is an antifungal antibiotic.
   (a) Acyclovir
   (b) Chloramphenicol
   (c) Artemisinin
   (d) Clotrimazole

   Ans. d

5. _____ is an antiviral antibiotic.
   (a) Azidothymidine (b) Rifampin
   (c) Streptomycin   (d) Flucytosine

   Ans. a

6. Aminoquinolines inhibit _____.
   (a) Viruses   (b) Bacteria
   (c) Protozoa (d) Fungi

   Ans. c

7. MIC stands for _____.
   (a) Minimum inhibitory concentration
   (b) Maximum inhibitory concentration
   (c) Medium inhibitory concentration
   (d) Mutual inhibitory concentration

   Ans. a

8. Alteration in antibiotic target site leads to _____.
   (a) Antibiotic susceptibility of the bacteria
   (b) Antibiotic resistance of the bacteria
   (c) Does not affect the antibiotic
   (d) None of the above

   Ans. b

SHORT NOTES

1. History of antibiotics
2. General properties of antibiotics
3. Classification of antibiotics
4. Antibiotic resistance
5. Antiviral drugs
6. Antifungal drugs
7. Antiprotozoal drugs

ESSAYS

1. Brief about the mode of action of antibacterial antibiotics.
2. Brief about the assay of antimicrobial activity.

14 Standard Safety Measures and Biomedical Waste Management

CHAPTER OBJECTIVES

14.1 Biomedical Waste Management
14.2 Risk Assessment
14.3 Standard Safety Measures

14.1 BIOMEDICAL WASTE MANAGEMENT

The term biomedical waste has a broad meaning and encompasses a large subject. Biomedical waste includes wastes arising from hospitals, diagnostic laboratories, teaching institutions, research institutions, and households. The term, being non-specific, has been used without much precision and difference. In general, biomedical waste comprises the following.

1. **Animal waste:** Animal corpses (dead animals), tissues, sectioned organs, blood and other body fluids, and bedding used for housing and caging.
2. **Hospital waste:** Wound dressings, surgical wastes, placental tissues, dissected organs and tissues, blood and body fluids (excluding non-pathogenic urine and faeces), autopsy specimens, swabs, and catheters.
3. **Wastes from diagnostic laboratories:** Diagnostic laboratories comprise microbiology, biochemistry, and pathology units. The wastes arising from procedures carried out for biological investigations, which can be a cause of infection, are termed as biomedical wastes (e.g., tissues for histopathology, microbial cultures, blood, body fluids, bandages, biopsy specimens, radioactive materials used in diagnosis).
4. **Wastes from educational and research institutes:** Institutions offering educational and research facilities in the areas of biomedical sciences generate biomedical waste. These comprise microbial cultures, animal cell cultures, experimental animal wastes (carcasses), microbial and animal cell culture broth and agar media, radioactive materials used for research, and so on.

5. **Sharps and other non-biodegradable items:** These include syringes, needles, precision knives, blades, scalpels, catheters, lancets, tubings, glass slides, vials, sample collection jars, broken or damaged glassware, and so on.
6. **Household wastes:** First-aid accessories (cotton, gauzes, plasters, and bandages), sanitary pads, and ear buds are some of the wastes generated by households.

Biomedical waste also includes the following:

1. **Biohazardous waste:** It refers to the wastes suspected of comprising infectious materials that possess threat by their physical and biological characters to human beings, animals, and the surrounding environment.
2. **Infectious waste:** It refers to the wastes that are contaminated with microbiological pathogens and can act as intermediate or temporary hosts by supporting the growth, enrichment, and spread of pathogens.

As biomedical wastes are a potential source of infection and spread of disease, precaution is of prime importance in their management. The following are the general guidelines to be followed without slackness starting from collection to treatment of these wastes.

14.1.1 Collection and Segregation of Wastes

The laboratory head holds the prime responsibility and the roles and responsibilities of all lab members needs to be precisely defined. No negligence should be shown by any member in carrying out his/her responsibilities. The laboratory should follow well-defined regulations on the collection and segregation of wastes. The regulations should be written such that it is clearly understood by all team members. Collection of biomedical wastes is of high importance, and all laboratory personnel who are associated with the unit need to be thoroughly trained on the different types of waste, their segregation, hygienic practices, and usage of personal protective equipments (gloves, masks, and protective glasses) during the handling process.

Segregation of biomedical wastes is the initial and one of the most important steps in the disposal and management of biomedical wastes. Good segregation practices reduce the burden of disposal and chances of disease transmittance to a great extent. Laboratories around the globe have adopted various methods of segregation, of which the most commonly adopted is the colour coding of bags carrying different types of waste. The colour code that is most adopted is red, yellow, blue, and black containers. Segregation of wastes needs to be done as shown in Table 14.1.

Table 14.1 Colour Code System for Segregation of Biomedical Waste

Yellow Bag	Red Bag	Blue Bag	Black Bag
Infectious waste, bandages, gauzes, cotton or any other materials in contact with body fluids, human body parts, surgical wastes, placental tissues, animal tissues, microbial cultures, and so on	Plastic wastes such as catheters, injections, syringes without needles, tubings, IV sets and so on	All types of glass bottle, broken glass articles, outdated and discarded medicines, and so on	Needles without syringes, blades, sharps, precision knives, and all other metal articles

Every laboratory personnel should follow the segregation procedure; if there has been an accidental mix in the segregated wastes, it has to be immediately reported to the person in charge, and the same has to be documented with the approval of the head. The mix-up has to be informed to the personnel engaged in the transportation and treatment of wastes.

14.1.2 Containment and Labelling

The choice of containers to store wastes is important in biomedical waste management. The containers should remain intact with no breaks, spillage, or exposure to the environment and should sustain mechanical and physical damages during handling and transportation. Based on the usage, the containers can be classified as single-use containers and multiple-use containers. Double jacketing should be preferred in case of suspicion of spillage or exposure to outside environment. Multiple-use or reusable containers need extreme care, should be made of metal or sturdy plastic, and should be regularly inspected for damages, cracks, and leaks. They should be cleaned thoroughly removing stains and odour and appropriately disinfected after every use.

Labelling of the containers is compulsory and of utmost importance. The container label should be precise giving full details about the type of waste packed or collected, indicated by the specific colour of the label; hazard type, that is, biohazardous, radioactive, cytotoxic, and so on, indicated by a clear logo; date and time of collection; place of collection; personnel in charge of collection in the laboratory and his/her full contact details; and name and designation of the office head. The label should be marked in bold as hazardous material with a caution logo, and complete contact details of the person to be contacted in case of emergency should be provided. The different hazard symbols used in labels are as represented in Figure 14.1.

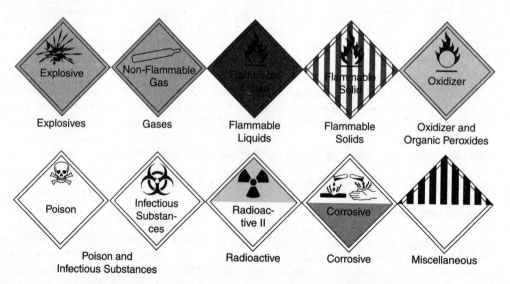

Figure 14.1 Biosafety Labels (See page 353 for the colour image)

Figure 14.2 is a sample of the label pasted on the containers carrying hazardous wastes.

HAZARDOUS WASTE

General Information

Name_____

Address_____

_____

HANDLE WITH CARE

Figure 14.2 Container Label (See page 353 for the colour image)

14.1.3 Transportation

The in-house movement of the waste once packed from the point of collection to that of storage or disposal should be as minimal as possible, with minimal or no exposure to laboratory personnel. The movement of these materials should always be through dirty corridors (non-sterile areas) and in no circumstances through clean corridors (sterile areas). The site of storage or pickup should be away from the main facility or building.

The transportation of biomedical waste should be time bound, and all the designated wastes as indicated by the colour codes should reach the appropriate treatment plants with no much delay. In case of delays in transportation, suitable measures, such as cold storage for decaying and infectious agents, to avoid the spread of infections and odour should be taken. The vehicle should be clearly marked with the biohazard symbol and the name and address of the treatment agency. The vehicle should halt only at designated collection points, and unnecessary stops should not be appreciated. The vehicle should also be thoroughly washed and disinfected appropriately at regular intervals. The driver and support staff also should follow hygienic practices.

14.1.4 Treatment of Biomedical Waste

Treatment of biomedical waste is another most important step in biomedical waste management. The treatment procedure employed should be appropriate for the material to be treated; the

Table 14.2 Biowaste and Their Appropriate Treatment Procedure

S. No.	Waste Type	Treatment Procedure
1	Microbiological waste	Autoclaving and chemical disinfection
2	Human and animal anatomical wastes	Incineration
3	Blood and body fluids	Incineration
4	Animal non-anatomical waste (bedding material)	Autoclaving
5	Plastics	Incineration and chemical disinfection followed by shredding
6	Sharps and metal instruments	Incineration and chemical disinfection followed by shredding

treatment procedure for a material should be changed only if necessary or else always carried out from start to finish. Under no circumstances, incomplete decontamination or disinfection procedure should be accepted. Various types of treatment procedure such as autoclaving, incineration, chemical disinfection, and so on should be employed depending on the type of waste. The different types of waste and treatment procedures employed are given in Table 14.2.

The treatment procedure, being the most important, should always follow a written standard operating procedure approved by competent authorities. All the personnel involved should have adequate training on the handling of different types of waste, instrumentation, and chemical treatment procedures dealt with. All the instruments should be maintained well and calibrated for proper functioning. Autoclaving should be done at the standard temperature and time ratios. Indicators (strips, colour bands, and spore suspension) of effective autoclaving need to be used to ensure accuracy. Autoclaving should not be carried out for human and animal anatomical wastes and blood and other body fluids. Cytotoxic compounds and chemotherapeutic drugs should not be autoclaved as they require higher temperature for disinfection. Incineration procedure should be followed for all such wastes.

Chemical treatments should be considered based on the microorganism and material involved, extent of contamination, efficacy of the disinfectant, concentration and dilutions to be prepared, time of contact between the material and disinfectant, pH, and temperature. Sodium hypochlorite (5.25%) is the most preferred reagent for chemical treatment.

After chemical treatment, sharps should be mechanically shredded. Sharps can also be treated in incinerators after shredding.

14.1.5 Disposal

After complete treatment, the treated biowaste has to be disposed according to the standard procedures. Incinerated anatomical wastes and shredded sharps should be landfilled. Autoclaved microbial cultures and chemical reagents can be sewer-drained only after ensuring complete decontamination. Cytotoxic and radioactive materials should not be landfilled and have to be disposed as per the guidelines from competent regulatory authorities.

14.1.6 Record Maintenance

A lucid record has to be maintained in all the facilities responsible for waste generation. The record should be maintained by the personnel solely responsible for waste disposal in the facility and should always be maintained to date. The record should comprise details such as the type of waste, hazard level, collection date, date and time of dispatch to treatment, quantity, mode of transportation (inclusive of driver and vehicle details), and treatment procedure employed.

14.2 RISK ASSESSMENT

The risk of getting infected through laboratory-associated and job-related activities is high in laboratory personnel. Assessment of such risks and adoption of appropriate safety and remedial measures ascertain the minimum spread of infection among the members inside and outside the laboratory. Risk assessment in any laboratory is dependent on the following factors.

14.2.1 Pathogenicity

The potentiality of the pathogen to cause infections is of high importance in determining the biosafety levels; more virulent the organism, higher are the chances of infection.

14.2.2 Route of Transmission

The route of transmission plays a very significant role in determining the safety standards. Aerosol infections have been the most prominent route of disease transmittance, followed by water, food, vector, and accidental injury through cuts and burns inside the laboratory.

14.2.3 Infectious Agent Stability

The viability, stability, and pathogenicity of a pathogen depend on its exposure to environmental factors such as desiccation, sunlight, UV light, and treatment by different chemical and physical procedures.

14.2.4 Infectious Dose

To cause a disease, any pathogen should infect a host at a dose where the human or animal immune system is unable to neutralize the pathogen and the effect of its toxins. The infectious dose of a pathogen depends on its virulence and can vary from one to several thousand units.

14.2.5 Susceptibility of the Host

The health of laboratory personnel is vital in assessing the risk of infections. A healthy individual can resist any infection, and his/her immunological status and power of resistance to infections determine the infectious dose for any pathogen to cause the disease. Pregnant women and individuals under treatment with immunosuppressants and steroids, with systemic infections, and who have undergone medical interventions or surgeries are more susceptible to infections in comparison with normal individuals.

14.2.6 Concentration and Volume of the Pathogen

Concentration is the number of infectious organisms in a unit volume. When the concentration and the volume of infectious organisms to be handled are higher, the risk of infection proportionately increases as the time and exposure of the personnel to the infectious organism also increase.

14.3 STANDARD SAFETY MEASURES

As the safety of a laboratory is an issue of paramount importance, biosafety measures hold high value. The principles of biosafety include containment, personal protection equipment, biological safety cabinets, facility, secondary barriers, and biosafety levels.

14.3.1 Containment

The term containment refers to the safe methods of managing infectious materials. The entire purpose of containment is to limit or eradicate the spread of and exposure to biohazardous material among the laboratory personnel or to the surrounding environment. Containment procedures involve three main facts, that is, the laboratory practices and techniques adopted for experimentation, safety equipments and measures adopted, and the design of the laboratory facility. The procedures vary according to the pathogen and the level of biohazard handled.

Containment is also classified into two stages based on the area of containment.

Primary Containment

Primary containment refers to the procedures adopted to control the spread of and exposure to infections with respect to the laboratory personnel and their immediate surroundings. Primary containment is achieved by personnel protective equipments (vaccinations, aprons, hand gloves, face masks, goggles, etc.), good microbiological practices (sterilization, inoculation, and disposal techniques), and proper usage of equipments (biosafety cabinets). Negligence in handling infectious samples and the usage of instruments has been the major cause of spread of infections among the laboratory personnel and thus is strictly recommended to follow the standard laboratory protocols.

Secondary Containment

Secondary containment refers to the prevention of exposure of infectious materials to outside environment external to the laboratory. Secondary containment can be achieved by the effective design of the facility comprising access controls, ventilation systems, air locks, sterile and non-sterile corridors, and biowaste management policies.

14.3.2 Personal Protective Equipment

During all microbiological processes, irrespective of the pathogenicity of the culture, laboratory personnel have to protect themselves against any possible infection. The first barrier against any infection is offered by the personal protective equipments worn by the laboratory personnel. Personal protective equipments include clean aprons, face masks, hand gloves, head masks or caps, shoe covers or laboratory shoes, face shields, and safety glasses. These personal protection equipments not only avoid direct exposure to microbial cultures but also offer considerable protection against accidental spillage of stains and laboratory reagents. They are a mandatory requirement in places where biosafety cabinets cannot be used and the laboratory personnel are in direct contact with infectious materials.

14.3.3 Biological Safety Cabinets

Safety equipments form an integral part of any biomedical laboratory working on pathogenic organisms. Safety equipments include biological safety cabinets, inoculation hoods, and closed cabinets used for microbiological purposes. The right combination of microbiological techniques and biosafety cabinets forms the basis of all safety procedures.

Biological safety cabinet, also referred to as biosafety cabinet, is the prime instrument used in any microbiological laboratory to curtail the spread of infectious materials. All microbiological and other cell culture procedures are supposed to be carried out within these instruments. Based on the usage and type of microorganisms handled, biological safety cabinets are classified into three categories.

Class I Biosafety Cabinet

These are open-fronted cabinets, usually referred to as inoculation hoods, and are used to handle non-pathogenic organisms (e.g., *Rhizobium* sp. and *Saccharomyces cerevisiae*). These cabinets in combination with good laboratory practices and personal protective equipments offer considerable level of protection. Presently, these have been replaced by Class II biosafety cabinets.

Class II Biosafety Cabinet

These are also open-fronted cabinets with glass or transparent doors, fitted with high-efficiency particulate air filters, and referred to as laminar air flow units. These instruments are used to handle moderately pathogenic organisms (e.g., *Bacillus anthracis*, *Escherichia coli*). These cabinets also provide product protection against external contamination during microbial cell culturing and preparation of microbial stock cultures. These units along with good laboratory practices and personal protective equipments offer significant level of protection and are an integral part of any microbiological laboratory.

Class III Biosafety Cabinet

These are airtight or gas-tight cabinets and are used to handle potentially pathogenic organisms (e.g., *Mycobacterium tuberculosis*). These cabinets provide the highest level of protection against microbial infections to the laboratory personnel and the surrounding environment. Personal protective equipments and good laboratory practices are mandatory during the usage of these cabinets.

14.3.4 Facility as a Barrier

The facility provides a significant level of protection to the people inside and outside the laboratory. A good design of the facility also helps in the prevention of spread of infection from the laboratory to the outside environment.

A good design of facility is usually a contribution of many professionals comprising biosafety engineers, Heating, Ventilation and Air Conditioning (HVAC) engineers, and animal-care professionals. A planned facility does have marked areas for all operations, such as work area, lounges, cafeteria, wash rooms, cleaning areas, sterile corridors, non-sterile corridors, animal facility, and biomedical waste collection points with no intermingling of activities. Instrumentations form an integral part of any facility. All the instruments irrespective of their frequency of usage should be periodically serviced and calibrated to ensure proper and thorough functioning.

The facility can be designed with primary and secondary barriers. Primary barriers refer to the walls, windows, doors, and compartments essential to differentiate areas. Depending on the type of microbiological or other biological research carried out, the facility is equipped with secondary barriers. Along with primary barriers, secondary barriers play an important role in reducing the spread of aerosol infections.

Secondary barriers include specialized ventilation units to ensure directional air flow, exhaust and air treatment filters for filtration of aerosols, controlled access doors, airlock systems, disinfecting shower systems, and separately constructed entire facility in the case of high-risk organisms.

14.3.5 Biosafety Levels

The laboratory facilities are classified into four biosafety levels based on the type, virulence, and pathogenicity of the organism; instrumentations; and infrastructure available to handle pathogens. The laboratory director is specifically and primarily responsible for assessing the risk and for effective implementation of the recommended biosafety levels.

Biosafety Level 1 (BSL-1)

These facilities handle non-pathogenic organisms that have been consistently not causing infections to the laboratory personnel (e.g., *Rhizobium* sp., *Aspergillus* sp., *Bacillus* sp.). These facilities follow basic safety precautions and require no special instrumentations or infrastructure.

Biosafety Level 2 (BSL-2)

These facilities handle pathogens of known aetiology that pose moderate risk if accidently inhaled, injected, swallowed, or exposed to skin (e.g., *Salmonella* sp., *B. anthracis*, *Cryptococcus neoformens*). The safety measures adopted in these facilities include limited access, personnel protective equipments, class I or II biosafety cabinets, and waste decontamination units, such as an autoclave.

Biosafety Level 3 (BSL-3)

These facilities handle pathogens that can be transmitted as aerosols and pose high risk with detrimental capacities (e.g., *M. tuberculosis*, yellow fever virus). These are specialized facilities with access-controlled double-door entry systems, class II or III biosafety cabinets, personnel protective equipments, clothing decontamination units, sealed windows, and specialized ventilation systems.

Biosafety Level 4 (BSL-4)

These are high-risk facilities handling life-threatening pathogens with unknown aetiology. These facilities handle aerosols and pathogens whose mode of transmittance is unknown (e.g., Ebola virus, smallpox virus, hemorrhagic fever viruses). These facilities are constructed in safe and high-security isolated zones with entirely separate buildings incorporated with all the facilities of a BSL-3 facility. In addition, all laboratory practices are carried out by individuals wearing positive pressure full-body suit. It is a mandatory procedure for all laboratory personnel to take decontamination shower before entry and exit from the facility.

MULTIPLE CHOICE QUESTIONS

1. Which of the following is not a biomedical waste?
   (a) Blood
   (b) Animal tissues
   (c) Microbiological cultures
   (d) Vegetables

   Ans. d

2. Yellow bag holds _____.
   (a) Needles
   (b) Anatomical wastes
   (c) Plastics
   (d) Glass items

   Ans. b

3. Label should not contain _____.
   (a) Waste type
   (b) Address of contact person
   (c) Advertisement
   (d) Colour code

   Ans. c

4. Blood and body fluids should be _____.
   (a) Autoclaved
   (b) Incinerated
   (c) Disposed directly
   (d) Chemically disinfected

   Ans. b

5. _____ are not personnel protective equipments.
   (a) Masks   (b) Gloves
   (c) Bangles   (d) Aprons

   Ans. c

6. Laminar air flow is a _____.
   (a) Class I biosafety cabinet
   (b) Class II biosafety cabinet
   (c) Class III biosafety cabinet
   (d) Class IV biosafety cabinet

   Ans. b

7. Positive pressure suits are worn in _____.
   (a) BSL-1 facility   (b) BSL-2 facility
   (c) BSL-3 facility   (d) BSL-4 facility

   Ans. d

SHORT NOTES

1. Types of biomedical waste
2. Collection and segregation of wastes
3. Labelling of biomedical wastes
4. Treatment of biomedical wastes

5. Risk assessment factors
6. Containment of infection
7. Facility as a barrier

ESSAYS

1. Explain in detail about biosafety cabinets.
2. Explain in detail about biosafety levels.

15 Hospital Acquired Infection and Hospital Infection Control Programme

CHAPTER OBJECTIVES

15.1 Introduction
15.2 Epidemiology of Nosocomial Infections
15.3 Infection Control Programmes: Protocols
15.4 Nosocomial Infection Surveillance

15.1 INTRODUCTION

Hospital-acquired infections are also referred to as 'nosocomial infections'. It is a new infection acquired by a patient (apart from what he/she is actually admitted for) on admission or visits to hospitals through hospital linen, equipment, and so on. It is important for hospitals to devote special care to the health and hygiene of the patients and hospital staff. There are many factors responsible for the occurrence of hospital-borne diseases, which are as follows:

1. Susceptible immune system of the host (patient)
2. Increased variations or modifications in medical techniques and procedures
3. Prevalence of infectious microorganisms
4. Poor health care practices resulting in the facilitation of transmission of drug-resistant bacteria

15.1.1 Occurrence of Infections

Hospital-acquired infections occur all over the world and are a major concern in both underdeveloped and developed countries. They are considered as a burden for both patients and the hospital population (staff and visitors) because of increased morbidity and hospital-acquired infections being the major cause of death among hospital patients. The nature and frequency of infections are influenced by a number of combinations of four major factors, which are as follows:

1. Patients with compromised resistance
2. Patient's contact with other infected persons

3. Contaminated environmental sites (patient's stay)
4. Drug-resistant endemic microorganisms

Infections occur through two main agents, namely endogenous and exogenous, which are discussed here.

Endogenous Infection

In this type, bacteria present in the normal flora cause infections because of transmission to sites outside the natural habitat (urinary tract), damage to tissue (wound), or inappropriate antibiotic therapy, which allows overgrowth.

Exogenous Cross-infection

In this type, bacteria are transmitted between patients in any one of the following ways:

1. Through direct contact between patients (hands, saliva droplets, or other body fluids)
2. Through the air (droplets or dust contaminated by a patient's bacteria)
3. Via staff contaminated during patient care
4. Via objects contaminated by the patient

In both the cases, the infecting organism might immediately attack the patient or might be introduced into the patient during surgical procedures, nursing procedures, or the handling of instruments.

Endemic or Epidemic Exogenous Environmental Infections: In the hospital environment, several types of microorganisms survive well in water, air, or certain damp areas. But occasionally, some organisms (e.g., *Pseudomonas*, *Acinetobacter*, and *Mycobacterium*) do survive in sterile products or disinfectants also.

The environment of the hospital, which acts as the chief source, comprises the following elements:

1. Contaminated air, foodstuff, water, and medicines
2. Equipment and instruments that are not hygienic
3. Soiled linens such as cotton, bed sheets, and garments
4. Biomedical waste

Patients get admitted to the hospital for various reasons, predominantly due to illness (infection), and they are subjected to medical investigations, which include contact with instruments and/or administration of drugs (antibiotics). Patients who are treated for malignancy are mostly post-operative patients who are on chemotherapy or radiotherapy. These factors make the immune defence mechanism of the host (patient) weak and subject them to the risk of infections. Antibiotic therapy makes the patients resistant to many drugs and also greatly affects and alters the normal flora. Implants and instruments may introduce various exogenous infections and trigger endogenous infections as well. Very rarely, auto-infections also arise.

15.1.2 Microbial Causes

A huge number of microorganisms are responsible for hospital infections, and they can be categorized into the following groups:

1. **Predictable pathogens:** These are organisms that can cause a disease in a healthy person who lacks specific immunity.
2. **Restrictive pathogens:** These are organisms that can cause a disease only in susceptible individuals who lack resistance to infections.
3. **Opportunistic pathogens:** These are organisms that can cause a widespread disease only in patients with greatly reduced immunity.

Obviously, it has to be understood that the above characteristics are not at all direct, and the category assigned to each group of pathogens can be challenged.

15.2 EPIDEMIOLOGY OF NOSOCOMIAL INFECTIONS

Worldwide studies have revealed that nosocomial infections are most important to the cause of morbidity and mortality. The poor quality of health care services is responsible for a high frequency of nosocomial infections. Factors responsible for the increase in the frequency of nosocomial infections are discussed below.

15.2.1 Nosocomial Infection Sites

The most common types of nosocomial infections are urinary tract infection, surgical-wound infection, pneumonia, and bloodstream infection. These infections are additional complications during patient care; at the same time, the normal defence mechanism of the host also gets damaged. About 80% of the nosocomial urinary tract infections are due to the use of indwelling urethral catheters. Nosocomial surgical-wound infections are due to the bacterial contamination of wounds during surgeries. Nosocomial pneumonia occurs more frequently in ICU patients with endotracheal intubation or mechanical intubation. Finally, nosocomial bloodstream infections are caused due to the use of indwelling central vascular catheters, and the duration of catheterization is also directly associated.

15.2.2 Urinary Tract Infections

The nature of infection can be assessed by a Gram staining technique done with the urine sample. This can help in the preliminary treatment before the culture results are available.

1. **Gram-negative infections:** Generally, antibiotics such as ceftazidime or cefoperazone are given in combination with aminoglycoside. In case of renal failure, aminoglycoside is replaced by fluoroquinolone.
2. **Gram-positive infections:** Generally, amoxicillin–clavulanic acid is given until the culture report is available.

The drug of choice for the infection caused by resistant Staphylococci is vancomycin. It is also used against Enterococci. For mixed Staphylococci and Gram-negative bacterial infections, gentamicin is used. Intravenous amphotericin B or intravenous fluconazole is administered for suspected *Candida* infection of the upper urinary tract.

15.2.3 Surgical Site Infections

Surgical site infections (SSIs) are dependent on the type of operation and the status of the patient who has been operated. They are very frequent. SSIs are classified into incisional SSIs and organ–space SSIs.

Around 33–67% of surgically infected wounds are cultured; among those, SSIs caused by *Staphylococcus aureus* account to around 15–20%, those caused by *Enterococcus* account to 15%, and the rest is caused by Gram-negative organisms and yeast. A single most significant intervention is the appropriate timing and use of peri-operative antibiotic prophylaxis. The risk of SSIs increases sixfold when the peri-operative antibiotic is administered after the incision; hence, the antibiotic should be administered 2 hours or less before the incision. In order to reduce nosocomial infections, reporting surgeon-specific SSI rates is important. SSIs are acquired usually during the operation itself. It might be exogenous, by the medical equipment, air, surgeons, or hospital staff, or endogenous, by the flora on the skin, flora in operative sites, or rarely from the blood used during surgery.

15.2.4 Nosocomial Pneumonia

About 1% of the admitted patient population develop pneumonia or bronchitis. The frequency increases in ICU patients. It is sensible to start empirical therapy as soon as a suitable sample of lower respiratory secretions is collected and sent for examination, without waiting for laboratory results. The antibiotic-sensitive organisms such as *Klebsiella* sp., *Pseudomonas aeruginosa*, and *Enterobacter* sp. reside in a hospital ICU. The choice of antibiotic used is dependent on the microbiological profile existing in the particular hospital's ICU.

15.2.5 Nosocomial Bloodstream Infections

Nosocomial bloodstream infections account to 14.2% of the total nosocomial infections. These infections are more common in ICUs and mainly include intravascular device-related infection and sepsis. There is a higher risk of getting bloodstream infections in a surgical ICU patient when compared with any other ICU patient. Nosocomial bloodstream infections can be grouped into the following:

1. Primary bloodstream infections: These are caused by unrecognized sources of infection.
2. Secondary bloodstream infections: These develop on the basis of documented infection with the same organism at some other anatomic site.

If not recognized and untreated, catheter-related sepsis results in increasingly severe sepsis. Before diagnosing catheter-related sepsis in a patient, it is important to rule out any other cause of sepsis. Catheter-related sepsis is characterized by frank pus at the entry site of the catheter in the presence of fever chills and tachycardia. The most common pathogen is coagulase-negative

Staphylococcus (31%). Other pathogens include *Staphylococcus aureus* (16%), *Enterococcus* sp. (9%), *Candida* sp. (8%), *Klebsiella pneumoniae* (5%), *Enterobacter* sp. (4%), and other organisms (27%).

15.2.6 Skin and Soft Tissue Infections

One of the major complicated situations in hospital-acquired skin and soft tissue infections is treating and controlling the burn wounds. Various agents such as 0.5% silver nitrate solution, 10.0% mafenide acetate cream, and silver sulphadiazine are used to treat topical wounds. Local antibiotics and prophylactic systemic antibiotic therapy constitute the best way for the treatment of burn wound infections. Immediately after burn injury, when the host's defences decline, systemic antibiotic therapy is recommended for the prevention of infections.

15.2.7 Other Nosocomial Infections

There are many other probable sites of infection, but the four most frequent and significant nosocomial infections are as follows:

1. Gastroenteritis, which is the most common nosocomial infection in children where rotavirus is the main pathogen
2. Other enteric infections and sinusitis
3. Endometritis and other infections of the reproductive organs during childbirth
4. Conjunctiva and the eye infections

15.3 INFECTION CONTROL PROGRAMMES: PROTOCOLS

The foremost duty of all individuals and services providing health care is the prevention of nosocomial infections. To reduce the risk of infection for patients and staff, everybody should ensure quality assurance, which includes personnel providing direct patient care, provision of material and products, training of health workers, management, and physical plant. Effective support at national and regional levels is a must.

15.3.1 National Programmes

To help the hospitals in reducing the risk of nosocomial infections, the responsible health authority must develop a national or regional programme. National programmes must set certain relevant objectives, which are consistent with other national health care objectives.

1. Guidelines for recommended health care surveillance, prevention, and practice should be developed and updated continually.
2. A national system to monitor selected infections and assess the effectiveness of interventions should also be developed.
3. Initial and continuing training programmes for health care professionals must be harmonized.
4. National or regional programmes must facilitate access to material and products essential for hygiene and safety.

15.3.2 Hospital Programmes

Effort must be taken to focus on the major preventions in hospitals and other health care facilities. To assess and promote good health care, sterilization, appropriate isolation, and other practices such as staff training and epidemiological surveillance should be developed in a yearly work plan.

1. **Infection control committee:** It is responsible for providing a team for multidisciplinary input and cooperation and information sharing.
2. **Infection control professionals or team:** The infection control team includes epidemiologists, nurses and physicians and the health care establishments must have access to all specialists.
3. **Infection control manual:** It is an important tool developed and updated by the infection control team, with review and approval by the committee. Nosocomial infection prevention manual must include recommended instructions and practices for patient care.

15.3.3 Infection-Controlling Responsibility

Responsibilities of Hospital Management

The responsibility of the hospital management is to provide leadership and adequate infrastructure and resources by supporting the hospital infection programme by the administration and medical management of the hospital.

Responsibilities of the Physicians

Their sole responsibility is the prevention and control of hospital infections by the following ways.

1. Giving direct patient care with the aid of practices that minimize infection
2. Maintaining appropriate practice of hygiene

Responsibilities of the Microbiologists

1. Microbiologists are responsible for handling staff and patient specimens to maximize the probability of a microbiological diagnosis.
2. They are responsible for developing appropriate guidelines for collection, transport, and handling of specimens.
3. Microbiologists ensure laboratory practices to be in appropriate standards.
4. They also ensure safer laboratory practices to prevent infections in staff.
5. They are responsible for performing antimicrobial susceptibility tests.

Responsibilities of the Hospital Pharmacist

The hospital pharmacist is responsible for obtaining, storing, and distributing pharmaceutical preparations by using practices through which transmission of infectious diseases is checked.

They participate in the hospital sterilization and disinfection practices. They also participate in the development of guidelines for antiseptics, disinfectants, and hand washing products.

Responsibilities of the Nursing Staff

The major role of the nurse is implementation of patient care practices in order to prevent infectious diseases. They should be familiar with the practice to prevent the occurrence of infection and check the spread of infection and should maintain appropriate hygiene for all patients throughout the duration of their hospital stay.

Responsibilities of the Central Sterilization Service

This department looks after the hospital areas, including operating suits. They are responsible for cleaning, decontaminating, testing, preparing for use, sterilizing, and storing aseptically all hospital equipment that have been sterilized.

Responsibilities of the Food Service

They should and must have knowledge in food safety, staff training, storage and preparation of foodstuffs, job analysis, and the use of equipment.

Responsibilities of the Laundry Service

One of the major roles of the laundry service is the selection of fabrics to use in different hospital areas. They also look after developing policies for working clothes and group of staff in each area and maintaining appropriate supplies.

Responsibilities of the Housekeeping Service

Their main role is the regular and routine cleaning of all surfaces to maintain high level of hygiene in the facility.

Responsibilities of the Maintenance

Maintenance gets collaborated with housekeeping, nursing staff, or other appropriate group in selecting equipment. They ensure early identification and prompt correction of any defects.

Responsibilities of the Infection Control Professionals

They are also called hospital hygiene service providers. Their major role is oversight and coordination of all infection control activities to ensure an effective programme.

1. An epidemiological surveillance programme is organized for nosocomial infections.
2. For supervising the use of anti-infective drugs, programmes have been developed with the participation of pharmacy.
3. Appropriate patient care practices are ensured to the level of patient risk.
4. The efficacy of the systems developed to improve hospital cleanliness and also the efficacy of the methods of disinfection and sterilization have been checked.

5. Development and provision of teaching programmes for medical personnel, allied health personnel, nursing staff as well as other categories of staff are organized.
6. Participation in the development and operation of regional and national infection control initiatives is ensured.
7. Assistance for smaller institutions is provided, and research (local, national, and international level) in hospital hygiene and infection control at the facility is undertaken.
8. In outbreak investigation and control, hospital hygiene service provides expert advice, analysis, and leadership.

15.4 NOSOCOMIAL INFECTION SURVEILLANCE

The excellence and safety of health care is indicated by the nosocomial infection rate in the patients in a health care facility. Monitoring the infection rate is an important step in the identification of local problems and priorities and the evaluation of the effectiveness of infection control activity by the progress of a surveillance process.

15.4.1 Objectives

The reduction in the occurrence of nosocomial infections and the costs involved in treating them is the ultimate aim of nosocomial infection surveillance. Some of the specific objectives undertaken by the surveillance programme are as follows:

1. The awareness of the clinical staff and other workers in the hospitals about nosocomial infections and antimicrobial resistance should be improved.
2. The frequency and distribution of nosocomial infections should be monitored.
3. The need for strengthened prevention programmes should be identified, and evaluation of the impact of preventive measures should be done.
4. Risk factor analysis must be performed in possible areas for improvement in patient care.

MULTIPLE CHOICE QUESTIONS

1. Hospital-acquired infections are caused by _____.
   (a) Lowered body defences in patients in the hospital
   (b) More sources of pathogenic organisms in the hospital
   (c) Immuno-compromized patients
   (d) All the above

   Ans. d

2. Patients who are immuno-compromized are likely to _____.
   (a) Acquire nosocomial infections very easily
   (b) Be monitored for infections from the hospital
   (c) Be treated as infectious
   (d) All the above

   Ans. a

3. The procedure from which a hospitalized patient acquires urinary tract infections is _____.
   (a) Urinary catheterization
   (b) Antibiotic therapy
   (c) Colonization of peri-urethral areas with hospital organisms
   (d) Immunosuppressive therapy

   Ans. a

4. Which of the following is not a common type of nosocomial infection?
   (a) Urinary tract infection
   (b) Surgical-wound infection
   (c) Skin infections
   (d) Nosocomial pneumonia

   Ans. c

5. The antibiotic-sensitive microorganisms such as *Klebsiella* sp., *Pseudomonas aeruginosa*, and *Enterobacter* sp. reside in _____.
   (a) Hospital linen
   (b) Hospital ICU
   (c) Hospital foodstuffs
   (d) Hospital equipment

   Ans. b

6. Which of the following is considered as an endogenous nosocomial infection?
   (a) Organisms infecting from another patient
   (b) Organisms infecting from a person's own gastrointestinal tract
   (c) Organisms inhaled from the surrounding environment
   (d) Organisms infecting from the surgical equipment

   Ans. b

7. What is the common vehicle transmission?
   (a) People infected getting exposed to the same pathogen
   (b) People getting infected by different sources
   (c) People getting infected by a single source
   (d) Infection occurring by all the ways mentioned above

   Ans. c

8. What kind of transmission is called faecal–oral transmission?
   (a) Pathogens affecting the gastrointestinal system
   (b) Pathogens ingested and then excreted in faeces
   (c) Food toxins produced before ingestion
   (d) Infected faeces leading to infections

   Ans. b

9. The survival of *Pseudomonas aeruginosa* in the hospital environment is aided by its capability to _____.
   (a) Survive the exposure to UV radiation
   (b) Resist a range of antibiotics
   (c) Grow even in some disinfectant solutions
   (d) Both (b) and (c)

   Ans. d

10. Methicillin-resistant strain of *Staphylococcus aureus* most commonly colonizes in the _____.
    (a) Liver   (b) Nose
    (c) Bladder   (d) Lungs

    Ans. b

11. After prolonged exposure to antibiotic treatment, which of the following opportunistic pathogen causes diarrhoea?
    (a) *Streptococcus pyogenes*
    (b) *Candida albicans*
    (c) *Clostridium difficile*
    (d) *Escherichia coli*

    Ans. c

12. By which of the following means is rotavirus commonly transmitted?

(a) Droplet transmission
(b) Faecal–oral route
(c) Sharp injuries
(d) Transmission by foodstuffs
   Ans. b

13. Which among the following patients is at most risk of getting infection easily?
    (a) A patient who has got severe and extensive burns
    (b) A patient with surgical wound
    (c) A patient with hepatitis B
    (d) A patient who is in the hospital for a long duration
       Ans. a

14. The removal of pathogens from an object is termed as _____.
    (a) Pasteurization
    (b) Sanitization
    (c) Sterilization
    (d) Disinfection
       Ans. c

15. Which of the following is the standard reason behind not taking precautions in hospitals to check nosocomial infections?
    (a) Not all patients in the hospital are a source of disease.
    (b) Special precautions are not necessary in hospitals as they will be clean.
    (c) Antibiotics treat almost all pathogens.
    (d) Infectious patients do not always show obvious symptoms and signs.
       Ans. d

16. The single most important preventive aetiology for infection control is _____.
    (a) Hand washing   (b) Isolation
    (c) Fumigation   (d) Bleaching
       Ans. a

17. An important cause of occupational blood-borne infection is _____.
    (a) HCV   (b) HIV
    (c) Influenza   (d) HBV
       Ans. c

18. The organism that is most commonly implicated in surgical site infections is _____.
    (a) *Staphylococcus aureus*
    (b) *Enterococcus* sp.
    (c) Gram-negative bacteria
    (d) All the above
       Ans. d

19. Coagulase-negative *Staphylococcus* is the most common organism responsible for _____.
    (a) Nosocomial urinary tract infection
    (b) Nosocomial bloodstream infection
    (c) Nosocomial pneumonia
    (d) Nosocomial surgical-wound infection
       Ans. b

20. Which among the following is an important patient factor responsible for getting nosocomial infection?
    (a) Diabetes mellitus
    (b) HIV infection
    (c) Age factor
    (d) All the above
       Ans. d

SHORT NOTES

1. Nosocomial infections
2. Primary source of infectious agents in hospitals

3. Most common types of nosocomial infection sites
4. Examples of Gram-negative and Gram-positive bacilli that cause nosocomial infection
5. Exogenous cross-infection
6. Endogenous infection

ESSAYS

1. Give an account on the epidemiology of nosocomial infections.
2. Explain the control measures taken by the hospital to reduce nosocomial infections.

Unit 4

PATHOGENIC ORGANISMS

Chapter 16 Bacteria	177
Chapter 17 Viruses	231
Chapter 18 Fungi	258
Chapter 19 Parasites	275
Chapter 20 Rodents and Vectors	285

16 Bacteria

CHAPTER OBJECTIVES

- 16.1 Staphylococcus
- 16.2 Streptococcus
- 16.3 Neisseria
- 16.4 Corynebacterium
- 16.5 Enterobacteriaceae
- 16.6 Mycobacterium
- 16.7 Vibrio cholerae
- 16.8 Spirochaetes
- 16.9 Mycoplasma
- 16.10 Rickettsia
- 16.11 Chlamydia

16.1 Staphylococcus
16.1.1 General Properties

Staphylococci are Gram-positive organisms that are spherical in shape and usually arranged in clusters (Figure 16.1). These organisms are a part of our normal flora. The genus *Staphylococcus* has 40 species; out of which, three are the most clinically important: *Staphylococcus aureus*, *Staphylococcus epidermidis*, and *Staphylococcus saprophyticus*. These organisms are omnipresent and cause localized suppurative lesions, pyogenic infections, food poisoning, and fatal septicemia in humans.

16.1.2 Cultural Characteristics

Staphylococci are non-motile, non-spore-forming, non-capsulated facultative anaerobes that are either single cocci or clusters in culture media. They grow readily on most nutrients, are metabolically active by fermenting number of sugars, and produce pigments. Their colonies are surrounded by a clear zone of haemolysis. *Staphylococci* are heat resistant, that is, they can withstand heat up to 60°C for 30 minutes.

Figure 16.1 Gram-positive Cocci (See page 354 for the colour image)

Their colonies grow at a pH of 7.4–7.6. Their colony morphology varies according to the type of medium used. In solid medium, they produce round and unblemished colonies. The colonies pick up the colour according to the pigment produced by the species of *Staphylococcus*.

16.1.3 Biochemical Properties

The pathogenic strains of *Staphylococcus* are catalase positive. They ferment mannitol without any gas production. The pathogenic strains of *Staphylococcus* are virulent mainly because of their ability to produce extracellular substances, which are toxins and enzymes. The exotoxins produced by these organisms are responsible for pore formation in white blood cells resulting in the lysis as well as the release of inflammatory mediators. They also generate enzymes such as coagulase, phosphatase, and DNAase, which are responsible for immunogenic responses in the host.

16.1.4 Mode of Transmission

Staphylococcal infections in humans are recurrent but localized at the entry by the normal host defences. They are normal flora of the body and are present in the nose, skin, and mucous membrane. The mode of infection is direct or indirect contact, especially through infected air droplets and dust and from heavily contaminated hospital environment. The carriers shed the bacteria through sneezing or coughing, and the infection may also spread through contaminated Intra Venous Syringes (IVS), catheters, and bandages. The entry of the cocci is through the skin and respiratory tract ending up in suppurative lesions and pneumonia.

16.1.5 Virulence Factor

Staphylococcus aureus is the most pathogenic organism, and they produce toxins and nontoxic aggressin, which are pathogenic to humans. The types of toxins produced are classified below:

1. **Cytotoxic toxin:** Cytotoxic toxins cause membrane damage ensuring pore formation on the outermost layers of the cell resulting in the lysis of the cell. There are four types of cytotoxic toxins, namely, α, β, γ, and δ. All these toxins are cellular destructive.
2. **Enterotoxin:** The enterotoxins produced by *Staphylococcus aureus* are superantigenic in nature. There are six types of enterotoxins produced by the organism, namely, enterotoxin A, B, C, D, E, and G. All these toxins are a leading cause for food poisoning. These toxins first act in the gut and end up in an emetic effect due to the action of toxins in the central nervous system resulting in vomiting and diarrhoea. An additional enterotoxin produced by *Staphylococcus*, that is, enterotoxin F, is responsible for 'toxic shock syndrome'.
3. **Exfoliative toxin:** These toxins are responsible for cutaneous deep infections of the skin. The 'staphylococcal skin scalded syndrome' is the common scarletiform eruption of the skin that occurs in children, and the virulence factors are exfoliative toxin A and B.
4. **Coagulase:** *Staphylococcus aureus* produces coagulase enzyme; it acts as a virulence factor by inhibiting phagocytosis and protecting the organism from the bactericidal essence present in the tissue. The enzyme protects the organism from internal barriers.

5. **Leukocidin:** This toxin acts as a destroyer of white blood cells in humans. They are of two types, namely, leukocidin S and F. They act as one of the major virulence factors for resistance-associated *Staphylococcus aureus*.

16.1.6 Pathogenesis

The potential virulence factors of *Staphylococcus* are surface proteins, which help the bacteria colonize into the tissues. They inhibit phagocytosis due to the presence of capsule. The toxins and enzymes secreted by *Staphylococcus* cause and spread infections in the host tissues causing diseases in human beings. The *Staphylococci* cause distinctive purulent lesions in humans. The types of infections are given below.

Cutaneous Infections

The host response to staphylococcal infection is inflammation, indurations, rise in temperature, and necrosis of tissues. They are responsible for skin infections that occur in the superficial region of the skin.

Symptoms: Staphylococcal infections cause boils and sepsis in wounds and burns. In serious conditions, skin infections may lead to pus formation, folliculitis (inflammation of hair follicles), furuncles, or impetigo (pus formation and eruption on the skin). Impetigo is formed due to close contact and is common in school children; hence, it is termed as school sores. The necrotic tissue heals after treatment.

Deep Infections

These cutaneous infections may enter and spread through the blood and lymphatics to other parts of the body.

Symptoms: The infection spreads via the bloodstream, and the bacteria get deposited inside the long bones leading to osteomyelitis. Staphylococcal infections in the bloodstream lead to septicemia, which in turn causes the spread of infections in the internal organs such as the lung, heart, skeletal muscle, or meninges resulting in pneumonia, endocarditis, meningitis, and other suppurative lesions in any organ.

Staphylococcal Skin Scalded Syndrome

It occurs mainly in children. A specific toxin produced by *Staphylococcus aureus*, that is, exfoliative toxin, acts on the skin causing 'erythematous cellulitis'. This syndrome is also termed as 'Ritter's disease'.

Symptoms: The symptoms are as follows:

1. Loss of the rigidity of the skin
2. Development of blisters (red rashes) on the layers of the skin
3. Skin getting peeled off with ensuing dermatitis

Toxic Shock Syndrome

It is a severe life-threatening syndrome generally caused by the toxic shock syndrome toxin-1. The toxins are multiorgan destructive in nature and lead to fatality.

Symptoms: The symptoms found are high fever, rashes, and hypotension. Many systems such as cardiovascular, mucosa, haematologic, hepatic, and central nervous system also get affected leading to organ failure.

Food Poisoning

Staphylococcal food poisoning is due to the toxins produced by the bacteria in the food. The infection starts its course of pathogenesis from the sixth hour of intake of contaminated food.

Symptoms: Common symptoms include vomiting, diarrhoea, and dryness occurring mainly due to the toxins (enterotoxins) produced by the bacteria. These symptoms stay for a week and then subside.

16.1.7 Laboratory Diagnosis

Specimen Collection

The collection of specimens is very important for clinical investigations. The samples are collected aseptically to avoid cross-contamination. The samples collected for diagnosis are as follows:

1. Swabs from infected areas
2. Pus
3. Blood
4. Sputum
5. Tracheal aspirates

Processing of Samples

As soon as the samples are collected, they are aseptically stored before processing. The specimens collected might be contaminated with mixed flora; therefore, they are cultured in 7.5% NaCl because salt inhibits the growth of other organisms. The process involves two steps, which are as follows:

1. **Direct smearing:** Direct smearing is done to perform a differential staining procedure called Gram staining. Examination of the stained smears of the sample will give an idea for further culturing and isolation of the organism.
2. **Culturing:** Gram staining categorizes the bacteria into Gram-positive or Gram-negative bacteria; then, the specific culture medium required to isolate the organism can be chosen accordingly.

A selective medium is one that has selective components for the enrichment of the organism is used for the isolation of the suspected organism from the samples obtained.

In the case of *Staphylococcus aureus*, blood agar and mannitol salt agar are the selective media used for isolation and identification. *Staphylococcus aureus* produces haemolysis in blood agar due to the presence of coagulase; hence, fibrinogen is converted to fibrin forming a 'zone of clearance'. Mannitol salt agar is a chromogenic medium for the isolation of *Staphylococcus aureus* because none of the other *Staphylococcus* species ferment mannitol.

Biochemical Analysis

The biochemical parameters of *Staphylococcus* are analyzed to differentiate the organism at the species level. Catalase and coagulase test are done for the confirmation of *Staphylococcus aureus* as these tests differentiate *Staphylococcus aureus* from other species of *Staphylococcus*.

Catalase Test: *Staphylococcus aureus* produces the enzyme cytochrome oxidase. A loopful of culture when introduced to a drop of hydrogen peroxide results in the formation of air bubbles indicating the release of oxygen due to the oxidation of H_2O_2 by the oxidase enzyme. This test acts as a confirmatory test for the identification of *Staphylococcus aureus*.

Coagulase Test: The pathogenic strain of *Staphylococcus* is coagulase positive. It produces clots when it reacts with human plasma. This test also acts as a confirmatory test for the identification of *Staphylococcus aureus*. Other non-pathogenic strains such as *Staphylococcus epidermidis* are coagulase negative.

Microbial Susceptibility: Susceptibility testing is very important in the case of staphylococcal infection because they are resistant to many antimicrobial agents such as penicillin and methicillin; hence, they are also called as Methicillin-Resistant *Staphylococcus aureus* (MRSA). The resistance to antimicrobial drugs is mainly due to the gene mecA, which binds to the proteins of antimicrobial agents and makes them inactive. The enzyme secreted by the organism binds to the consistent structure of antibiotics and inactivates the drug (β-lactam ring of penicillin gets inactivated by the enzyme secreted by *Staphylococcus aureus*). The MRSA strains should be tested for susceptibility before rendering treatment.

16.1.8 Prevention and Treatment

1. Proper hygiene among health care professionals is important.
2. Close contact with infected persons should be avoided.
3. The use of tampons should be avoided.
4. Care should be taken during the preparation of food, which will decrease the contamination of food.
5. Intravenous vaccines available for skin scalded syndrome can be used.
6. The MRSA strains are the cause of communicable diseases; hence, effective antibiotics should be tested for susceptibility in the laboratory before being prescribed.
7. Vancomycin, mupirocin (impetigo), and beta-lactamase-stable drugs (nafcillin and oxacillin) are commercially available for MSRA strains.

16.2 Streptococcus

16.2.1 Introduction

Streptococcus sp. are Gram-positive aerobes and facultatively anaerobic in nature. They are cocci in chains and are part of our normal flora. Certain species of *Streptococcus* are pathogenic in nature causing upper respiratory tract infection. The most important species is *Streptococcus pyogenes*, which causes pyogenic infection leading to suppurative and non-suppurative lesions. These species are classified on the basis of haemolysis on blood agar.

Alpha haemolysis or partial haemolysis produces green-coloured pigmentation on blood agar, and the bacilli belonging to this group are called as viridans streptococci or *Streptococcus viridians*. They cause upper respiratory tract infection.

Haemolytic streptococci produce beta haemolysis or complete haemolysis on blood agar. Most of the pathogenic strains belong to this group, and they are also grouped from A to V according to the carbohydrate moiety present on the cell wall. *Streptococcus pyogenes* belongs to group A; these groups are named after the scientist Rebecca Lancefield and hence termed as Lancefield groups.

In gamma-haemolytic or non-haemolytic *Streptococci*, no haemolysis is produced. Faecal *Streptococci* such as *Enterococcus* belong to this group.

16.2.2 General Properties

Streptococcus pyogenes are group A *Streptococcus* (non-motile, Gram-positive in chains, non-spore-forming, and facultatively anaerobic bacilli). This bacilli was identified by the scientist named Billroth in 1874 from an infected wound. These bacilli are recurrent pathogens affecting the upper respiratory tract, genitals, skin, and soft tissues.

16.2.3 Cultural Characteristics

They are facultatively anaerobic in nature and grow well at a temperature of 37°C. They grow in a normal basal medium, but the growth can be enhanced by the addition of carbohydrates and blood to the medium. The colonies formed are small, translucent, and opaque. On blood agar, they produce complete haemolytic colonies, which occur due to the complete clearance of the media colour. Haemolytic colonies are formed indicating the property of beta haemolysis (which is due to the presence of hyaluronidase and streptolysin, which lyse the haeme present in blood agar). This is one of the confirmatory tests followed in laboratories.

16.2.4 Biochemical Properties

They grow well in anaerobic conditions and are very sensitive to high temperature, hence stored at 4°C. They ferment sugars producing acid but no gas. *Streptococci* are catalase negative. These bacilli are sensitive to bacitracin.

16.2.5 Mode of Transmission

These pathogens are airborne; hence, they enter the respiratory tract of humans through inhalation. Then, they attach onto the epithelial cells of the throat and start invading the tissues with the help of the surface antigens and toxins they produce [M protein, surface antigens, haemolysin, exotoxin (pyrogenic), hyaluronidase, and streptokinase]. Each toxin and enzyme of the list mentioned above acts as a virulence factor for enhancing pathogenicity. After localization, these bacilli spread to the surrounding tissues leading to severe complications.

16.2.6 Pathogenesis

Streptococcus pyogenes causes both suppurative and non-suppurative infections in humans.

Suppurative Infections

Respiratory Infections: Respiratory infections such as sore throat, tonsillitis, pharyngitis, adenitis, and otitis media in severe cases lead to meningitis.

Skin Infections: Skin infections in wounds and burns lead to lymphangitis and cellulites in severe cases, which may in turn lead to septicemia. These infections of the skin are called as erysipelas in older patients and as impetigo in children. In such infections, suppurative lesions are formed, which affect the subcutaneous muscles leading to toxic shock syndrome.

Genital Infections: Genital infections are generally caused by group B *Streptococcus* affecting the genital tract of the mother during childbirth or miscarriage. This leads to puerperal sepsis leading to septicemia and death.

Non-suppurative Infections

Acute Rheumatoid Fever: It is the pharyngitis infection that leads to the production of cross-reactive antibodies, which in turn target the heart, brain, and joints leading to the degeneration of heart valves and myocardial lesions.

Glomerulonephritis (Inflammation of the Kidney): It leads to the loss of appetite, nausea, and bloody urine due to the inflammation of the Glomeruli of the kidney. It is caused by the cross-reactive antigens leading to the destruction of self cells. Untreated cases may result in death.

16.2.7 Laboratory Diagnosis

The samples collected are respiratory aspirates, pus from lesions, blood, urine, and genital swabs. The samples are processed for the isolation of the organism using selective and enriched media. A preliminary test, that is, Gram staining, is performed; for further confirmation of streptococcal infections, serological tests such as anti-streptolysin O titre and slide agglutination test are done.

16.2.8 Treatment

Streptococcal bacilli are sensitive to penicillin, amoxicillin, and erythromycin groups. Continuous antimicrobial treatment should be provided in order to avoid rheumatic fever. Prevention is better than cure, so it is always better to prevent mild infections with proper antimicrobial therapy.

16.3 Neisseria

16.3.1 Introduction

The genus *Neisseria* comprises non-motile, non-sporulating, facultative aerobic Gram-negative diplococci. They are commensals of the mucous membrane. *Neisseria* comprises 11 species; out of which, the following two are pathogenic to humans: *Neisseria gonorrhoeae* and *Neisseria meningitidis*. Both the species share 70% DNA homology but are differentiated by the type of diseases they cause and certain specific biochemical characteristics such as oxidation of carbohydrates.

16.3.2 Neisseria gonorrhoeae

General Properties

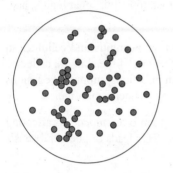

Neisseria gonorrhoeae is a non-motile, fastidious, Gram-positive diplococci (Figure 16.2). This organism causes venereal infections in humans. It was first discovered by Albert Ludwig Neisser in 1879. These cocci are intracellular pathogenic organisms and are therefore found predominantly within the polymorphonuclear cells. They carry certain virulence factors that safeguard them from phagocytosis.

Cultural Characteristics

Figure 16.2 Gram-negative Cocci (See page 354 for the colour image)

They are aerobic, non-motile, non-sporulating organisms but can grow well in anaerobic conditions. They grow best at an optimum pH of 7.2, temperature of 37°C, and 5%–10% CO_2. Thayer Martin agar is the medium of choice for the isolation of *Neisseria* sp. This medium contains a basic supplement of minerals and vitamins including haemoglobin (which is lysed by heating) for the enrichment of the bacteria. This medium also contains antibiotics such as vancomycin, lincomycin, and nystatin, which inhibit the growth of other organisms, and hence serves as a selective medium for the isolation of *Neisseria* sp. The colonies of gonococci are small, curved, translucent, non-pigmented, non-haemolytic, and emulsifiable.

Biochemical Properties

Gonococci oxidize glucose by producing acid but no gas, and hence, the carbohydrate patterns obtained due to the oxidation are useful in identifying the species. Gonococcus is oxidase positive, and the oxidase test serves as the confirmatory test for the identification of the isolate.

Virulence Factors

Gonococci are antigenically heterogeneous and hence are capable of modifying their exterior structures to evade host defence mechanisms.

Pili (fimbriae): These are hair-like appendages used for the attachment of organisms to host cells. They are made up of a protein called pilin, which contains hydrophobic amino acids. These proteins enhance the attachment of organisms to host cells and also resist phagocytosis.

The pilin protein is antigenically different in almost all strains of *Neisseria*, and every single strain of *Neisseria* has the ability to express a distinct form of pilin.

OPA Proteins: These are outer membrane proteins that help to facilitate the interaction with host cells. They act as receptors for host cells facilitating bacterial attachment, incursion, and response by a particular host. Three to four types of OPA proteins are present on the outer layer, which are coded by 12 or more genes.

POR (Porin Protein P II): These proteins present in the cell membrane are responsible for the penetration and intracellular attack on human cells. These proteins provide variable resistance to gonococci against human serum by binding to complement components. Each strain of *Neisseria* comprises two or three outer membrane POR proteins that are antigenically variable.

Rmp (Protein III): Reduction-modifiable protein is an outer membrane protein that complexes with POR, OPA, and lipooligosaccharide (LOS). Its function is to block the antibodies produced against the organism.

Lipooligosaccharide: The toxic effects of gonococcal infections are due to the presence of LOS in the cell wall of the organism. The LOS secretes two enzymes, namely, protease and phospholipase, which mediate the mucosal damage of host tissues. The gonococcal LOS structurally resembles the glycosphingolipids of the human cell membrane, thereby masking themselves from the immune response of host cells. They provide resistance to gonococci from being killed by the bactericidal activity of normal human serum.

Other Proteins: Other proteins such as Lip (H8), Fbp (ferric-binding protein), and IgA1 protease also participate and enhance the pathogenesis of the organism.

Mode of Transmission

The transmission is through sexual contact with an infected person. The bacteria are transmitted through the mucosal exudates of the infected person from vagina or seminal fluids. Gonococcal conjunctivitis occurs in neonates from the infected mother's vaginal canal during birth. The incubation period of the infection may extend for months if it is asymptomatic; else, it is usually 2–7 days from the onset of infection. In females, the infection starts from the cervix and ascends into the uterus and fallopian tubes. Females are mostly asymptomatic when compared with males. Males show symptoms after the onset of the infection. Occasionally, they enter into the bloodstream ending up in bacteremia, where the bacterial cells get lysed because of the bactericidal effects of the serum and endotoxins released due to the free Lipo oligosaccharide (LOS) result in circulation.

Pathogenesis

The pilated virulent bacteria attack the mucosal layers ending up in urethritis, salpingitis, and chronic gonococcal cervicitis in women. In men, the infection leads to urethritis. The symptoms in males and females are as follows:

Symptoms and Diseases in Females: The symptoms are as follows:

1. Inflammation of the urethra
2. Pain or burning sensation during urination

3. Frequent urination
4. Yellow or bloody vaginal discharges
5. Chronic pelvic pain in untreated cases
6. Pelvic inflammatory diseases (PIDs)
7. Tubular infertility
8. Ectopic pregnancy
9. Nausea and vomiting
10. Abdominal cramps
11. Fever

Symptoms and Diseases in Males: The symptoms are as follows:

1. Painful pus discharge from penis
2. Pain or blazing sensation during urination
3. Distended testicles

Symptoms and Diseases in Neonates: The symptoms are as follows:

1. Infection of eyes
2. Conjunctivitis
3. Blindness in the case of progressing infection

Other Symptoms: The other symptoms include the following:

1. Bacteremia
2. Infection in rectum resulting in itching and painful bowel movement
3. Skin infection due to bacteremia
4. Bone infections
5. Haemorrhagic pustules in limbs, hands, forehead, and legs.

Laboratory Diagnosis

Specimens Collected:

1. Pus
2. Urine
3. Vaginal discharges
4. Swabs from cervix and conjunctiva
5. Blood sample in the case of bacteremia
6. Fluids from arthritic joints (suitable for culturing)

Processing of Samples: For microscopic identification, smears are prepared from the swabs and discharges before culturing. Gram staining is the standard method followed for the differentiation of the organism as Gram-positive or Gram-negative. The stained smears from the sample show intracellular Gram-negative kidney bean-shaped diplococci in polymorphonuclear leukocytes. This staining is also used as a confirmatory test after culturing the isolates in selective media.

Culture Technique: The primary inoculation is done in chocolate agar and selective media that contain the antimicrobial agents that inhibit the growth of the bacterial and fungal contaminates in the clinical isolates and enhance the growth of gonococcus. The isolates from the specimens collected are inoculated onto the selective media and incubated at 37°C with 5% CO_2. After 24 hours of incubation, colonies are selected and confirmed using Gram staining.

Biochemical Investigation: The isolates are confirmed by performing certain biochemical tests such as oxidase test and catalase test. The confirmation of isolates is also done using advanced tools including nucleic acid methods such as real-time PCR-based assays, which are sensitive and less time-consuming due to automation.

Prevention and Treatment

Due to the widespread use of penicillin, gonococci have developed resistance against penicillin. Fluoroquinolones were the recommended drugs for gonococcal infections. The Center for Disease Control and Prevention has recommended ceftriaxone to be administered intravenously as a single dose in the case of chronic infections. The other antibiotics prescribed for sexually transmitted diseases are as follows:

1. Cephalosporin
2. Cefixime
3. Azithromycin

16.3.3 *Neisseria meningitidis*

General Properties

Neisseria meningitidis are non-motile, fastidious, Gram-positive bean-shaped diplococci. This organism causes meningitis in humans. These cocci are found in the upper respiratory tract in humans, where they invade the mucous membrane causing meningitis. Both gonococci and meningococci share the same homology and cultural characteristics but meningococci can be differentiated by the presence of polysaccharide capsule, which is absent in gonococci. Meningococci oxidize glucose and maltose by producing acid but no gas, and the carbohydrate patterns obtained after oxidation are useful in identifying the species.

Virulence Factor

The capsular polysaccharide acts as an antigen and is found in circulation during the vigorous infection. Pili present in meningococci enhance the attachment and evade phagocytosis. The cytolytic enzyme acts as the major virulence-causing mediator in the host. The POR and OPA proteins present resemble that of gonococci and share the similar pathogenesis pattern caused in the host.

Mode of Transmission

The transmission is through aerosols, droplets, or human secretions, and the infection is spread due to the inhalation of the infected droplet nuclei. The infected droplet, which carries the organism, attaches with the help of pili, and the organism invades the host mucosal membrane. The isolates are most frequently found in the oropharynges or nasopharynges. The pathogen may also enter the bloodstream to cause bacteremia.

Pathogenesis

The bacteria invade the mucosal area of the upper respiratory tract causing the following symptoms:

1. High fever
2. Rashes
3. Headache
4. Vomiting
5. Rigid muscle
6. Intravascular clotting
7. Circulatory failure
8. Thrombosis of blood vessels (in organs and the brain)
9. Infiltration of polymorphonuclear leukocytes (purulent exudates covering the brain)

Laboratory Diagnosis

Specimens Collected:

1. Blood
2. Nasopharyngeal swab
3. Cerebrospinal fluid

Processing of Samples: Smears are prepared from the swabs and cerebrospinal fluid, and Gram staining is performed. The stained smears show intracellular Gram-negative bean-shaped diplococci in polymorphonuclear leukocytes. This staining is also used as a confirmation test after culturing the isolates in selective media.

Culturing and Confirmatory Test: The sample from the cerebrospinal fluid is inoculated onto the chocolate agar and selective media such as Thayer Martin agar, which contains the antimicrobial agents that enhance the growth of meningococci. The colonies are then picked up to confirm the organism by Gram staining and biochemical oxidase test, which is the positive test for the identification of *Neisseria meningitidis*.

Treatment

Penicillin G is the only drug of choice; however, in the case of penicillin allergies, chloramphenicol or cephalosporin can be given orally. Polysaccharide vaccines are available for administration. These trigger the immune system to produce T cell-dependent response to the vaccine; hence, the immunity against the infections lasts long.

16.4 Corynebacterium
16.4.1 General Properties

Corynebacterium sp. are non-motile, Gram-positive pleomorphic rods (Figure 16.3). These bacilli were first cultivated by Loffler in 1884 but were observed by Klebs, hence called as Klebs–Loffler bacilli. They are club-shaped bacilli and have metachromatic granules (which can be stained with aniline dye). Following are the few species of *Corynebacterium* that are medically important: *Corynebacterium diphtheriae, Corynebacterium pseudodiphtheriticum, Corynebacterium minutissimum, Corynebacterium xerosis*, and *Corynebacterium striatum*. These bacilli have a characteristic feature of producing a pseudomembrane on the diseased site, hence called as diphtheros (leather-like appearance).

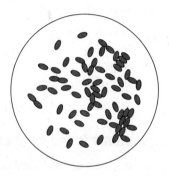

Figure 16.3 Gram-positive Bacilli (See page 354 for the colour image)

16.4.2 Cultural Characteristics

These bacilli are non-spore-forming, non-capsulated, non-motile pleomorphic bacilli. They grow well in aerobic conditions but are facultatively anaerobic in nature. They grow well in media enriched with blood or serum. An example is Loeffler's serum slope, which comprises bovine serum, peptic digest, egg powder, sodium chloride, and agar. The pH of this medium is 7.6. The ingredients enhance the proteolytic activity of the organisms and stimulate their growth. Another medium used for their cultivation is blood tellurite agar; the tellurite present in the medium inhibits the growth of other organisms.

16.4.3 Biochemical Properties

Corynebacterium sp. ferment sugar without gas formation. They are catalase positive and oxidase negative.

16.4.4 Mode of Transmission

They are transmitted through infected droplets. They enter the respiratory tract, localize, and invade the tissues causing pseudomembrane formation resulting in a contagious disease called diphtheria. The exotoxin produced by the organism forms the pseudomembrane (which is nothing but the dead cells of the invaded tissues).

16.4.5 Pathogenesis

Subsequent to invasion by the pathogen, it takes about 2–5 days for colonization (incubation time). Then, the exotoxin, that is, diphtheria toxin, () produced by the organism forms a pseudomembrane on the invaded surfaces such as tonsils, pharynx, and even wounds. Thereafter, the toxin enters the circulation and invades other vital organs.

Symptoms

The symptoms are as follows:

1. High fever
2. Pharyngitis
3. Tonsillitis
4. Swelling of the neck
5. Inflammation in the wound areas

Symptoms in the case of Critical Conditions

The symptoms under critical conditions are as follows:

1. Myocarditis (leading to heart failure)
2. Neuritis resulting in diaphragmatic paralysis

16.4.6 Laboratory Diagnosis

Throat swabs, exudates from the lesions, mucopulurent discharges, and blood products are collected for the isolation of diphtheria bacilli. Gram staining and culturing in Loeffler's serum slope are the preliminary tests done for the isolation of the bacilli. Confirmatory tests such as *in vivo* and *in vitro* test for toxicity are also done.

The most important *in vitro* test done for the confirmation of the bacilli is the Elek's gel precipitation test. Here, the diphtheria antitoxin is incorporated into a serum-containing medium. The medium is then inoculated with the bacilli and incubated. The toxin produced by the bacilli reacts with the antitoxin in the medium resulting in a precipitation reaction indicating the presence of bacilli in the sample.

In vivo test involves the inoculation of the grown culture from the medium subcutaneously into guinea pigs. If the culture is positive for the bacilli, the pigs die within 4 days. This invasive procedure is presently not practised.

16.4.7 Treatment

Immunization is the most important mode of treatment for the disease. Active immunization is given in the form of triple vaccine, which is a combination of diphtheria toxoid, tetanus toxoid, and live pertussis, also called as DTP. The first three doses are given at regular intervals, and the fourth dose is given after a year. These doses are given to enhance permanent immunity in the host. Passive immunity is enhanced by the administration of Anti Diphtheria Serum (ADS) to the patient who is affected with chronic diphtheria. Antimicrobial drugs such as penicillin, erythromycin, and other higher penicillins can be administrated to the patient.

16.5 Enterobacteriaceae
16.5.1 Introduction

Enterobacteriaceae comprises the organisms that are localized in the gastrointestinal tract of humans and animals. All the organisms belonging to this family are Gram-negative rods (Figure 16.4) and are facultatively anaerobic in nature, but they have the capacity to grow well in the presence of oxygen. The family includes many organisms, namely, *Escherichia, Proteus, Salmonella, Shigella, Enterobacter, Klebsiella,* and *Serratia*. Some organisms are a part of the normal flora in human beings, and some are pathogenic to the human population. Most members of *Enterobacteriaceae* share a common characteristic, that is, they are motile with peritrichous flagella, and some are non-motile. They ferment glucose producing gas, are catalase positive and oxidase negative, and reduce nitrates to nitrites. These organisms are classified into the following three groups according to their capability to ferment lactose: lactose fermenters, non-lactose fermenters, and late-lactose fermenters.

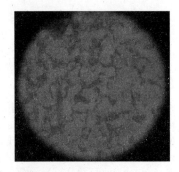

Figure 16.4 Gram-negative Bacilli (See page 355 for the colour image)

The members of *Enterobacteriaceae* can also be classified according to the antigens found in the cytoplasm, cell membrane, cell wall (O antigen), enterobacterial common antigen, flagella (H antigen), capsule (K antigen), and fimbriae (F antigen). The pathogenicity of the organism is enhanced by these factors, and hence, they are used for identifying the genus and species. The *Enterobacteriaceae* bacilli have a thin layer of mucopeptide and a thick outer layer. The thick outer layer is composed of a complex made of lipopolysaccharide and lipids. These complexes form the fundamental part of the Gram-negative cell wall. The outer layer acts as a pathological agent and is responsible for secreting endotoxins in humans.

16.5.2 Escherichia

General Properties

Escherichia are motile, non-spore-forming, rod-shaped Gram-negative organisms. They are facultative anaerobes that ferment carbohydrates with gas production. *Escherichia coli* was first identified by a German scientist Theodore Escherich in 1885; he isolated these bacilli from the intestinal flora of an infant. This organism has wide potential in being used as a vector for cloning purposes and also as a control organism for testing the competence of antimicrobial agents and other chemical substances. As *Escherichia coli* is found in the normal intestinal flora of humans and animals, it is excreted in the faeces; it has the ability to survive in the environment. *Escherichia coli* strains are usually motile with peritrichous flagella, and they produce different types of pili, which are antigen specific. The cells are characterized by the presence of a capsule (slimy layer of acidic polysaccharide).

Cultural Characteristics

Escherichia coli produces acid on exposure to lactose and hence is referred to as a lactose fermenter. It produces a metallic sheen on differential media such as eosin methylene blue

(EMB) agar. It is motile and is found in the intestinal tract of warm-blooded animals, hence termed as faecal coliform. It has the capacity to grow rapidly (with a doubling time of 20 minutes), hence called as rapid grower. The optimum temperature required for its growth is 37°C.

Biochemical Properties

Escherichia coli is indole and methyl red positive, and this characteristic is used for its identification. *Escherichia coli* is a lactose fermenter and produces pink-coloured colonies on MacConkey agar (This colour change is due to the fermentation of sugar.). It produces acid and gas on exposure to sugars such as glucose and mannitol. It is positive for 4-methylumbelliferyl-β-D-glucuronide (MUG) test because of its tendency to produce the enzyme β-glucuronidase; hence, this is used as a confirmatory test for identifying *Escherichia coli* from clinical isolates.

Mode of Transmission

The organism belongs to the normal flora of the intestinal tract of human beings. Generally, it does not cause infections but becomes pathogenic when it invades sites other than the intestine. The transmission is through the ingestion of contaminated water or food containing the toxins produced by the organism. The transmission may also occur through contaminated catheters in the case of hospital-acquired infections. The route of entry and invasion is the urinary tract, gastrointestinal tract, and bloodstream (in severe conditions).

Pathogenesis

Urinary Tract Infections: The most common infection caused by *Escherichia coli* in young children and women is the urinary tract infection (UTI). The infection spreads from the urethra and affects the kidney leading to cystitis (inflammation of the bladder), pyelonephritis (inflammation of the kidneys), and ascending infections.

Symptoms: The symptoms are as follows:

1. Pain during urination (dysuria)
2. Low-grade fever
3. Frequent urination
4. Red blood cells in the urine (haematuria)
5. Urine containing pus cells (pyuria)

The type of UTI depends on the type of antigen produced by the organism. O and K antigens cause the mild type of infection: cystitis is caused by O type antigens. And severe infections such as pyelonephritis and upper UTI are caused by pili antigens or K antigens, which are more virulent when compared with capsular O antigens.

Diarrhoeal Diseases: These are often caused by the toxins produced by the organism. Hence, they differ in the characteristic virulence produced and are as follows:

1. **Enteropathogenic *Escherichia coli* (EPEC):** This strain is responsible for causing diarrhoea in infants and children resulting in fever, nausea, vomiting, and watery

diarrhoea without bloody stool. The transmission is by contaminated food and improper hygiene. The pathogenic *Escherichia coli* attacks the microvillus by adhering to the surface causing distinctive lesions. The virulence factor accountable for the disease is caused by the O and H antigens. The disease can be controlled by effective antibiotic administration.

2. **Enterotoxigenic *Escherichia coli* (ETEC):** They are otherwise called as 'traveler's diarrhoea'. Two types of toxins are produced, namely, the heat-labile exotoxin and the heat-stable toxin. These toxins resemble the cholera toxin but are mild in comparison. Watery stools, fever, and nausea are the symptoms caused by Enterotoxigenic *Escherichia coli*. Watery stools are mainly due to the fluid accumulation in the gut lumen as a result of the production of increased amounts of cAMP leading to the secretion of chlorides and water. Rehydration and antibiotic therapy are successful prophylaxis measures for Enterotoxigenic infections.

3. **Enteroinvasive *Escherichia coli* (EIEC):** It resembles *Shigella* dysentery. This strain is a non-lactose fermenter and is non-motile. The epithelial cells of the intestinal mucosa are invaded resulting in dysentery in mostly children and people who travel.

4. **Enteroaggregative *Escherichia coli* (EAEC):** Rigorous and chronic diarrhoeal diseases are caused by this strain. They are mainly transmitted by contaminated food. The toxin produced by this strain is called as *Shigella* toxin. They are associated with haemorrhagic colitis, haemorrhagic uremic syndrome, anaemia, and thrombocytopaenia. The only preventive measure to be followed is proper cooking of the food products.

Sepsis: *Escherichia coli* enters into the blood causing sepsis. This secondary infection occurs if the primary infection is not treated properly.

Meningitis: Meningitis caused by *Escherichia coli* resembles the infection caused by group B streptococci as in both cases, capsular K antigens attack the meninges leading to meningitis.

Laboratory Diagnosis

Specimen Collection: Different samples are collected according to the severity of the infection, which are as follows:

1. Urine
2. Blood
3. Stool
4. Spinal fluid

Processing of Samples: The samples are processed in two ways. They are smeared for staining and cultured in blood agar and differential media for further biochemical processing. They are further confirmed by inoculating into special media such as EMB to see their characteristic appearance.

Treatment

There are no specific treatments available. Replacement of fluids lost due to dehydration and a combination of drugs can be administered in chronic cases. The drugs prescribed are as follows:

1. Sulphonamides
2. Ampicillin
3. Cephalosporin
4. Fluoroquinolones
5. Tetracycline
6. Bismuth subsalicylate (which inactivates the endotoxins produced by *Escherichia coli*)

These drugs are given in various concentrations and ratios. Antimicrobial testing is done before prescribing drugs as plasmid-mediated drug resistance is possible. In certain conditions, surgical corrections are also done, such as in the case of urinary tract obstruction due to infections and so on. Proper food sanitation and hygiene should be maintained to avoid contamination.

16.5.3 Klebsiella

General Characteristics

Klebsiella sp. are non-motile capsulated rods that are commensals of the intestine. They are also widely distributed in the soil and water. This bacillus was first identified by a German bacteriologist Edwin Klebs. However, the causative agent of pneumonia, that is, *Klebsiella pneumoniae*, was identified by Friedlander in 1883; hence, it is also termed as Friedlander's bacilli. Many different species have been identified according to the similarities in DNA homology; out of which, the following three species have been identified as human pathogens: *Klebsiella pneumoniae*, *Klebsiella ozaenae*, and *Klebsiella rhinoscleromatis*. The pathogenicity of the organisms is characterized by two types of antigens present on the surface of the bacteria, namely, cell surface antigen (lipopolysaccharide), that is, O antigen, and capsular K antigen. These pathogens are more prevalent in hospital areas and generally affect hospitalized patients 48 hours after admission; hence, the infections caused are also termed as nosocomial infections or hospital-acquired infections.

Cultural Characteristics

The optimum temperature for their growth is 37°C. They are non-motile and ferment sugars with the production of acid and gases. They are capsulated and hence produce mucoid colonies on media. The capsules are identified by performing a special technique called capsular staining or Gram staining (in which an aura around the bacilli is shown).

Biochemical Properties

IMViC reaction ($- - + +$) is used for the differentiation and identification of the bacteria from other *Enterobacteriaceae* species. Acid and gases are produced at 44°C on inoculation into a lactose-containing medium, hence termed as lactose fermenters. They also produce the enzyme urease.

Mode of Transmission

The transmission is by contaminated ventilators, food substances, catheters, medical equipment, and blood products. Transmission may also be due to infected hospital staff. As soon as the patients come in contact with the contaminated materials in the hospital, they get infected. The incubation period is about 48 hours, which ends up in nosocomial infections.

Pathogenesis

Klebsiella sp. causes different kinds of hospital-acquired infections. They are as follows:

Pneumonia: Pneumonia is caused by *Klebsiella pneumoniae*, which enters through contaminated medical devices or procedures done to the patients. The organism colonizes the mucosal region, attacks the epithelial cells of the upper respiratory tract, and starts eliciting enterotoxins resulting in pneumonia. Immense inflammation in the lung region leads to the formation of mucus exudates resulting in leucocytosis and in fatal cases necrosis and abscess formation in the lungs.

Symptoms: The symptoms are as follows:

1. Fever
2. Chills
3. Cough
4. Bloody sputum
5. Lung necrosis (fatal cases)

Urinary Tract Infections: *Klebsiella* sp. are a common cause of UTIs in hospitals. Most commonly, *Klebsiella ozaenae* is the common cause of UTIs in hospitals. Most strains are resistant to antibiotics, and if not treated, infections may lead to septicemia and meningitis.

Symptoms: The symptoms are as follows:

1. Irritation during urination
2. Fever
3. Abdominal pain
4. Bacteriuria
5. Cystitis
6. Hepatic diseases (liver abscess) in severe cases

Bacteremia: Here, the bacteria spread in the blood due to the spread of toxins in the blood.

Ozena: It is a disease caused by *Klebsiella ozaenae* in the nasal mucosa. These capsulated organisms invade the nasal cells resulting in foul-smelling mucopurulent discharges. They are also responsible for various other infections such as UTIs and ear infections.

Rhinoscleroma: It is caused by *Klebsiella rhinoscleromatis*, which is an intracellular pathogen causing unceasing granulomatous respiratory infection that affects the nose.

Laboratory Diagnosis

Samples such as sputum, nasal swabs, blood, urine, and lesions in the infected regions are collected. Gram staining and motility and biochemical tests are carried simultaneously with the culturing of the clinical isolates in their selective and differential media. Antibiotic sensitivity testing is done because of the characteristic of drug resistance of the pathogens.

Treatment and Control

Ampicillin, sulphonamides, cephalosporins, tetracycline, nalidixic acid, ciprofloxacin, and gentamicin are prescribed after antibiotic sensitivity testing is done in the clinical isolates. Proper hygiene, sterilization of the hospital equipment in a periodic pattern, and health education and awareness given to the hospital staff are necessary in controlling the diseases.

16.5.4 *Proteus*

General Properties

These bacilli are motile, Gram-negative pleomorphic rods and are facultatively anaerobic in nature. *Proteus* sp. is a normal commensal of the intestine but on entry into the urinary tract or respiratory tract may cause infections in humans. They are classified into three important genera, which are as follows:

1. *Proteus*
2. *Provindencia*
3. *Morganella*

All these strains are strict urease producers resulting in stone formation. The genus *Proteus* has two medically important species that cause UTIs in humans (which is hospital acquired, that is, nosocomial), namely, *Proteus mirabilis* and *Proteus vulgaris*, which can be differentiated by their biochemical properties.

Biochemical Characteristics

Proteus mirabilis is indole negative, methyl red positive, and Voges–Proskauer negative. However, *Proteus vulgaris* is indole positive; this differentiates the two species. They reduce nitrate without producing gas. They are catalase positive, oxidase negative, phenyl deaminase positive, and urease positive. The urease enzyme produced by the organisms hydrolyses urea into ammonia and carbon dioxide.

Virulence Factors

The flagellar H antigens and fimbriae act as virulence factors producing a characteristic swarming growth, which is due to the peritrichous flagella. This type of bacilli with swarming growth are called 'Hauch' bacilli (meaning that they grow like mist). The pili and flagella act as the mode of transport and attachment to the urinary tract even after flushing and affect humans causing UTIs, cystitis, and pyelonephrilis. These bacilli are pleomorphic in nature: they are swimmer cells in liquid media and have a swarming growth in solid media. The outer membrane protein acts as a virulence factor and possesses immunogenic properties.

Mode of Transmission

The transmission is through contact (inoculation). As they are the normal flora of the intestine, they are also found in the soil and water. They swim to the host and on contact invade, colonize, and cause infections. They have the ability to form biofilms after colonization mainly on the instruments such as catheters, which on insertion into the patients cause urinary tract infestation producing nosocomial infections.

Pathogenesis

Proteus mirabilis causes UTIs in human beings, which is hospital acquired, and is also related to kidney infections such as the formation of stones due to ascending infections. The organism produces two types of UTIs, namely, systemic and ascending infections.

Symptoms: The symptoms are as follows:

1. Low-grade fever
2. Discomfort during urination
3. Inflammation of the bladder

Symptoms due to Stone Formation: The symptoms owing to stone formation are as follows:

1. Unbearable pain in the pelvic region (renal colic)
2. No bladder control
3. Restlessness
4. Haematuria
5. Nausea
6. Vomiting

Proteus vulgaris is also an important nosocomial pathogen causing UTIs and respiratory and eye infections from infected catheters, ventilators, and retinoscopes that are not disinfected properly.

Provindencia sp. (*Providencia rettgeri*, *Providencia alcalifaciens*, and *Providencia stuartii*) cause hospital-acquired UTIs and also burn infections.

Morganella sp. are a part of the normal flora of the intestine, but certain strains are antibiotic resistant and cause hospital outbreaks.

Laboratory Diagnosis

The samples collected for processing are urine and faeces. Samples are also collected from the instruments used for patients. The collected samples are subjected to culturing and biochemical tests for the confirmation of the species.

Treatment

Proper sanitation and disinfection of the hospital equipment and proper administration of drugs after antibiotic sensitivity testing will stop the outbreaks in hospitals. The antibiotics prescribed are as follows:

1. Ampicillin
2. Sulphonamides

3. Cephalosporin
4. Nitrofurantoin
5. Gentamicin
6. Amikacin

16.5.5 Shigella

General Properties

Shigella causes dysentery (characterized by blood-stained mucopurulent stools) in human beings. This bacillus was identified by a Japanese scientist Dr. Kiyoshi Shiga in 1897 during an outbreak in Japan. The *Shigella* bacilli are classified according to their antigenic structure and are of the following four types:

1. *Shigella sonnei*
2. *Shigella boydii*
3. *Shigella flexneri*
4. *Shigella dysenteriae*

These are non-motile, Gram-negative facultatively anaerobic bacilli belonging to the *Enterobacteriaceae* family. These bacilli attach to the tissue, invade the cells, and produce toxins causing dysentery.

Cultural Characteristics

These bacilli are facultatively anaerobic in nature but grow well in aerobic conditions also. Their optimum growth temperature is 37°C. They are non-lactose fermenters except for *Shigella sonnei* and grow readily even in a basal medium. The selective medium used for the isolation of *Shigella* sp. is deoxycholate citrate agar, in which red-coloured smooth and translucent colonies are produced. These organisms are sensitive to heat and acids. *Shigella sonnei* is more resistant when compared with other *Shigella* species.

Biochemical Properties

Shigella sp. are urease, catalase, and oxidase negative. They reduce nitrates. They ferment glucose by producing acid with gas. Other than *Shigella dysenteriae*, the remaining three species of *Shigella* ferment mannitol.

Mode of Transmission

The transmission is through the ingestion of contaminated food and improper hygiene. The consumption of contaminated food results in the invasion of the bacilli on the epithelial cells of villi, where they multiply in large numbers producing three different toxins, namely, enterotoxin, neurotoxin, and cytotoxin (Shiga toxin), resulting in the haemorrhage of the tissues.

Pathogenesis

The intense infection caused by *Shigella* sp. is bacillary dysentery. The incubation period for the infection ranges from 3 to 7 days, and 1- to 100-bacilli load is enough for the inception of the infection. The symptoms include the following:

1. Low-grade fever
2. Vomiting
3. Abdominal cramps
4. Bloody stool
5. Mucopulurent discharges
6. Haemolytic uremic syndrome (in severe cases)

Laboratory Diagnosis

Samples Collected: Faeces, blood, and discharges are collected, and the bacilli are isolated using the selective medium. As these bacilli are delicate and are killed on contact with acid and heat, they are to be inoculated into the transport medium that contains mucus and has a pH of 7.0–7.4 (which is optimal for their growth). On inoculation into the selective medium followed by incubation, colonies are isolated for additional tests such as motility and biochemical test for further confirmation. The slide agglutination test is used for the isolation of the species using polyvalent sera.

Treatment and Prevention

Antibiotics such as ampicillin and fluoroquinolones are used. In severe cases, oral rehydration is given to avoid dehydration due to the loss of fluids. Proper hygiene and sanitation are required. Proper cooking of food and avoiding the consumption of reheated foods will minimize the cause of infections.

16.5.6 *Salmonella typhi*

General Characteristics

Salmonella bacilli belong to *Enterobacteriaceae* family. These bacilli are Gram-negative, motile, and facultatively anaerobic in nature. They are intestinal parasites causing enteric fever and gastroenteritis disorders in humans. *Salmonella typhi* is the causative organism of typhoid fever, which is mainly transmitted from the faeces of infected individuals through contaminated food and water. This bacillus was isolated by a pathologist named Georg Theodor August Gaffky and was confirmed by Eberth in 1880; hence, the bacillus has been named after the two scientists as Gaffky–Eberth bacillus.

Cultural Characteristics

Salmonella typhi are facultatively anaerobic in nature but grow well even in aerobic conditions. The optimum temperature required for their growth is 37°C. They are non-lactose fermenters but produce H_2S and form black-coloured colonies on selective media such as Wilson and Blair bismuth sulphite medium.

Biochemical Properties

Many subtypes of *Salmonella* sp. cause enteric fever. They are grouped according to their biochemical properties and antigenic structure. The four important bacilli causing typhoid infections are *Salmonella typhi*, *Salmonella paratyphi A*, *Salmonella paratyphi B*, and *Salmonella paratyphi C*. All these bacilli ferment glucose forming acid and gas except for *Salmonella typhi* (which produces acid but no gas). They are citrate positive, indole negative, urease negative, and do not hydrolyse gelatin. H_2S is produced by all the subtypes except for *Salmonella paratyphi A*.

Antigenic Structure

The three important antigens are the flagellar H antigen, surface Vi antigen, and somatic O antigen. These antigenic structures cause variation in between the species and group them. These act as virulence factors by inhibiting phagocytosis and bacterial lysis and complement activation.

Mode of Transmission

The infection is transmitted by the ingestion of contaminated food and water. The infectious dose to cause the disease in humans ranges from 1000 to 10,00,000 bacilli. These bacilli enter, attach themselves to the microvilli of the mucosa, and undergo penetration, where they are attacked by phagocytic cells. The virulent bacteria tend to multiply inside these cells and use them as a vehicle to move and infect various other organs and finally enter the bloodstream causing septicemia.

Pathogenesis

Enteric fever or typhoid fever is caused by *Salmonella typhi*. This bacillus grows well inside the gall bladder as bile is a good medium for growth. They proliferate rapidly and a significant bacterial load enters the intestine causing ulceration, intestinal perforation, and haemorrhage.

Symptoms: The symptoms are as follows:

1. High fever
2. Abdominal pain
3. Malaise
4. Fatigue
5. Weakness
6. Dizziness
7. Rose spots on the chest region
8. Hepatosplenomegaly

Salmonella paratyphi A and *B* cause paratyphoid, which is milder when compared with the typhoid infection caused by *Salmonella typhi*.

Laboratory Diagnosis

The samples collected for the isolation of the bacilli are blood products, stool, and bone marrow aspirates. These samples are processed for culturing using appropriate media, and the isolates undergo confirmation by staining, motility test, and biochemical test. The widely used test for the confirmation and grouping of the bacilli according to severity is the Widal test. This is a serological test performed to quantify the H and O agglutinins for the typhoid and paratyphoid bacilli from the patients' sera. Agglutination reactions are observed in positive samples, and the agglutination titre indicates the severity of the disease. However, this test is time-consuming. Recent improvements in the field of science have introduced various other tests such as ELISA, CIE, and haemagglutination test, which are very effective and less time-consuming.

Prevention and Treatment

The typhoid fever can be controlled by good sanitation, clean water supply, and hygienic food. Vaccines are available; for example, Typhoid-paratyphoid A and B (TAB) vaccine is given in two doses for the prevention of typhoid. The individuals who are infected must be treated using proper oral hydration methods. Antibiotics such as ciprofloxacin, amoxicillin, and cefixime are used in combination due to the characteristic of multi-drug resistance of the bacterium.

16.6 Mycobacterium
16.6.1 General Properties

Mycobacterium species are aerobic, non-motile, non-spore-forming, non-capsulated slender rod-shaped organisms (Figure 16.5). They structurally (occasionally) resemble branched filamentous mycelium and hence are termed as *Mycobacterium* meaning fungus-like bacteria. They have a special cell wall that retains the primary stain; they resist decolourization by acids and alcohol and therefore are referred to as 'acid-fast bacilli'.

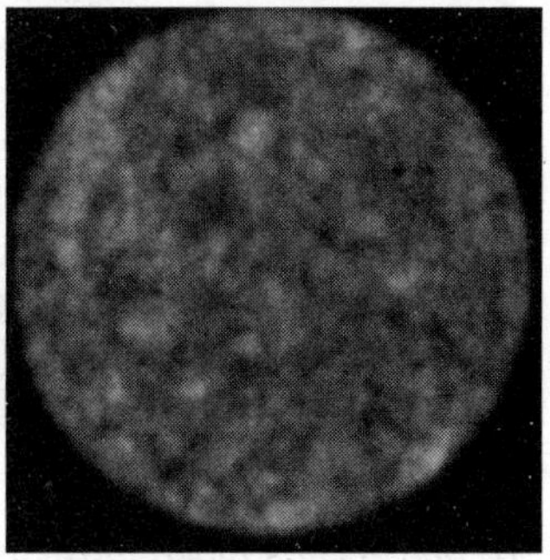

Figure 16.5 *Mycobacterium* (See page 355 for the colour image)

16.6.2 *Mycobacterium tuberculosis*

Tuberculosis is an infectious disease caused by *Mycobacterium tuberculosis*. In 1882, Robert Koch, a German physician and scientist, presented his discovery of the *Mycobacterium* that causes tuberculosis (TB). He also discovered a staining technique that enabled him to identify the bacteria under the microscope. In 1883, two scientists Ziehl and Neelsen modified Koch staining technique and gave better visibility to the bacterium. Since then, the Ziehl–Neelsen acid-fast technique has been used as the 'gold standard' method for the identification of the bacterium.

General Properties

Mycobacterium tuberculosis is an acid-fast bacillus. Upon Gram staining, they retain the primary stain and appear Gram-positive under the microscope. They are non-motile pleomorphic rods. They are obligate aerobes and grow well in tissues with oxygen (e.g., lungs). They are intracellular pathogens (usually affect the mononuclear phagocytes, e.g., macrophages). Their generation time is about 12–18 hours, and hence, they are slow growers. They lack the outer cell membrane and are hydrophobic in nature (due to the high lipid content of the cell wall).

Cultural Characteristics

Mycobacterium tuberculosis is an obligate aerobe and is eugenic in nature. Its generation time is 12–18 hours, and it requires a specific type of medium having selective ingredients to facilitate its growth. It grows well at an optimum temperature of 37°C and requires a pH of 6.4–7.0 for enhanced growth.

The selective media used for its growth are Middlebrook's medium and Lowenstein–Jensen medium. These are agar-based selective media. Lowenstein–Jensen medium has specific ingredients, namely, asparagine, mineral salts, malachite green, and coagulated egg. Malachite green inhibits the growth of other organisms in the medium. The addition of glycerol to the medium enhances the growth of *Mycobacterium tuberculosis* but inhibits the growth of other organisms. The colonies are small, raised, irregular, and velvety white when they appear on the media and become buffy yellowish later. They are slow growers and take around 5 weeks to form visible colonies on the media.

Mycobacterium tuberculosis is termed as acid-fast bacillus because of its cell wall properties. This bacillus has a complex cell wall, most of which is made up of lipids. The cell wall of *Mycobacterium tuberculosis* comprises three important components: mycolic acid, cord factor, and wax D. These components are the major determinants of virulence for the bacterium. This bacterium resists decolourization even after the addition of acid or alkali, which is due to the high lipid content of the cell wall (which makes it impermeable to the stains and resists decolourization). This impermeability protects the bacterium from external damage and osmotic lysis and makes it survive inside macrophages.

Biochemical Properties

Mycobacterium tuberculosis is differentiated from the other species of mycobacterium family by two important biochemical tests, namely, niacin accumulation test and nitrate reduction test, which are positive in the case of *Mycobacterium tuberculosis* and negative in the other species. They are aerophilic in nature and produce eugonic colonies.

Niacin Accumulation Test: *Mycobacterium tuberculosis* produces niacin in the culture media because of its inability to convert niacin ribonucleotide into nicotinamide adenine dinucleotide. The niacin thus secreted by the bacterium in the medium reacts with cyanogen bromide to produce a cannery yellow colour.

Nitrate Reduction Test: *Mycobacterium tuberculosis* produces the enzyme nitroreductase. It catalyses the reduction of nitrate into nitrite, which is indicated by the change in colour from pink to red, indicating a positive result.

Mode of Transmission

The transmission is through the inhalation of droplets generated by infected persons. The droplet nuclei carry viable organisms from infected persons to others. The entry of more than 10 bacilli is enough to infect a healthy person. The incubation period of the bacilli after localization is 3–6 weeks depending upon the person's immune competence.

Pathogenesis

The TB infection sets on as soon as the pathogen enters the lungs through inhalation and localizes. During localization, the bacteria are engulfed by alveolar macrophages. As the bacterial membrane is impermeable, the lysozyme activity is neutralized. The bacteria replicate within the macrophages and subsequently lyse and kill them.

The replicated bacteria use the macrophages as a vehicle to reach the lymph nodes causing 'Ghon focus', otherwise called primary infection or primary complex, in children. The infection in the upper parts of the lungs is referred to as 'pulmonary infection'. This spreads from the lungs to different parts of the body ending in extrapulmonary infection in children and immunosuppressed individuals. The cellular immune response slows down the spread of the infection by arresting the bacteria within the macrophages. Therefore, the bacteria are not allowed to disintegrate out, and the immune cells such as T cells and B cells form an aggregate with the macrophage resulting in granuloma formation. This response is termed as 'granulomatous inflammatory response'.

The activated immune cells actually mask the growing bacteria limiting further replication and the spread of the tubercle bacilli. Most bacteria are either killed or arrested. However, the pathogen is not completely eradicated in some individuals leading to latent infections.

Symptoms: The symptoms are as follows:

1. Cough persisting for more than 3–4 weeks
2. Cough with sputum and spotting of blood
3. Persistent fever and chills
4. Weight loss
5. Fatigue
6. Loss of appetite
7. Discomfort in breathing with chest pain
8. Swollen lymph node

Laboratory Diagnosis

Specimen Collection and Processing: The samples collected for the isolation of the bacteria are sputum, blood, lung aspirates, pleural effusion, urine, and stool. The samples are collected aseptically in sterile containers. The sputum samples are collected early in the morning because only then they show best results during staining. The golden standards for the isolation and identification of the bacteria are Petroff's deposition method, culturing, and staining methods.

Petroff's Method: On prolonged exposure, sodium hydroxide is toxic to many pathogens including tubercle bacilli. The sputum samples are processed with 4% NaOH and then subjected to centrifugation. This eliminates other contaminants before culturing.

Culturing: The Petroff's sputum deposits are used for inoculation into Lowenstein–Jensen medium. Visible colonies appear on the culture medium in about 2–4 weeks. The colonies are small, raised, irregular, and velvety white when they appear on the media and become buffy yellowish later.

Microscopic Examination: The golden standards for the identification of the tubercle bacilli are acid-fast staining and fluorochrome staining using fluorescence microscopy. Both the methods are very sensitive and rapid. The interpretation of acid-fast-stained bacilli is shown in Table 16.1.

Table 16.1

S.No.	No. of Bacilli per Field	Result
1.	More than 10 bacilli per field	3+ (highly positive)
2.	1–10 bacilli per field	2+ (strongly positive)
3.	10–99 bacilli per 100 fields	1+ (positive)
4.	1–9 bacilli per 100 fields	Scanty
5.	No bacilli seen in 100 fields	Negative

Mantoux Tuberculin Skin Test: This is a type of delayed hypersensitive reaction often done to screen people who are at high risk of tuberculosis. The purified protein product of the tubercle bacilli is inactivated and inoculated on the superficial layer of the skin. If the person is infected, an induration is observed at the site of inoculation indicating a positive result.

Recent Diagnostic Technique: Rapid radiometric culture methods are used for the rapid identification of TB bacilli. Autoanalysers and strip tests are also being used for the quick identification of the pathogens.

Prevention and Control

The World Health Organization plays a very important role in controlling and eradicating this deadly disease in most developing countries. The prevention and treatment modules such as vaccines and treatments using many drugs through directly observed therapies are followed all over the world. The BCG (Bacillus Calmette–Guérin) vaccine is administered to all children soon after birth.

Directly observed therapy: This is the most effective method in the current scenario. Direct supervision is done by health personnel for 6 months of drug treatment given to the patients in terms of recording their everyday intake of drugs and maintaining documents effectively. The disadvantage of this method is its labour intensiveness.

Treatment

Multidrug regimes are prescribed for patients, which include drugs such as the following:

1. Rifampicin
2. Ethambutol
3. Pyrazinamide

Many such combinations of drugs are being given for the effective treatment and eradication of tuberculosis.

16.6.3 Mycobacterium leprae

Leprosy, which is considered to be a disease of severe deformity and disfiguration, is caused by *Mycobacterium leprae*. This disease was always considered as a dreadful disease in the past, and the people suffering from this condition were considered as 'unclean' and were isolated from others. This disease-causing organism was first identified by Armauer Hansen in 1874 in Norway; hence, from then, it is called as Hansen's disease.

General Properties

Mycobacterium leprae are non-motile, Gram-positive straight or slightly curved rods. These bacilli are acid-fast but require only 5% acid or alkali for decolourization. They are obligate intracellular pathogens seen either as single bacilli or as cluster fragments outside the cell. The clustered bacilli are bound together like a bundle and have a 'cigar bundle' appearance, which is mainly due to the lipid-like substance or waxy outer layer (which attaches the bacilli together).

Cultural Characteristics

Mycobacterium leprae are non-motile, non-spore-forming pleomorphic bacteria. These bacilli are obligate intracellular pathogens that cannot be grown *in vitro*. They require an optimum temperature of 27°C–30°C for growth in the footpad of mouse and 32°C–37°C for growth in armadillo. This was first attempted by a scientist named Shepard in 1960. He successfully inoculated the bacilli isolated from a patient lesion onto the footpad of the mouse. Since then, this method is being followed as the standard protocol for studying leprosy bacilli. This bacterium is slow-growing, and it takes about 12–14 days to grow in the footpad of mouse. These bacilli are identified by Ziehl–Neelsen staining technique, and this is one of the golden standards for the identification of the bacteria. *Mycobacterium leprae* is less acid-fast when compared with *Mycobacterium tuberculosis*.

Mode of Transmission

The mode of transmission of *Mycobacterium leprae* is still vague, but most of the research concluded saying that it is mainly due to the inhalation of the infected droplet nuclei by healthy individuals. Hence, the factors influencing the spread of infection are environmental factors, exposure, and susceptibility of the person.

Pathogenesis

Mycobacterium leprae is the causative organism of leprosy, which is a persistent granulomatous disease of human beings principally involving the skin and nervous system. This bacillus targets the neural cells, namely, the glia cells, consequently affecting the nervous system causing muscle paralysis. The infection induces both humoral and cell-mediated immunity in humans.

The two most acute forms of the disease are lepromatous and tuberculoid leprosy, which are discussed overleaf.

Tuberculoid Leprosy: The progress of tuberculoid type infection leads to the formation of granulomatous lesions on the skin by few bacilli (paucibacillary disease), which affect the nervous system leading to deformities in hands and feet. They have very good cellular immunity and hence show a positive result for the lepromin test (skin test).

Lepromatous Leprosy: The lepromatous type is caused by the most virulent bacilli that affect the upper respiratory tract; hence, a large number of bacilli (multibacillary disease) are shed in oral secretions. The cell-mediated immunity is poor in this type. They affect the eyes, kidney, and bones, and septicemia is common. In the case of lepromatous type, the bacilli carry autoantibodies, and hence, the antibody titre is high. The lesions on the skin are abundant due to the heavy loads of bacilli consequently seen within the cells (intracellular pathogen). They show a negative result for the lepromin test.

Indeterminate Leprosy: This type is intermediate to tuberculoid and lepromatous leprosy. The lesions are transitional and heal spontaneously.

Dimorphic Leprosy: This is otherwise termed as borderline tuberculoid or lepromatous leprosy and hence possesses the characteristics of both types.

Symptoms: The general symptom is skin lesions. The common symptoms are given below:

1. Light red patches on the skin surface
2. Numbness on the skin
3. Muscle flaw and paralysis (in severe cases), especially in hands and feet.
4. Skin rigidity
5. Eyes getting affected leading to blindness
6. Sensory loss
7. Disfiguration

Laboratory Diagnosis

Collection and Processing of Samples: The samples collected are skin biopsy specimens, nasal secretions, and pus from skin lesions. The nasal discharges and the tissue fluids from skin lesions are smeared onto the slide, and acid-fast staining technique is performed on the smears. These samples are then examined and graded based on the number of bacilli seen per field (which is considered as the bacillary index). The interpretation of stained bacilli is shown in Table 16.2.

Table 16.2

S.No.	No. of Bacilli per Field	Result
1.	1–10 bacilli in 100 fields	1+
2.	1–10 bacilli in 10 fields	2+
3.	1–10 bacilli per field	3+
4.	10–100 bacilli per field	4+
5.	100–1000 bacilli per field	5+
6.	More than thousand bacilli, clumps, and globi in every field	6+

The bacillary index is calculated considering the morphology of the stained bacilli. The bacilli that are stained intensely and consistently are considered while counting and the rest are not because they are regarded as dead bacilli.

The skin biopsy, that is, slit skin test, is performed after the confirmation of the bacillary index. The biopsy specimens are fixed on the slides, and a histological study is performed for the confirmation of the type of *Mycobacterium leprae* infection. Commercially available protein antigens are used for the early serological diagnosis of *Mycobacterium leprae*.

Lepromin Test: This is done to determine the type of leprosy in patients. The leprosy bacilli are inactivated by heating and injected underneath the skin. The response to the inoculation can be either of the two types of reactions, which are as follows:

1. **Fernandez reaction:** This is the immediate reaction (which occurs in 24–48 hours) developing indurations and erythema at the site of injection. This reaction is similar to tuberculin test.
2. **Mitsuda reaction:** This is the late reaction (which occurs in 3–4 weeks). The reaction developed here is the same as mentioned above but the indurations get ulcerated.

The lepromin test does not have any diagnostic value but is mainly employed for classifying the type of leprosy (i.e., tuberculoid or lepromatous).

Prevention and Control

Earlier, sulphones such as diaminodiphenyl sulphone and dapsone were given as treatment but due to the resistance developed by the bacilli, a combination of drugs is prescribed for patients to control the infection. The antibiotics prescribed along with Dapsone are as follows:

1. Rifampicin
2. Clofazimine
3. Ofloxacin
4. Prothionamide

This multidrug treatment is given until the patients show a negative result in the smear.

Prophylaxis

BCG vaccines are given to children as soon as they are born in order to elicit an immunological response. This vaccine not only gives protection against tuberculosis but also confers immunity for leprosy. However, research is still going on whether BCG vaccine can be used for the prophylaxis of leprosy.

16.7 *Vibrio cholerae*
16.7.1 *General Properties*

Vibrio cholerae are Gram-negative, motile, facultatively anaerobic bacilli. These bacilli are highly motile due to the presence of polar flagella (flagella at one pole); hence, they have vibratory or darting movement (Figure 16.6). This bacillus was first identified by Robert Koch in 1883. Upon staining, they appear like curved rods. *Vibrio cholerae* causes cholera, which is

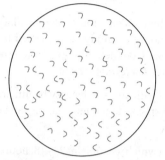

Vibrio cholerae
Gram-Negative Bacilli [Oil Immersion]

Figure 16.6 Vibrio-illustration

a discerning diarrhoeal disease transmitted mainly through contaminated food and water. In untreated cases, it leads to profuse rice watery stools and vomiting leading to death in 24 hours.

16.7.2 Cultural Characteristics

This bacillus grows well in aerobic conditions and requires an optimum temperature of 37°C and an alkaline pH (8.5). They are late-lactose fermenters and produce alpha-haemolytic colonies in blood agar. They grow well in simple media; when salt is added to the media, it enhances the growth of the organisms by inhibition of the contaminants. They liquefy gelatin and produce powdery deposit on liquid cultures. The colonies are moist, translucent, and pigmented.

16.7.3 Biochemical Properties

These bacilli are positive for catalase and oxidase. They ferment sugars producing acid but no gas. They are indole positive. They reduce nitrates to nitrites; due to this characteristic feature, the bacilli produce a derivative compound, namely, nitroso indole, producing a red precipitate indicating the presence of both indole and nitrite in the media. This reaction is called as cholera red reaction.

16.7.4 Mode of Transmission

The infection is transmitted through the consumption of contaminated food (seafood) and water. The bacteria enter the gastrointestinal tract and produce cholera toxins. These toxins attack the mucosal cell membrane and secrete fluids containing more of potassium and bicarbonates in huge quantities. This results in the profuse loss of electrolytes from the body leading to dehydration. The cholera toxins are heat-labile toxins, which are potent toxigenic agents produced by a specific gene called the tox gene. These genes are attached to the pili of the bacterium; hence, throughout the episode of infection, they tend to attach to the mucosal cell and cause infection rather than invading the cell.

16.7.5 Pathogenesis

Humans are the reservoirs of these bacilli. A heavy load of bacilli is required to cause the infection in humans. The most distinctive feature of cholera infection is the painless watery diarrhoea, that is, 'rice watery stools' leading to dehydration and death in 24 hours in untreated cases.

Symptoms

The symptoms are as follows:

1. Watery diarrhoea
2. Vomiting

3. Septicemia
4. Muscular cramps
5. Dehydration

The patients who are treated for the infection might still show symptoms for 4 days. In this duration, they lose maximum amount of body electrolytes. Replacement of fluids and electrolytes should be done, and intravenous treatment should be provided at the appropriate time.

16.7.6 Laboratory Diagnosis

Stools and rectal swabs are the only samples that can be collected for the isolation of the organism. The bacilli can be isolated before the administration of antibiotics. As the bacilli are fragile and die in a few minutes if exposed outside, during sample collection, they have to be transported via a transport media. The specimens collected should be processed immediately or else they have to be stored at 4°C in the transport medium. The preliminary tests such as motility test and inoculation into selective media (Thio Sulfate-Citrate Bile Salts (TCBS) agar) are done after the isolation of the bacilli. Confirmatory tests such as gelatin stab culture method, string test, and serological test are also performed.

16.7.7 Prophylaxis and Treatment

The only preventive method is the provision of hygienic food and water. Environmental sanitation is necessary for the prevention of such communicable diseases. As far as treatment is concerned, lots of oral fluids and electrolytes must be supplemented to compensate for the loss of excessive fluids from the body. Vaccines are widely used in the case of endemic outbreaks. Broad-spectrum antibiotics are prescribed for bringing down the load of bacteria. However, the strains have become multidrug resistant; hence, vaccination is the only source of control, which is yet to be commercialized.

16.8 *Spirochaetes*

16.8.1 Introduction

Spirochaetes are motile helical bacteria. They belong to the order Spirochaetales, which consists of the following two families:

1. **Spirochaetaceae:** This family includes *Spirochaetes* that are anaerobic and facultatively anaerobic and not hooked. It includes the following four genera:
   (a) *Spirochaeta*
   (b) *Cristispira*
   (c) *Treponema*
   (d) *Borrelia*
2. **Leptospiraceae:** This family includes hooked and aerobic *Spirochaetes*. It includes the following two genera:
   (a) *Leptospira*
   (b) *Leptonema*

16.8.2 Size and Structure

Spirochaetes vary in size from 5 to 500 μm in length. *Spirochaetes* have a Gram-negative cell wall. They have varying numbers of fine fibrils that are linked sub-terminally at each pole of the cell and extend towards the opposite pole. These are referred to as 'endoflagella' owing to their similarity with other bacterial flagella (Figure 16.7).

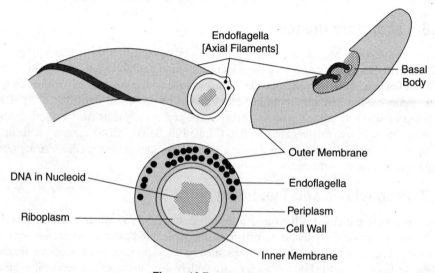

Figure 16.7 *Spirochaetes*

The spiral shape and serpentine motility of *Spirochaetes* depend on the structural integrity of the endoflagella. Motility is either active or sluggish, and its types include the following:

1. Flexion and extension
2. Corkscrew-like rotator movement
3. Translatory motion

The number of endoflagella at the end of the cell is a morphological characteristic of each and every species. This characteristic of some species is given in Table 16.3.

Table 16.3

Species	Number of Endoflagella
Treponema	3, occasionally 4
Borrelia	15–20
Leptospira	1

16.8.3 Habitat

Most of them are free-living, while few of them are obligate parasites. Genus *Spirochaeta* includes free-living *Spirochaetes* whose natural habitat is water. Genus *Cristispira* includes *Spirochaetes* found in the digestive tract of molluscs as commensals. The differentiating features of *Treponema*, *Borrelia*, and *Leptospira* are given in Table 16.4.

Table 16.4

Character/feature	*Treponema*	*Borrelia*	*Leptospira*
Size	6–14 × 0.2 µm	8–30 × 0.2–0.5 µm	6–20 × 0.1 µm
Spirals	6–12 regular, closed spirals with pointed ends at 1-µm intervals	3–10 loose, open spirals at 3-µm intervals	Tightly coiled regular spirals with hooked ends
Number of endoflagella	3–4 at each pole	15–20 at each pole	1 at each pole
Motility	Actively motile exhibiting flexion and extension, translatory, and corkscrew-like rotator motility; having a tendency to bend at right angles near its midpoint	Flexion and extension, corkscrew-like rotatory, and translatory motility	Rotation around the long axis, forward and backward, and bending and flexion
Staining	Can be stained with Giemsa and silver impregnation stains; not stained by routine stains	Readily stained by routine methods; Gram-negative	Can be stained with Giemsa and silver impregnation stains; not stained by routine stains
Mode of infection	Sexual contact, intimate cutaneous contact, blood transfusion, and transplacental	Ticks and lice	Water contaminated with rodent urine
Incubation period	10–90 days	2–14 days	7–14 days
Pathogenicity	Syphilis, yaws, pinta, bejel	Relapsing fever, Lyme disease	Weil's disease

16.9 *Mycoplasma*

16.9.1 Introduction

Mycoplasma (along with *Ureaplasma*) is classified under the class Mollicutes (*Mollis* means 'soft' and *Kutis* means 'skin') and order Mycoplasmatales. This order contains the following four families:

1. Mycoplasmataceae
2. Acholeplasmataceae

3. Spiroplasmataceae
4. Anaeroplasmataceae

1. The family Mycoplasmataceae contains the following genera:
   (a) *Mycoplasma*: This genus contains over 90 species, and they are known to exist as commensals, parasites, and pathogens of plants and animals. Six *Mycoplasma* species are found in humans. These are as follows: *Mycoplasma pneumoniae*, *Mycoplasma salivarium*, *Mycoplasma orale*, *Mycoplasma hominis*, *Mycoplasma genitalium*, and *Mycoplasma fermentans*.
   (b) *Ureaplasma*: This genus contains five species; of which, only one species *Ureaplasma urealyticum* may be recovered from human genital and respiratory tracts.
2. The family Acholeplasmataceae contains only one genus *Acholeplasma*. It comprises 10 species; of which, only one species *Acholeplasma laidlawii* may be found in the specimens from the human oral cavity and respiratory and genital tracts.
3. The family Spiroplasmataceae contains the genus *Spiroplasma*. It includes helical motile organisms.
4. The family Anaeroplasmataceae contains the genus *Anaeroplasma*, which are strict anaerobes.

16.9.2 Morphology and General Characteristics

1. *Mycoplasma* and *Ureaplasma* are very small organisms measuring 0.2–0.3 μm in diameter. They can pass through bacterial filters.
2. They differ from other bacteria in that they lack a rigid cell wall. These organisms are bound only by a single trilaminar cell membrane that contains a sterol (Figure 16.8).
3. *Mycoplasma* cannot synthesize its own cholesterol and require it as a growth factor in the culture medium.
4. Because of the fact that they lack a cell wall, they are extremely pleomorphic varying in shape from coccoid to filamentous to other undefined shapes.
5. In addition, due to the fact that they lack the bacterial cell wall (containing peptidoglycan), these organisms are insensitive to cell wall-active antibiotics such as penicillins and cephalosporins.
6. They possess a small genome ($5-10 \times 10^8$ Daltons molecular weight). Therefore, they have limited biosynthetic capabilities.
7. They multiply by binary fission.
8. They do not possess flagella or pili.
9. They stain poorly by Gram stain and are Gram-negative. They stain well with Giemsa and Diene's stain.
10. Most species on solid media form characteristic small 'fried egg' colonies.

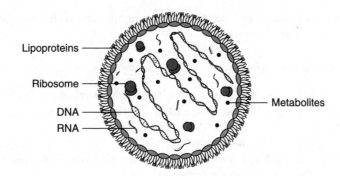

Figure 16.8 *Mycoplasma*

16.9.3 Cultural Characteristics

With the exception of *Anaeroplasma*, all the organisms of the order Mycoplasmatales are aerobes and facultative anaerobes. For primary isolation, an atmosphere of 95% N_2 and 5% CO_2 is preferred. They can grow within a temperature range of 22°C–41°C. For fermentative organisms, the initial pH of the medium should be 7.3–7.8. The medium widely used for the isolation of *Mycoplasma* consists of bovine heart infusion broth (Pleuro Pneumonia like Organism (PPLO) broth) to which 20% horse serum and 10% fresh yeast are added along with glucose and phenol red (as pH indicator). This medium solidifies by the addition of agar. Antibiotics such as penicillin, ampicillin, and polymyxin B are added to inhibit contaminating bacteria. Colonies appear after 2–3 days of incubation. Initially, the *Mycoplasma* cells multiply within the agar to form an opaque ball-shaped colony that grows up to the surface of the agar and spreads around it forming a translucent peripheral zone. Such a colony presents a 'fried egg' appearance. Colony size varies from 200 to 500 μm for 'large colony' *Mycoplasma*.

Colonies may be seen with a hand lens. However, visualization is facilitated by the application of Diene's stain directly to the agar surface. *Mycoplasma* with the fried egg colonies appears highly granular and stain with a dark blue centre and a light blue periphery.

16.9.4 Biochemical Reactions

Mycoplasma pneumoniae, *Mycoplasma fermentans*, *Mycoplasma genitalium*, and *Acholeplasma laidlawii* utilize glucose and other carbohydrates. *Mycoplasma salivarium*, *Mycoplasma orale*, *Mycoplasma hominis*, and *Mycoplasma fermentans* utilize arginine as a major source of energy. The carbohydrate-fermenting species catabolize glucose to produce lactic acid, resulting in a shift to an acidic pH. Arginine-metabolizing species produce ammonia, carbon dioxide, and adenosine triphosphate. The production of ammonia results in a shift to an alkaline pH.

16.9.5 Susceptibility to Physical and Chemical Agents

Mycoplasma is killed by heating at 56°C for 30 minutes. They can be preserved at −70°C or in liquid nitrogen. *Mycoplasma* is believed to be resistant to ultraviolet rays and the photodynamic action of methylene blue. Antiseptic solutions such as chlorhexidine and cetrimide inhibit the growth of *Mycoplasma*.

16.9.6 Antigenic Properties

Mycoplasma pneumoniae possesses cell membrane-bound glycolipids and proteins that act as haptens. Glycolipids induce antibodies that react in complement fixation test and other *in vitro* antigen–antibody reactions. Glycolipids with a similar antigenic structure have been found in the human brain. Therefore, antibodies to *Mycoplasma pneumoniae* glycolipids may cross-react with the brain cells leading to cell damage. This may account for the neurological manifestations of *Mycoplasma pneumoniae* infection.

16.9.7 Pathogenicity

Mycoplasma is an opportunistic infectious agent. Therefore, the infections caused by these organisms occur more frequently in patients with hypogammaglobulinemia, Hodgkin's disease, lymphoma, leukaemia, AIDS, and organ transplant.

Mycoplasma pneumoniae

Mycoplasma pneumoniae causes tracheobronchitis, pharyngitis, sinusitis, and primary atypical pneumonia accompanied by the formation of the following:

1. Cold haemagglutinins that agglutinate human red blood cells in cold
2. *Streptococcus* MG agglutinins
3. Antibodies giving biological false-positive Wassermann reaction

Mycoplasma pneumoniae may also cause extrapulmonary lesions such as arthritis, hepatitis, meningoencephalitis, cerebral ataxia, myocarditis, and pericarditis. *Mycoplasma pneumoniae* is transmitted from one person to another through airborne droplets containing the organisms. They attach to the epithelial cells in the respiratory tract and multiply. *Mycoplasma pneumoniae* possesses a membrane-associated 169-kDa protein called PI, which is known to mediate its adherence to the host cells. It is a surface parasite and does not penetrate the epithelial cells of the respiratory tract but remains localized; and invasion of the bloodstream is rare. Incubation period is 1–3 weeks. The onset is gradual. Patients develop fever with chills, malaise, headache, sore throat, nasal congestion, and non-productive cough. The disease is self-limited, with recovery occurring in 3–10 days without antimicrobial therapy.

Mycoplasma hominis

Mycoplasma hominis may cause salpingitis, tubo ovarian abscess, pelvic abscess, septic abortion, puerperal infection, post-operative wounds, septic arthritis, peritonitis, and brain abscess.

Mycoplasma genitalium

Mycoplasma genitalium has been associated with some cases of non-gonococcal urethritis and PIDs.

16.9.8 Laboratory Diagnosis

Laboratory diagnosis of *Mycoplasma* infections may be carried out by the following methods:

1. Culturing
2. Serological tests

Culturing

Mycoplasma pneumoniae may be recovered from throat swabs, sputum, throat washings, bronchoalveolar lavage, and tracheal aspirates. Genital *Mycoplasma* may be isolated from urethral, vaginal, and cervical samples; semen; prostatic secretions; urine; and blood. Colonies of *Mycoplasma pneumoniae* may be identified by the following tests:

1. **Haemadsorption test:** Colonies growing on the surface of the agar are flooded with 2 ml of 0.2%–0.4% of washed guinea pig erythrocytes suspended in *Mycoplasma* broth medium. The plate is then incubated at 35°C for 30 minutes. The colonies are examined after washes with the broth medium.
2. **Tetrazolium reduction test:** *Mycoplasma pneumoniae* has the ability to reduce the colourless compound triphenyltetrazolium to a red compound formazan. The agar surface bearing the colonies is flooded with 2-*p*-iodophenyl-3-*p*-nitrophenyl-5-phenyl tetrazolium chloride and incubated at 35°C for 1 hour. Colonies of *Mycoplasma pneumoniae* appear reddish after an hour and may appear purple to black after 3–4 hours.
3. **Serological techniques:** These include the inhibition of colony development around discs impregnated with specific antiserum or the fluorescence of colonies treated with such antiserum labelled with a fluorochrome.

Serological Tests

There are two types of serological tests, which are as follows:

1. **Detection of antigen and nucleic acids:** The detection of antigen can be performed using direct immunofluorescence, countercurrent immunoelectrophoresis, immunoblotting, and ELISA. Specific DNA detection is done using dot blot hybridization and PCR methods.
2. **Detection of antibody:** Cold agglutinins can be detected by the agglutination of O Rh-negative erythrocytes at 4°C. Complement fixation test is the most widely used serological test for the diagnosis of *Mycoplasma pneumoniae* infection. Apart from this, ELISA using IgM, IgG, and IgA antibodies has been developed.

16.9.9 Treatment and Prophylaxis

Mycoplasma is resistant to penicillins and cephalosporins, which act on the cell wall, and sensitive to tetracyclines and erythromycin, which inhibit protein synthesis. Therefore, tetracyclines and erythromycin are the drugs of choice for the treatment of diseases caused by *Mycoplasma*. The best prevention technique is to avoid contact with infected persons. At present, no vaccine against *Mycoplasma* is available.

16.10 Rickettsia
16.10.1 General Characteristics

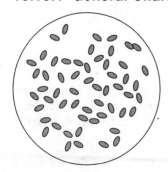

Figure 16.9 *Rickettsia* (See page 356 for the colour image)

1. The organisms of this genus are small Gram-negative bacilli. They stain red with Gimenez and Macchiavello and purple with Giemsa (Figure 16.9).
2. They are obligate intracellular parasites.
3. They require an arthropod vector to complete their lifecycle and are transmitted to humans through blood-sucking arthropods.
4. They multiply by binary fission.
5. They possess a trilaminar cytoplasmic membrane and a cell wall of bacterial type.
6. They are sensitive to lysozyme and antibiotics.
7. They possess both DNA and RNA.

16.10.2 Cultural Characteristics

Rickettsiae are unable to grow on cell-free media. Growth generally takes place in the cytoplasm of infected cells, but in the case of spotted fever *Rickettsiae*, growth may take place in the nucleus as well. Embryonated hen's egg inoculated into the yolk sac during the 5th or 6th day of development is highly susceptible to infection. *Rickettsiae* grow in the cells of the membrane surrounding the yolk. The inoculated eggs are incubated for 8 days at 35°C. After incubation, the eggs are harvested and the yolk sac membrane is removed. They also grow on mouse fibroblasts and cell lines such as HeLa and Hep-2. They may also be propagated in arthropods.

16.10.3 Pathogenesis

Rickettsiae possess three types of antigens, which are as follows:

1. A group-specific soluble antigen present on the surface of the organism
2. A species-specific or strain-specific antigen associated with the cell wall of the organism
3. An alkali-stable polysaccharide found in some *Rickettsiae*

The diseases caused by various species of *Rickettsia* are given in Table 16.5.

Table 16.5

Group	Disease	Organism	Insect Vector	Mode of Transmission
Typhus fevers	Epidemic typhus	*Rickettsia prowazekii*	Human body louse	Louse faeces scratched into the skin
	Brill–Zinsser disease	*Rickettsia prowazekii*	–	–

(continued)

Table 16.5 (Continued)

Group	Disease	Organism	Insect Vector	Mode of Transmission
	Murine typhus	*Rickettsia typhi*	Rat fleas	Rat flea faeces scratched into the skin
Spotted fevers	Rocky mountain spotted fever	*Rickettsia rickettsii*	Ixodid ticks	Tick bite
	Boutonneuse fever	*Rickettsia conorii*	Ixodid ticks	Tick bite
	Australian tick typhus	*Rickettsia australis*	Ixodid ticks	Tick bite
	Siberian tick typhus	*Rickettsia sibirica*	Ixodid ticks	Tick bite
	Rickettsialpox	*Rickettsia akari*	Mites	Mite bite
Scrub typhus	Scrub typhus	*Orientia tsutsugamushi*	Trombiculid mites	Mite bite

16.10.4 Laboratory Diagnosis

Diagnosis is mediated by the following:

1. Isolation of *Rickettsiae* in laboratory animals, fertile hens' eggs, and cell cultures
2. Direct detection of the organism and its antigens in clinical specimens using direct staining, immunofluorescence, and DNA probe
3. Serological tests by Weil–Felix reaction, complement fixation, and ELISA.

16.10.5 Treatment

Rickettsial infections may be treated with tetracyclines or chloramphenicol.

16.11 *Chlamydia*

16.11.1 Classification

Order: Chlamydiales
Family: Chlamydiaceae
Genus: *Chlamydia*
Species: *Chlamydia trachomatis*, *Chlamydia psittaci*, and *Chlamydia pneumoniae*

16.11.2 General Characteristics

The organisms of this genus possess the following characteristics:

1. They are small, obligate, intracellular Gram-negative bacteria.
2. They are different from true bacteria in the sense that they lack peptidoglycan. However, they posses RNA, DNA, and ribosomes like Gram-negative bacteria.

3. They are truly intracellular, and they use the ATP of the host as they cannot synthesize their own ATP. They multiply in the cytoplasm of the host cells forming microcolonies or inclusion bodies, which wrap around the nucleus like a mantle (chlamys meaning 'mantle').
4. They divide and multiply by binary fission.
5. They are non-motile.
6. They stain poorly with Gram stain. They stain blue with Castaneda and red with Macchiavello and Giemsa. Giemsa staining is preferred for staining inclusions in the cell culture.
7. Inclusion bodies of *Chlamydiae* are basophilic in nature.
8. They can also be viewed by direct immunofluorescence staining.
9. They infect vertebrate hosts such as birds and mammals including humans.
10. They are susceptible to antibiotics such as tetracyclines, erythromycin, and rifampicin.

16.11.3 Developmental Cycle

Chlamydiae consists of two morphologically distinct forms, namely, the elementary body (EB) and the reticulate body (RB) as shown in Figure 16.10.

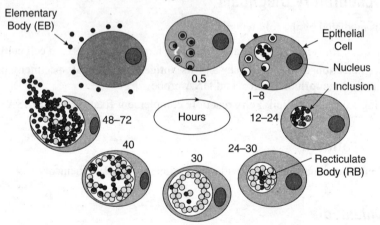

Figure 16.10 Developmental Cycle of *Chlamydia trachomatis*
(See page 356 for the colour image)

Elementary Body

The EB is an extracellular infectious particle. It has an irregular electron-dense central nucleoid and is capable of extracellular survival. The rigidity of the cell wall is due to the disulphide cross-linking (even in the absence of peptidoglycan).

Reticulate Body

The RB is the intracellular metabolically active form that divides by binary fission to form EBs. It is 800–1200 mm in diameter, and its cell wall lacks disulphide cross-linking.

Infection is initiated by the attachment of infectious EB to the susceptible host cell. After attachment, the organism enters the host cell within a vesicle. By 9 hours of infection, the EB within the vesicle loses its dense DNA core, its cell wall becomes less rigid due to the breaking of the disulphide bonds, and it increases in size and differentiates into RB. By 18 hours, within the enlarging vesicle, the RB divides by binary fission to yield pleomorphic organisms. By 24 hours, there is no condensation of DNA within the RB, disulphide bonds are formed in the outer membrane proteins, and new EBs develop within the vesicle. The developing chlamydial microcolony within the vesicle is termed as inclusion body, which is typically perinuclear and may contain 100–500 EBs. By 40–70 hours, infectious EBs are released from the cell by rupture of the inclusion.

16.11.4 Cultural Characteristics

Chlamydiae can be isolated by the following ways:

1. **Intranasal, intraperitoneal, or intracerebral inoculation into mice:** The mice die within 10 days and smears of the lung, peritoneal exudates, spleen, or brain show EBs.
2. **In the yolk sac of 6- to 8-day-old chick embryo:** In the yolk sac, the organisms multiply in the endothelial cells and can be detected in smears stained by Giemsa.
3. **In cell cultures:** Cells that have been irradiated or treated with a metabolic inhibitor are used for the isolation of *Chlamydiae*. Mouse fibroblast cell lines can be used for the isolation of *Chlamydiae*. The presence of the organism in cell culture is detected by staining for inclusions or EBs.

16.11.5 Pathogenesis

Chlamydiae produce infections of the eyes, genitals, and respiratory tract in humans (Table 16.6).

Table 16.6

Site of Infection	Disease	Symptoms	Organism
Eyes	Trachoma	Follicles, papillary hyperplasia, pannus formation	Chlamydia trachomatis
	Inclusion conjunctivitis	Intense hyperaemia, mucopurulent discharge, scarring, corneal lesions	Chlamydia trachomatis
	Ophthalmia neonatorum	Swelling of eyelids, hyperaemia, purulent infiltration of the conjunctiva	Chlamydia trachomatis
Genital tract			
Male	Urethritis, epididymitis, proctitis	Infection of the specified organ or organ parts	Chlamydia trachomatis

(continued)

Table 16.6 (Continued)

Site of Infection	Disease	Symptoms	Organism
Female	Urethritis, cervicitis, salpingitis, infertility, proctitis	Infection of the specified organ or organ parts	*Chlamydia trachomatis*
	Abortion, stillbirth		*Chlamydia psittaci*
Male and female	Lymphogranuloma venereum	Papule or ulcer on penis (males), infection of fourchette (females)	*Chlamydia trachomatis*
Respiratory tract	Pneumonitis of infants	Malaise, fever, anorexia, headache	*Chlamydia trachomatis*
	Pharyngitis, pneumonia	Malaise, fever, anorexia, headache	*Chlamydia pneumoniae*
	Psittacosis	Mucopurulent respiratory discharge	*Chlamydia psittaci*

16.11.6 Laboratory Diagnosis

Ocular, vaginal, and cervical specimens are collected for the diagnosis of chlamydial infections. The specimens are processed by the following methods:

1. **Light microscopy:** This method involves the staining of the inclusion bodies with Giemsa and other dyes. It has low sensitivity.
2. **Immunofluorescence:** Staining of the smears is done with Fluorescein isothiocyanate (FITC)-labelled antibodies (monoclonal) against species-specific or genus-specific antigens. The method is highly sensitive and specific.
3. **ELISA:** It involves the detection of soluble antigens using enzyme-labelled antibodies.
4. **DNA probes:** It involves direct detection using DNA hybridization.
5. Chemiluminescence assay
6. Polymerase chain reaction

16.11.7 Treatment

The various medications used for the treatment are listed in Table 16.7.

Table 16.7

Disease condition	Drug
Trachoma	Tetracycline
Trachoma in children	Erythromycin

(continued)

Table 16.7 (Continued)

Disease condition	Drug
Inclusion conjunctivitis	Tetracycline or erythromycin
Ophthalmia neonatorum	Erythromycin
Genital infections	Tetracycline

MULTIPLE CHOICE QUESTIONS

1. Toxic shock syndrome caused by *Staphylococcus aureus* is due to _____.
   - (a) Enterotoxin A
   - (b) Enterotoxin D
   - (c) Enterotoxin E
   - (d) Enterotoxin F

   Ans. d

2. The mode of transmission of respiratory infections is _____.
   - (a) Inhalation
   - (b) Ingestion
   - (c) Inoculation
   - (d) Insects

   Ans. a

3. Which of the following is the confirmative test for *Staphylococcus aureus*?
   - (a) Catalase test
   - (b) Pigmentation
   - (c) Mannitol fermentation
   - (d) Haemolysis of blood agar

   Ans. a

4. MRSA means _____.
   - (a) Methicillin-resistant *Staphylococcus aureus*
   - (b) Methicillin recombinant *Staphylococcus aureus*
   - (c) Methanol-resistant *Staphylococcus aureus*
   - (d) Methicillin-restricted *Staphylococcus aureus*

   Ans. a

5. The enzyme produced by *Staphylococcus* that is responsible for immunogenic responses in the host is _____.
   - (a) Coagulase
   - (b) Mannase
   - (c) Dehydrogenase
   - (d) Lipase

   Ans. a

6. M protein present in *Staphylococcus* is responsible for _____.
   - (a) Resistance
   - (b) Pathogenicity
   - (c) Antigenicity
   - (d) Absorption

   Ans. b

7. A chromogenic medium used for the isolation of *Staphylococcus aureus* is _____.
   - (a) Mannitol agar
   - (b) Mannitol sucrose agar
   - (c) Mannitol salt agar
   - (d) Mannitol sugar agar

   Ans. c

8. The infection of the long bones is termed as _____.
   - (a) Osteoporosis
   - (b) Osmosis
   - (c) Osteomyelitis
   - (d) Arthritis

   Ans. c

9. Staphylococcal skin scalded syndrome is the common scarletiform eruption of the skin causing _____.
   - (a) Ritter's disease
   - (b) Goiter

(c) Wool-sorter's disease
(d) Impetigo
Ans. a

10. Food poisoning caused by *Staphylococcus aureus* is due to _____.
    (a) Enterotoxin   (b) Endotoxin
    (c) Exotoxin   (d) Serotoxin
    Ans. a

11. Which of the following is Gram-positive cocci in chains?
    (a) *Streptococcus* sp.
    (b) *Staphylococcus* sp.
    (c) *Neisseria* sp.
    (d) *Pneumococcus* sp.
    Ans. a

12. The enrichment media used for the growth of *Streptococcus* sp. is _____.
    (a) Nutrient agar
    (b) MacConkey agar
    (c) Blood agar
    (d) Tellurite media
    Ans. c

13. Pyogenic infection is caused by _____.
    (a) *Streptococcus pyogenes*
    (b) *Clostridium sordellii*
    (c) Pneumococci
    (d) *Mycobacterium tuberculosis*
    Ans. a

14. Beta-haemolytic *Streptococcus* produces _____.
    (a) Complete haemolysis
    (b) Partial haemolysis
    (c) No haemolysis
    (d) Lysis
    Ans. a

15. *Neisseria gonococci* was discovered by _____.
    (a) Albert Ludwig Neisser
    (b) Albert John Neisser
    (c) Albert von Neisser
    (d) John Neisser
    Ans. a

16. *Neisseria gonococci* oxidizes _____ by producing acid without gas.
    (a) Glucose   (b) Lactose
    (c) Maltose   (d) Fructose
    Ans. a

17. The toxic effect is due to the occurrence of _____ in the cell wall of gonococcus.
    (a) Lipopolysaccharide
    (b) Lipooligosaccharide
    (c) Peptidoglycan
    (d) POR protein
    Ans. b

18. The mode of transmission of *Neisseria meningitidis* is _____.
    (a) Inhalation   (b) Ingestion
    (c) Inoculation (d) Invasion
    Ans. a

19. Neonatal conjunctivitis is caused by _____.
    (a) *Neisseria meningitidis*
    (b) *Neisseria gonorrhoeae*
    (c) *Staphylococcus aureus*
    (d) *Treponema*
    Ans. b

20. In a Gram-stained smear, gonococci are seen inside _____.
    (a) T cells   (b) B cells
    (c) Neutrophils   (d) Macrophages
    Ans. d

21. The virulence of *Neisseria meningitidis* is due to _____.
    (a) Cell membrane
    (b) Cytolytic enzyme
    (c) Pili
    (d) Capsule
    Ans. b

22. Which of the following inhibits the growth of *Neisseria*?
    (a) Heated blood   (b) Haemin
    (c) Animal proteins (d) Fatty acids
    Ans. d

23. The selective media used for the growth of *Neisseria* is _____.
    (a) Thayer Martin medium
    (b) Blood agar
    (c) Muller–Hinton agar
    (d) Mannitol salt agar
    Ans. a

24. Penicillin G is the drug of choice for treating _____.
    (a) *Neisseria meningitidis*
    (b) *Neisseria gonorrhoeae*
    (c) *Kingella*
    (d) *Simonsiella*
    Ans. a

25. The toxin produced by *Corynebacterium diphtheriae* is _____.
    (a) Exotoxin   (b) Endotoxin
    (c) Enterotoxin   (d) Neurotoxin
    Ans. a

26. *Corynebacterium diphtheriae* was first cultivated by _____.
    (a) Loffler in 1884
    (b) Loffler in 1886
    (c) Thomas Escherich in 1885
    (d) Robert Koch in 1885
    Ans. a

27. Pseudomembrane formation is caused by _____.
    (a) *Corynebacterium diphtheriae*
    (b) *Klebsiella rhinoscleromatis*
    (c) *Streptococcus pneumoniae*
    (d) Enteroaggregative *Escherichia coli*
    Ans. a

28. Klebs–Loffler bacilli are _____.
    (a) *Klebsiella rhinoscleromatis*
    (b) *Corynebacterium diphtheriae*
    (c) *Klebsiella ozaenae*
    (d) *Klebsiella oxytoca*
    Ans. b

29. Capsular antigens are _____.
    (a) H antigens   (b) K antigens
    (c) O antigens   (d) F antigens
    Ans. b

30. *Escherichia coli* was first identified by a German scientist _____.
    (a) Theodore Escherich in 1885
    (b) Robert Escherich in 1886
    (c) Thomas Escherich in 1885
    (d) Robert Koch in 1885
    Ans. a

31. Traveler's diarrhoea is a disease caused by _____.
    (a) Enterotoxigenic *Escherichia coli* (ETEC)
    (b) Enteropathogenic *Escherichia coli* (EPEC)
    (c) Enteroinvasive *Escherichia coli* (EIEC)
    (d) Enteroaggregative *Escherichia coli* (EAEC)
    Ans. a

32. *Klebsiella rhinoscleromatis* causes _____.
    (a) Rhinoscleroma
    (b) Scleroma
    (c) Sinusitis
    (d) Respiratory infections
    Ans. a

33. Friedlander's bacilli is _____.
    (a) *Klebsiella rhinoscleromatis*
    (b) *Klebsiella pneumoniae*
    (c) *Klebsiella ozaenae*
    (d) *Klebsiella oxytoca*
    Ans. b

34. Swarming motility is found in _____.
    (a) *Escherichia coli*
    (b) *Vibrio cholerae*

(c) *Proteus mirabilis*
(d) *Klebsiella* sp.
Ans. c

35. Kidney stones are produced by the ascending urinary tract infection caused by _____.
    (a) *Proteus mirabilis*
    (b) *Providencia rettgeri*
    (c) *Providencia alcalifaciens*
    (d) *Proteus vulgaris*
    Ans. a

36. Bacilli in which swarming growth, which is due to peritrichous flagella, is observed are called _____.
    (a) Koch bacilli   (b) Hauch bacilli
    (c) Haunch bacilli   (d) Notch bacilli
    Ans. b

37. Active motility of *Proteus* sp. is due to _____.
    (a) Fimbriae
    (b) Pili
    (c) Brownian movement
    (d) Flagella
    Ans. d

38. *Shigella* sp. was first identified by _____.
    (a) Robert Koch   (b) Kiyoshi Shiga
    (c) Komag Shiga   (d) Kansa Shiga
    Ans. b

39. Bacillary disease is caused by _____.
    (a) *Shigella dysenteriae*
    (b) *Escherichia coli*
    (c) *Vibrio cholerae*
    (d) *Amoeba*
    Ans. a

40. The selective medium used for the isolation of *Shigella* sp. is _____.
    (a) Nutrient agar
    (b) Muller–Hinton agar

(c) Deoxycholate citrate agar
(d) Blood agar
Ans. d

41. Which one of the following is a late-lactose fermenter?
    (a) *Shigella sonnei*
    (b) *Shigella boydii*
    (c) *Shigella flexneri*
    (d) *Shigella dysenteriae*
    Ans. a

42. Gaffky-Eberth bacillus is otherwise called as _____.
    (a) *Salmonella typhi*
    (b) *Salmonella* Paratyphi A
    (c) *Salmonella* Paratyphi B
    (d) *Salmonella* Paratyphi C
    Ans. a

43. Enteric fever is caused by _____.
    (a) *Salmonella typhi*
    (b) *Corynebacterium diphtheriae*
    (c) *Escherichia coli*
    (d) *Salmonella choleraesuis*
    Ans. a

44. Widal test is the confirmatory test for identifying _____.
    (a) Dengue fever   (b) Enteric fever
    (c) Malarial fever   (d) Jaundice
    Ans. b

45. The selective medium used for the isolation of *Salmonella typhi* is _____.
    (a) Wilson–Blair bismuth sulphite medium
    (b) Nutrient agar
    (c) Bile salt agar
    (d) MacConkey agar
    Ans. a

46. The mode of transmission of tuberculosis is _____.

(a) Inhalation   (b) Ingestion
(c) Inoculation (d) Insects
                                    Ans. a

47. The pathogen that is a more frequent cause of death worldwide is _____.
    (a) *Mycobacterium tuberculosis*
    (b) *Mycoplasma pneumoniae*
    (c) *Legionella pneumophila*
    (d) *Bordetella pertussis*
                                    Ans. a

48. Which one of the following is referred to as acid-fast bacilli?
    (a) *Mycobacterium tuberculosis*
    (b) *Salmonella typhi*
    (c) *Staphylococcus aureus*
    (d) *Pseudomonas*
                                    Ans. a

49. Mantoux tuberculin skin test is characterized by _____.
    (a) Delayed hypersensitivity
    (b) Immediate hypersensitivity
    (c) Ag–Ab-mediated hypersensitivity
    (d) Allergic reaction
                                    Ans. a

50. Ghon focus, that is, the primary infection of tuberculosis in children is referred to as _____.
    (a) Primary complex
    (b) Secondary complex
    (c) Tertiary complex
    (d) Secondary infection
                                    Ans. a

51. Leprosy is a disease caused by _____.
    (a) *Mycobacterium leprae*
    (b) *Mycobacterium tuberculosis*
    (c) *Mycobacterium lepraemurium*
    (d) Atypical *Mycobacterium*
                                    Ans. a

52. The organism causing leprosy was first identified by _____.
    (a) Robert Koch
    (b) Armauer Hansen
    (c) Edward Jenner
    (d) Pasteur
                                    Ans. b

53. Leprosy bacilli are identified microscopically by _____.
    (a) Ziehl–Neelsen staining
    (b) Gram staining
    (c) Spore staining
    (d) Capsular staining
                                    Ans. a

54. Lepromatous leprosy is termed as _____.
    (a) Paucibacillary disease
    (b) Multibacillary disease
    (c) Borderline disease
    (d) Bacillary disease
                                    Ans. b

55. The leprosy bacilli target the neural cells, namely, _____.
    (a) Glia cells
    (b) Myocytes
    (c) White blood cells
    (d) Germ cells
                                    Ans. a

56. *Vibrio cholerae* are _____.
    (a) Gram-positive bacilli
    (b) Gram-negative bacilli
    (c) Gram-variable bacilli
    (d) Acid-fast bacilli
                                    Ans. b

57. The name *Vibrio* is derived from _____.
    (a) Non-motility
    (b) Vibratory motility
    (c) Brownian movement
    (d) Sluggish motility
                                    Ans. b

58. The selective medium used for the isolation of cholera bacilli is _____.
    (a) Thiosulphate citrate bile salt agar
    (b) Trypticase soy agar
    (c) Trypsin agar
    (d) MacConkey agar
    Ans. a

59. The toxin produced by *Vibrio cholerae* is _____.
    (a) Exotoxin   (b) Shiga toxin
    (c) Enterotoxin   (d) Aflatoxin
    Ans. c

60. Which of the following genera belongs to family Spirochaetaceae?
    (a) *Spirochaeta*
    (b) *Treponema*
    (c) *Borrelia*
    (d) All the above
    Ans. d

61. Which of the following genera does not belong to family Spirochaetaceae?
    (a) *Leptospira*   (b) *Spirochaeta*
    (c) *Treponema*   (d) *Borrelia*
    Ans. a

62. Three to four endoflagella are present in each pole in _____.
    (a) *Leptospira*   (b) *Spirochaeta*
    (c) *Treponema*   (d) *Borrelia*
    Ans. c

63. *Spirochaetes* demonstrate _____.
    (a) Flexion and extension motility
    (b) Corkscrew-like rotatory motility
    (c) Translatory motion
    (d) All the above
    Ans. d

64. Weil's disease is caused by _____.
    (a) *Leptospira*   (b) *Spirochaeta*
    (c) *Treponema*   (d) *Borrelia*
    Ans. a

65. Colonies with fried egg appearance are seen in _____.
    (a) *Mycoplasma*   (b) *Pseudomonas*
    (c) *Haemophilus* (d) *Bordetella*
    Ans. a

66. Which of the following tests can help in the laboratory diagnosis of primary atypical pneumonia?
    (a) Cold agglutinin test
    (b) *Streptococcus* MG agglutination test
    (c) Culturing in *Mycoplasma* broth medium
    (d) All the above
    Ans. d

67. Which of the following drugs is the most suitable for the treatment of *Mycoplasma* infections?
    (a) Penicillins
    (b) Cephalosporins
    (c) Tetracyclines
    (d) Nalidixic acid
    Ans. c

68. The cell wall is absent in _____.
    (a) *Mycoplasma*
    (b) *Actinomyces*
    (c) *Corynebacterium*
    (d) *Brucella*
    Ans. a

69. Colonies of *Mycoplasma pneumoniae* can be identified by _____.
    (a) Haemadsorption test
    (b) Tetrazolium reduction test
    (c) Fluorescence detection
    (d) All the above
    Ans. d

70. What is the shape of *Mycoplasma* cells?
    (a) Coccoid   (b) Filamentous
    (c) Bizarre   (d) All the above
    Ans. d

71. Which of the following antibiotics is administered to treat rickettsial infections?
    (a) Tetracycline
    (b) Chloramphenicol

(c) Both (a) and (b)
(d) None of the above
                                    Ans. c
72. Which of the following stains is used to stain *Rickettsiae*?
    (a) Giemsa   (b) Gimenez
    (c) Macchiavello (d) All the above
                                    Ans. d
73. Epidemic typhus is transmitted by _____.
    (a) Body lice   (b) Fleas
    (c) Ticks   (d) Mites
                                    Ans. a
74. Rocky mountain spotted fever is caused by _____.
    (a) *Rickettsia prowazekii*
    (b) *Rickettsia rickettsii*
    (c) *Rickettsia sibirica*
    (d) *Rickettsia australis*
                                    Ans. b
75. Brill–Zinsser disease is caused by _____.
    (a) *Rickettsia prowazekii*
    (b) *Rickettsia rickettsii*
    (c) *Rickettsia sibirica*
    (d) *Rickettsia australis*
                                    Ans. a
76. The morphological forms of *Chlamydia* observed in its developmental cycle are _____.
    (a) Elementary body
    (b) Reticulate body
    (c) Both (a) and (b)
    (d) None of the above
                                    Ans. c
77. *Chlamydiae* cause infections of the _____.
    (a) Eyes   (b) Genital tract
    (c) Respiratory tract (d) All the above
                                    Ans. d
78. Which one of the following is mostly responsible for infections?
    (a) *Chlamydia trachomatis*
    (b) *Chlamydia psittaci*
    (c) *Chlamydia pneumoniae*
    (d) All the above
                                    Ans. a
79. Which of the following antibiotics is administered to treat chlamydial infections?
    (a) Tetracycline
    (b) Erythromycin
    (c) Both (a) and (b)
    (d) None of the above
                                    Ans. c
80. Which of the following stains is used to stain *Chlamydiae*?
    (a) Giemsa   (b) Castaneda
    (c) Macchiavello (d) All the above
                                    Ans. d

SHORT NOTES

1. Toxins produced by *Staphylococcus aureus*
2. Toxic shock syndrome
3. Skin scalded syndrome
4. Biochemical properties of *Staphylococcus aureus*
5. Laboratory diagnosis of *Staphylococcus aureus*
6. Pathogenesis of *Staphylococcus aureus*

7. Microbial susceptibility of *Staphylococcus aureus*
8. General properties of *Staphylococcus*
9. Laboratory diagnosis of *Streptococcus pyogenes*
10. Pathogenesis of *Streptococcus pyogenes*
11. Pathogenesis of *Neisseria gonorrhoeae*
12. Pathogenesis of *Neisseria meningitidis*
13. Laboratory diagnosis of *Neisseria gonorrhoeae*
14. Virulence factors of *Neisseria gonorrhoeae*
15. Laboratory diagnosis of *Neisseria meningitidis*
16. Cultural characteristics and biochemical properties of *Neisseria gonorrhoeae*
17. General properties of *Neisseria*
18. Prevention of and treatment for diseases caused by *Neisseria gonorrhoeae* and *Neisseria meningitidis*
19. Pathogenesis of *Corynebacterium diphtheriae*
20. Laboratory diagnosis of *Corynebacterium diphtheriae*
21. Active immunization for *Corynebacterium diphtheriae*
22. Pathogenesis of *Klebsiella pneumoniae*
23. Pathogenesis of *Escherichia coli*
24. Laboratory diagnosis of *Escherichia coli*
25. Laboratory diagnosis of *Klebsiella* sp.
26. Pathogenesis of *Proteus* sp.
27. General characteristics of *Proteus* sp.
28. Pathogenesis of *Shigella* sp.
29. Laboratory diagnosis of *Shigella*
30. Pathogenesis of *Salmonella typhi*
31. Laboratory diagnosis of *Salmonella typhi*
32. Pathogenesis of *Mycobacterium tuberculosis*
33. Laboratory diagnosis of *Mycobacterium tuberculosis*
34. Cultural characteristics of *Mycobacterium tuberculosis*
35. Pathogenesis of *Mycobacterium leprae*
36. Laboratory diagnosis of *Mycobacterium leprae*
37. Cultural characteristics of *Mycobacterium leprae*
38. Pathogenesis and mode of transmission of *Vibrio cholerae*

39. Laboratory diagnosis of *Vibrio cholerae*
40. Structure and pathogenicity of *Treponema*
41. Structure and pathogenicity of *Borrelia*
42. Structure and pathogenicity of *Leptospira*
43. Motility of *Spirochaetes*
44. Endoflagella of *Spirochaetes*
45. Classification of *Spirochaetes*
46. Cold agglutinin test
47. Morphology and general characteristics of *Mycoplasma*
48. Classification and order of Mycoplasmatales
49. Cultural characteristics of *Mycoplasma*
50. Cultural characteristics of *Rickettsia*
51. Pathogenesis of *Rickettsia*
52. Laboratory diagnosis and treatment of diseases caused by *Rickettsia*
53. Developmental cycle of *Chlamydia*
54. Pathogenesis of *Chlamydia*
55. Laboratory diagnosis and treatment of diseases caused by *Chlamydia*

ESSAYS

1. Explain in detail about the pathogenicity and laboratory diagnosis of *Staphylococcus aureus*.
2. Describe the morphology, cultural characteristics, and biochemical properties of *Staphylococcus aureus*.
3. Explain the laboratory analysis and treatment of *Staphylococcus aureus* infections.
4. Explain in detail about the biochemical properties, pathogenicity, and laboratory diagnosis of *Streptococcus pyogenes*.
5. Explain in detail about the sexually transmitted diseases caused by *Neisseria gonorrhoeae*.
6. Describe the morphology, pathogenesis, and laboratory diagnosis of bacteria causing pyogenic meningitis.
7. Describe the morphology, virulence factors, and laboratory diagnosis of *Neisseria gonorrhoeae*.
8. Describe the morphology, pathogenesis, and laboratory diagnosis of bacteria causing diphtheria.

9. Describe the morphology, pathogenesis, and laboratory diagnosis of bacteria causing enteric diseases.
10. Give a brief account on the pathogenesis, diagnosis, and preventive methods of diseases caused by *Klebsiella* sp.
11. Describe the morphology, pathogenesis, and laboratory diagnosis of *Proteus* sp.
12. Give a brief account on the pathogenesis, diagnosis, and preventive methods of diseases caused by *Shigella* sp.
13. Give a brief account on the pathogenesis, diagnosis, and preventive methods of diseases caused by *Salmonella typhi*.
14. Describe the morphology, pathogenesis, and laboratory diagnosis of *Mycobacterium tuberculosis*.
15. Describe the morphology, pathogenesis, and laboratory diagnosis of *Mycobacterium leprae*.
16. Describe the cultural characteristics, pathogenesis, and laboratory diagnosis of bacteria causing cholera.
17. Explain in detail the classification of *Spirochaetes* and add a note on their size, structure, and habitat.
18. Explain in detail the differentiating features of *Treponema*, *Borrelia*, and *Leptospira*.
19. Discuss the pathogenicity and laboratory diagnosis of *Mycoplasma* infections.
20. Discuss the classification and cultural characteristics of *Mycoplasma*.
21. Explain in detail the cultural characteristics and pathogenesis of *Chlamydia* and the laboratory diagnosis and treatment of diseases caused by *Chlamydia*.
22. Explain in detail the general characteristics and developmental cycle of *Chlamydia*.

17 Viruses

CHAPTER OBJECTIVES

17.1 General Properties of Viruses
17.2 Herpesviridae
17.3 Picornaviridae
17.4 Rhabdoviridae
17.5 Retroviridae

17.1 GENERAL PROPERTIES OF VIRUSES

17.1.1 Introduction

Viruses are the simplest form of life and the smallest known infectious agents. They are intracellular obligatory parasites. The main properties of viruses that differentiate them from other microorganisms are as follows:

1. **Small size:** Their size varies from 10 to 300 nm. They can pass through bacterial filters and cannot be observed under the light microscope. Viruses can be measured by passage through membrane filters, ultracentrifugation, and electron microscopy.
2. **Genome:** Viruses carry their own genetic material in the form of DNA or RNA (but not both). The genome may be single stranded or double stranded, circular or linear, and segmented or unsegmented.
3. **Metabolically inert:** Viruses have no metabolic activity outside their susceptible hosts. They do not possess the machinery for translation and hence can multiply only inside living cells. Therefore, they are referred to as 'intracellular obligatory parasites'.

17.1.2 Structure of Viruses

Viruses consist of a nucleic acid core surrounded by a protein coat called capsid. The capsid with the enclosed nucleic acid is known as nucleocapsid. The capsid is composed of repeating protein units called capsomers.

Functions of Capsid

1. It protects the viral genome from physical destruction and enzymatic inactivation.
2. It serves as a vehicle of transmission from one host to another.
3. It facilitates the assembly and package of viral genetic information.
4. It is antigenic and specific for each virus type.
5. It provides structural symmetry to the virus particle.

Virus Symmetry

The capsid shows three types of symmetry based on the arrangement of capsomers around the nucleic acid. The three types are as follows:

1. **Icosahedral symmetry:** The capsomers are arranged as if they lay on the faces of an icosahedron that has 20 equilateral triangular faces and 12 corners or apices. Viruses with icosahedral symmetry have a rigid structure.
2. **Helical symmetry:** The nucleic acid and capsomers are wound in the form of a helix.
3. **Complex symmetry:** Viruses (e.g., poxviruses) that do not show icosahedral or helical symmetry due to the complexity of their structure have complex symmetry.

Envelope

DNA viruses replicate and are assembled in the nucleus, and RNA viruses are assembled in the cytoplasm. The final assembly of some viruses occurs in the nuclear or cytoplasmic membrane. As the virus particle moves from the nucleus to the cytoplasm or passes from the cytoplasm to the extracellular space, an external lipid-containing envelope (host origin) with virus-coded polypeptides or virus-specific glycoproteins is added to the nucleocapsid. In mature virus particles, the glycoproteins often appear as projecting spikes on the outer surface of the envelope. These are known as peplomers. Many peplomers mediate the attachment of the virus to the host cell receptors to initiate the entry of the virion into the cell. Some viral glycoproteins attach to receptors of RBCs and cause haemagglutination.

A summary of general properties of viruses is provided in Figure 17.1.

17.1.3 Replication of Viruses

The genetic information necessary for viral multiplication is contained in the viral nucleic acid but the biosynthetic enzymes are lacking. Hence, viruses replicate by taking over the biochemical machinery of the host cell and redirecting it to the manufacture of virus components. The viral replication is divided into six stages, which are as follows:

1. **Adsorption:** It is the attachment of the virus particle to the cell surface. Virions come into contact with cells by random collision, but adsorption or attachment is specific and is mediated by the binding of the virion surface structures known as ligands to receptors on the cell surface.

Characteristics	Viral Family	Important Genera
Single-Stranded DNA Non-enveloped 18–25 nm	Parvoviridae	Human parvovirus B19
Double-Stranded DNA Non-enveloped 70–90 nm	Adenoviridae 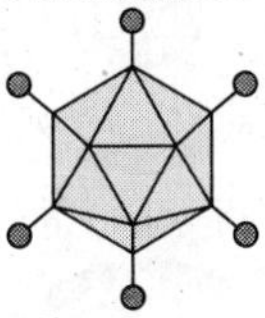	Mastadenovirus
40–57 nm	Papovaviridae 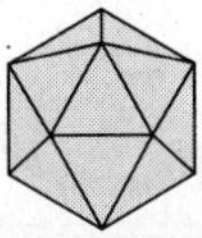	Papillomavirus [Human Wart Virus] Polyomavirus
Double-Stranded DNA enveloped 200–350 nm	Poxviridae 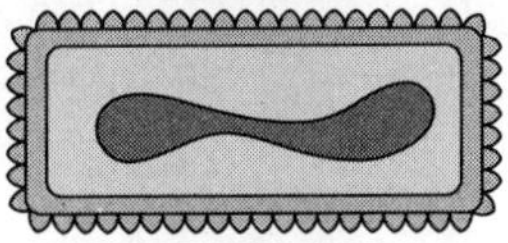	Orthopoxvirus [Vaccinia & Small Pox viruses] Molluscipoxvirus
150–200 nm	Herpesviridae 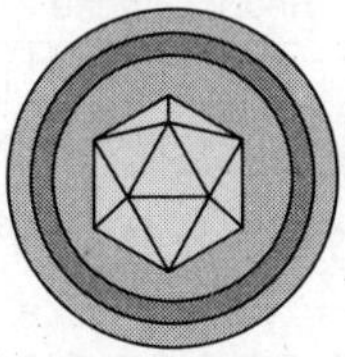	Simplexvirus [HHV-1 & 2] Varicellovirus [HHV-3] Lymphocryptovirus [HHV-4] Cytomegalovirus [HHV-5] Roseolovirus [HHV-6] HHV-7 Kaposi's Sarcoma [HHV-8]
Double-Stranded DNA enveloped 22–42 nm	Hepadnaviridae 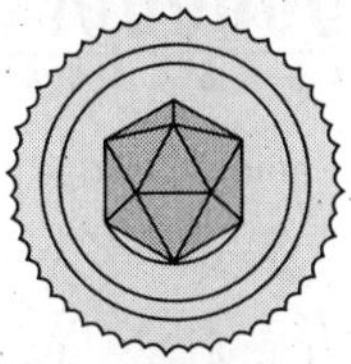	Hepadnavirus [Hepatitis B Virus]
Single-Stranded RNA, + Strand Non-enveloped 28–30 nm	Picornaviridae 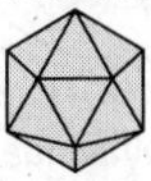	Enterovirus Rhinovirus [Common Cold Virus] Hepatitis A Virus
35–40 nm	Caliciviridae 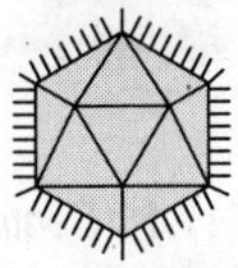	Hepatitis E Virus Norovirus

Figure 17.1 General Properties of Viruses (See pages 357, 358 and 359 for the colour image)

Characteristics	Viral Family	Important Genera
Single-Stranded RNA + Strand enveloped 60–70 nm	Togaviridae 	Alphavirus Rubivirus [Rubella Virus]
40–50 nm	Flaviviridae 	Flavivirus Pestivirus Hepatitis Virus
Nidovirales 80–160 nm	Coronaviridae 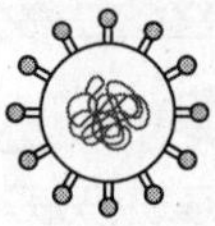	Coronavirus
Mononegavirales - Strand, One Strand of RNA 70–180 nm	Rhabdoviridae 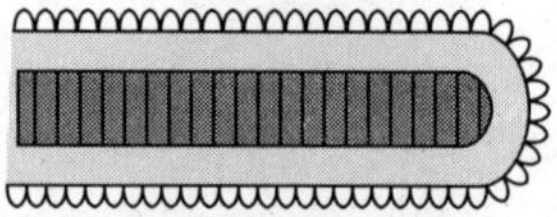	Vesiculovirus [Vesicular stomatatis Virus] Lyssavirus [Rabies Virus]
80–14,000 nm	Filoviridae 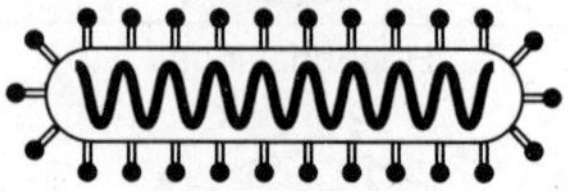	Filovirus
150–300 nm	Paramyxoviridae 	Paramyxovirus Morbillivirus [Measles-Like Virus]
- Strand one Strand of RNA 32 nm	Deltaviridae 	Hepatitis D

Figure 17.1 (*Continued*)

Characteristics	Viral Family	Important Genera
- Strand, Multiple Strands of RNA 80–200 nm	Orthomyxoviridae 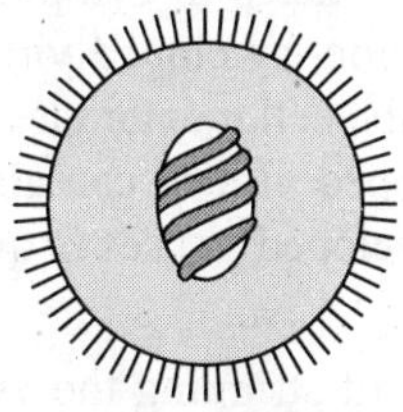	Influenza Virus A, B, and C
90–120 nm	Bunyaviridae 	Bunyavirus [Cailfornia Encephalitis Virus] Hantavirus
110–130 nm	Arenaviridae 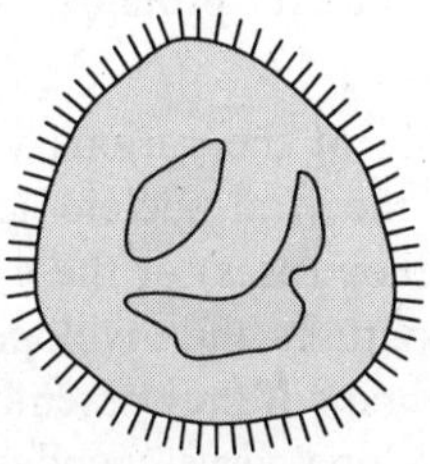	Arena Virus
Produce DNA 100–120 nm	Retroviridae 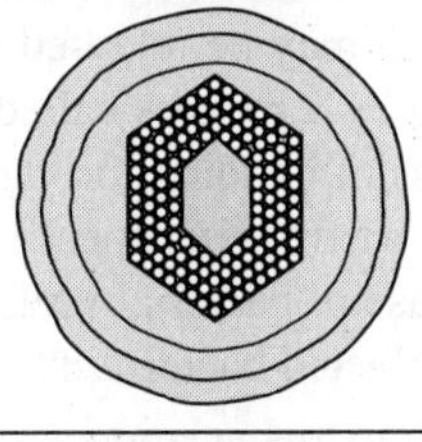	Oncoviruses Lentivirus [HIV]
Double-Stranded RNA Non-Enveloped 60–80 nm	Reoviridae 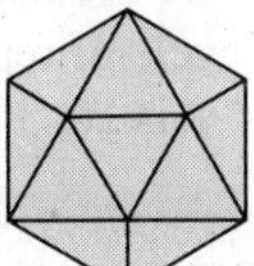	Reovirus Rotavirus

Figure 17.1 (*Continued*)

2. **Entry into the host cells:** Viruses enter the cells by one of the following methods:
   (a) **Non-enveloped viruses:** Non-enveloped viruses enter the cells by endocytosis or engulfment, which is by the invagination of a section of the plasma membrane with the accumulation of virus particles in cytoplasmic vesicles (phagosomes). This is known as viropexis. Some non-enveloped viruses enter by the translocation of the whole virus particle across the cell membrane.
   (b) **Enveloped viruses:** Enveloped viruses can also enter by endocytosis of the virion but they differ from non-enveloped viruses in that their envelopes fuse with the membranes of endosomes.
3. **Uncoating:** This is the process of stripping the virus of its outer layers and capsid so that the nucleic acid is released into the cell.
4. **Biosynthesis:** This phase includes the synthesis of the following: the viral nucleic acid; capsid protein; enzymes necessary in the various stages of viral synthesis, assembly, and release; and regulatory proteins required to direct the sequential production of viral components. Biosynthesis consists of the following steps:
   (a) The transcription of mRNA from the viral nucleic acid
   (b) The translation of mRNA into early proteins or non-structural proteins
   (c) The replication of viral nucleic acid
   (d) The synthesis of late or structural proteins, which constitute the daughter virion capsids
5. **Virion assembly:** The assembly of the various viral components into virions occurs shortly after the replication of the viral nucleic acid and may take place in either the nucleus (herpesviruses and adenoviruses) or the cytoplasm (picornaviruses and poxviruses). In the case of enveloped viruses, the envelope is derived from the nucleus (herpesviruses) and from plasma membrane if they assembled in the cytoplasm of the host cell (orthomyxoviruses). In this envelope, virus-encoded peplomers are also embedded.
6. **Release:** The release of completed viruses is the final step in virus multiplication. Viruses that exist as naked nucleocapsids may be released by the lysis of the host cell (polioviruses), or they may be extruded by a process called as reverse phagocytosis. Enveloped viruses are released by a process of budding through special areas of the host cell membrane (cytoplasmic or nuclear), where virus-specific transmembrane glycoproteins (peplomers) are embedded. In the case of bacterial viruses, the release of the progeny virions takes place by the lysis of the infected bacterium.

From the stage of penetration to the appearance of mature daughter virions, the viruses cannot be demonstrated inside the host cell. This period is known as eclipse phase. The time taken for a single cycle of replication in the case of animal viruses is 5–30 hours, and for bacteriophages, it is 15–30 minutes.

17.1.4 Classification of Viruses

Viruses are classified as DNA and RNA viruses based on their genetic material. Examples are listed in the following page:

DNA Viruses Infecting Humans

1. Poxviridae
2. Herpesviridae
3. Adenoviridae
4. Parvoviridae

RNA Viruses Infecting Humans

1. Orthomyxoviridae
2. Paramyxoviridae
3. Rhabdoviridae
4. Picornaviridae

17.2 HERPESVIRIDAE

The family Herpesviridae has been divided into three subfamilies (Table 17.1). The virions of this family have the capacity to establish lifelong latent infections from which the virus may be reactivated.

Table 17.1 Human Herpesviruses

Subfamily	Common Name	Scientific Name
Alphaherpesvirinae	Herpes simplex virus type 1	Human herpesvirus 1
	Herpes simplex virus type 1	Human herpesvirus 2
	Varicella zoster virus	Human herpesvirus 3
	Simian herpes virus	Cercopithecine herpesvirus 1
Gammaherpesvirinae	Epstein–Barr virus	Human herpesvirus 4
	Kaposi's sarcoma-associated herpesvirus	Human herpesvirus 8
Betaherpesvirinae	Human cytomegalovirus	Human herpesvirus 5
		Human herpesvirus 6
		Human herpesvirus 7

17.2.1 Morphology

Herpesviruses are 120–200 nm in diameter. They comprise four distinct structural elements: envelope, tegument, capsid, and core (Figure 17.2). Envelope is the outermost and is composed of lipid with numerous small glycoprotein peplomers. Tegument is the electron-dense material present between the envelope and the capsid. It contains several proteins. Next to the tegument is the icosahedral capsid of 100 nm diameter. It has a total of 162 capsomers. Core, which is

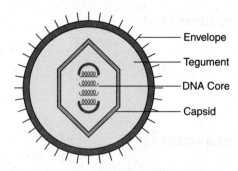

Figure 17.2 Herpesviridae (See page 360 for the colour image)

inside the capsid, consists of a double-stranded 124- to 235-kb DNA. With the exception of Epstein–Barr virus (EBV), members of the family Herpesviridae can be cultivated in cell cultures and produce giant cells and intranuclear inclusion bodies in infected cells.

17.2.2 Herpes Simplex Virus

There are two types of herpes simplex virus (HSV): herpes simplex virus type 1 (HSV-1) and herpes simplex virus type 2 (HSV-2). Herpes simplex virus type 1 primarily infects the mouth, eyes, and central nervous system (CNS). Herpes simplex virus type 2 infects the genital and anal regions. The infections caused by HSVs can be divided into primary infection, latent infection, reactivation, and recrudescence.

Primary Infections

Herpes simplex virus is transmitted only by contact. The portal of entry in primary infections is the damaged skin or mucosa, and the classic lesion is a vesicle beneath the keratinized squamous epithelial cells. The infection of epithelial cells is cytolytic; the cells lose adhesion, occasionally become multinucleate as a result of virus-induced cell fusion, and contain eosinophilic intranuclear inclusions. The vesicle and the surrounding tissue contain a dense infiltrate of inflammatory cells, mostly mononuclear. The vesicle drains and the lesion crusts before healing occurs, sometimes with residual scarring, and draining lymph nodes are commonly enlarged during this process. Recurrent lesions are morphologically and histologically similar.

Herpes Simplex Virus Type 1

HSV-1 causes the following infections:

1. **Acute gingivostomatitis:** It refers to the acute painful ulcers coated with a grayish slough inside the mouth and buccal mucosa and on the gums.
2. **Herpetic whitlow:** It is an occupational hazard of doctors and nurses who acquire the infection through the saliva and respiratory secretions of patients.
3. **Keratoconjunctivitis:** It involves the infection of the eyes, painful ulceration of the cornea, and vesiculation of the eyelids.

4. **Eczema herpectium:** It is the superinfection of eczematous skin.
5. **Encephalitis:** It is the infection of the temporal lobe of the CNS.

Herpes Simplex Virus Type 2

HSV-2 causes the following infections:

1. **Genital herpes:** It is a sexually transmitted disease. It involves the development of painful vesicles on the genitalia or anal regions with fever, malaise, and tender swollen lymph nodes. In females, lesions may occur on the perineum, vagina, cervix, or vulva. In males, the lesions may occur on the glans, prepuce, or shaft of the penis.
2. **Aseptic meningitis:** It occurs as a complication of HSV-2 genital infection.
3. **Neonatal infection:** It is acquired by neonates usually from their mothers during the passage through an infected birth canal.

Latent Infection

During primary infections, the virus travels from the site of infection in the mouth to the trigeminal and probably other cranial and cervical ganglia. In genital herpes, HSV-2 travels to sacral ganglia. Within the sensory ganglia, the viral DNA exists as a free circular episome.

Reactivation and Recrudescence

Reactivation of the virus is provoked by various stimuli such as common cold, fever, pneumonia, menstruation, and stress. Reactivation recurs sporadically, sometimes often, throughout life.

Laboratory Diagnosis

Specimens include vesicle fluid, skin swab, saliva, conjunctival fluid, corneal scrapings, brain biopsy, and cerebrospinal fluid (CSF). The diagnosis of HSV infection can be done by the direct examination of clinical specimens by electron microscopy (EM), fluorescence microscopy, and light microscopy. Herpes virions may be demonstrated in the negatively stained smear of the specimen by EM. Herpes simplex virus can be isolated on cell lines. The differentiation of HSV-1 and HSV-2 can be made by the use of monoclonal antibodies in immunofluorescent staining or by neutralization test. Primary infections can be diagnosed serologically by the detection of virus-specific IgM or of a rising IgG titre by complement fixation, neutralization, immunofluorescence, or ELISA.

17.2.3 Varicella Zoster Virus

Varicella zoster virus (VZV) causes varicella (chickenpox) in children and zoster (shingles) in adults and immunocompromised patients. Varicella follows primary infection in a non-immune individual, whereas zoster is the reactivation of the latent virus when immunity has fallen to ineffective levels.

Varicella

It is one of the common childhood exanthemata. The portal of entry of the virus is the respiratory tract. The incubation period is about 2 weeks. The earliest manifestation is a maculopapular rash that progresses within a few hours to the vesicular stage. Vesicles characteristically are surrounded by a red rim. The lesions then rupture and crust or may become secondarily infected and pustular before healing.

Pocks are centripetal in distribution, that is, they are more profuse on the trunk followed by the neck and proximal areas of limbs. Successive crops of vesicles appear over 2–5 days, and as a result, at any one time, the patient will have lesions at various stages of development on the same area of the skin.

Infections of adults are generally more severe than that of children; the vesicles heal more slowly, secondary bacterial infection and scarring are more common, and the accompanying fever is higher and more prolonged. Varicella zoster virus can cross the placenta following viraemia in pregnant women and infect the foetuses. A syndrome of congenital malformations with hypoplasia of limbs, chorioretinitis, and scarring of skin associated with maternal varicella in the first trimester may develop.

Zoster

Zoster or shingles is an endogenous reactivation of the virus that has remained latent in one or more sensory ganglia following primary varicella many years before. The virus travels down the sensory nerves to produce painful vesicles in the affected area of the skin. Thoracic nerves supplying the chest wall are most often affected. When the ophthalmic nerve of trigeminal ganglion is affected, the rash is distributed on the scalp and forehead.

Laboratory Diagnosis

1. Direct examination of the vesicle fluid by EM can be done.
2. Stained smears from the base of the lesion from biopsy tissue show multinucleate giant cells containing acidophilic intranuclear inclusion bodies.
3. Monoclonal antibody-based fluorescent antibody technique is used.
4. ELISA can be performed using specimens such as the vesicle fluid.
5. PCR-based assay for viral antigens can be carried out.

Acyclovir and vidarabine given intravenously are effective in the treatment of severe varicella and zoster. Live attenuated vaccine is available, which protects 90% of recipients for several years.

17.2.4 Epstein–Barr Virus

Epstein–Barr virus has been named after the virologists Epstein and Barr, who first observed it under the electron microscope in cultures of lymphocytes from Burkitt's lymphoma. Epstein–Barr virus replicates in epithelial cells of the nasopharynx and salivary glands, especially the parotid, lysing them and releasing infectious virions into the saliva. B lymphocytes appear to become infected when they infiltrate infected nasopharyngeal mucosa. Epstein–Barr virus has oncogenic properties, and a proportion of infected B lymphocytes undergo transformations and transform cells *in vitro*. Activated B lymphocytes secrete immunoglobulins, especially IgM.

Epstein–Barr virus possesses the envelope glycoprotein gp350/220, which mediates the attachment of the virus to CD21 receptors present on the susceptible cells. Most shedding of the virus takes place in the oral cavity, and therefore, the transmission of the virus requires salivary contact.

Infections Caused by EBV

Infectious mononucleosis (glandular fever) is a primary EBV infection seen mainly in the 15–25 years age group. Epstein–Barr virus is commonly transmitted by infected saliva and initiates infection in the oropharynx. Viral replication occurs in epithelial cells of the pharynx and salivary glands. Following the replication in epithelial cells, the virus infects B cells where it persists in a latent state.

Epstein–Barr virus-infected B cells synthesize immunoglobulins. Autoantibodies are typical of the disease. After an incubation period of 4–7 weeks, patients present with sore throat due to exudative tonsillitis, generalized lyphadenopathy, fever, malaise, headache,and sweating.

EBV-Associated Malignancies

1. Burkitt's lymphoma
2. Nasopharyngeal carcinoma
3. B cell lymphoma
4. Oral hairy leucoplakia

Laboratory Diagnosis

The laboratory diagnosis of EBV includes the following:

1. Differential white blood cell count
2. Paul–Bunnell heterophile antibodies
3. EBV-specific antibodies
4. Virus isolation

17.2.5 Cytomegalovirus

Cytomegalovirus is the most common agent to cause intrauterine infections and prenatal damage to the foetus leading to congenital abnormalities.

Perinatal Infection

It is acquired from infected maternal genital secretions or from breastfeeding.

Post-natal Infection

It may be acquired through saliva during sexual intercourse or artificial insemination, blood transfusion, and organ transplantation.

Laboratory Diagnosis

Cytomegalovirus can be isolated from urine, saliva, stool, breast milk, semen, cervical secretions, and blood leucocytes. When stained, multinucleated (post culture on cell lines) giant cells containing acidophilic inclusions in the nuclei and cytoplasm can be observed. Cytomegalovirus DNA in the specimen can be amplified by PCR. Cytomegalovirus-specific IgM can be detected in the patient serum by ELISA.

17.3 PICORNAVIRIDAE

The family Picornaviridae comprises a large number of very small RNA viruses (pico means small, and rna refers to RNA). Their diameter ranges from 27 to 30 nm (Figure 17.3). The capsid is a naked icosahedron made up of 60 protein subunits. The genome consists of a single linear molecule of single-stranded RNA of positive polarity with 7–8 kbp. Picornaviridae is divided into five genera, three of which, namely, *Enterovirus*, *Rhinovirus*, and *Hepatovirus*, possess human pathogens (Table 17.2). The other two genera, namely, *Aphthovirus* and *Cardiovirus*, cause foot and mouth disease and meningoencephalomyelitis, respectively, in mice.

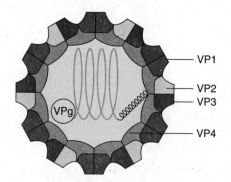

Figure 17.3 Picornaviridae (See page 360 for the colour image)

17.3.1 Enteroviruses

They are the most stable viruses. They can remain viable for years even at a temperature of −20°C or −70°C.

Polioviruses

Polioviruses are divided into three serotypes based on neutralization tests, which are as follows:

1. Type 1 (common epidemic type)
2. Type 2 (associated with endemic infections)
3. Type 3 (occasionally causes epidemics)

Table 17.2 Human Picornaviruses

Genus	Species	Major Disease
Enterovirus	Polioviruses	Paralytic poliomyelitis
	Coxsackieviruses	Aseptic meningitis
	Echoviruses	Aseptic meningitis
	Enteroviruses	Conjunctivitis
Rhinovirus	Rhinoviruses	Common cold
Hepatovirus	Hepatitis A virus	Hepatitis

These viruses have affinity for nervous tissue and a narrow host range. Only humans and some primates are susceptible. Natural infection occurs only in humans. The virus is spread from person to person by the faecal–oral route, and as early multiplication occurs in both the oropharynx and the intestinal mucosa, the virus spreads by pharyngeal secretions during the first week of illness. On entering the body of a new host, the virus multiplies in the tonsils and Peyer's patches of the ileum. The spread of the infection to the regional lymph nodes (cervical and mesenteric) leads to viraemia, enabling the virus to become disseminated throughout the body including the spinal cord and brain.

Clinical Features: There are four types of polio infection, which are as follows:

1. **Inapparent infection:** The patient does not have any symptom, but the virus may be isolated from the throat or stool. This type is seen in 90–95% individuals.
2. **Minor illness:** The patient develops mild, transient 'influenza-like' illness. This type is seen in 4–8% cases.
3. **Non-paralytic poliomyelitis:** When the patient develops headache, neck stiffness, and back pain in addition to minor illness, it may indicate aseptic meningitis. This illness lasts 2–10 days with rapid and complete recovery. This type is observed in 1–2% individuals.
4. **Paralytic poliomyelitis:** The patient develops paralysis during the course of the illness. On the basis of the site of involvement, the paralysis may be spinal, bulbar, or bulbospinal.

Laboratory Diagnosis: The virus can be isolated from faeces and throat early in the disease but not from CSF. It can be cultivated on cell lines, and cytopathic effect is usually seen in the cells within 48 hours. The identification of the serotype is carried out by neutralization tests.

Prophylaxis: Two effective vaccines are available, which are as follows:

1. Inactivated polio vaccine (Salk)
2. Live attenuated oral polio vaccine (Sabin)

Coxsackieviruses

These viruses were so named as they were first isolated from Coxsackie village in New York. Coxsackieviruses are classified into two groups, namely, group A and group B (Table 17.3).

Table 17.3 Coxsackieviruses

	Group A	Group B
Pathological changes induced by the inoculation of suckling mice	Generalized myositis, flaccid paralysis, death within a week	Patchy focal myositis, spastic paralysis, localized lesions in the liver, pancreas, myocardium, brain, and brown fat pads

Pathogenesis: Like other enteroviruses, coxsackieviruses inhabit the alimentary canal primarily and are spread by the faecal–oral route. The conditions caused by them are mentioned below.

Group A viruses:

1. Aseptic meningitis
2. Herpangina (vesicular pharyngitis; fever and sore throat)
3. Hand-foot-and-mouth disease (painful stomatitis and rash on hands and feet)

Group B viruses:

1. Epidemic myalgia (Bornholm disease; stitch-like pain in the muscles of the chest)
2. Myocarditis and pericarditis
3. Aseptic meningitis

Laboratory Diagnosis: Specimens from the lesions or faeces may be inoculated by intracerebral and/or intraperitoneal route into the newborn. Histological examination of the sacrificed mice is performed. The isolate in the animal or cell culture can be identified by means of neutralization test with reference antisera.

17.3.2 Rhinoviruses

Rhinoviruses are small RNA viruses that are morphologically and biochemically similar to other members of the family. They are acid labile, and their optimal temperature of replication is 33°C. They are inactivated below pH 5. They are stable in the temperature range of 20°C–37°C. On the basis of type-specific antigen in their capsid, rhinoviruses have been subdivided into more than 100 serotypes.

Pathogenesis

Rhinoviruses are the major cause of common cold. They are transmitted by the inhalation of droplets expelled from the nose of a patient during sneezing and coughing. During the acute phase of illness, high concentrations of the virus are present in nasal secretions. After an incubation period of 2–4 days, the patient develops profuse watery discharge with nasal obstruction, sneezing, sore throat, cough, headache, and malaise.

17.3.3 Hepatovirus (Hepatitis A Virus)

Hepatitis A virus (HAV) is a small, non-enveloped, 27-nm icosahedral virus containing a linear single-stranded RNA of 7.5 kb length and positive polarity. It has only one serotype. Hepatitis A virus is one of the most stable viruses infecting humans.

Pathogenesis

Hepatitis A virus is shed early in the stools of infected individuals, 1–2 weeks prior to the onset of symptoms, and persists for the first several days after the transaminase levels peak. Hepatitis A infection is an acute self-limiting disease with an incubation period of 2–6 weeks. The onset is abrupt with fever, malaise, anorexia, and nausea. Hepatomegaly due to cell necrosis causes the blockage of biliary excretions resulting in jaundice. It may also produce pain in the right upper abdominal quadrant. Transmission during birth by the exposure to maternal faeces or by breastfeeding has been reported.

Laboratory Diagnosis

1. Serum levels of both alanine and aspartate aminotransferase are markedly raised.
2. Virus particles can be demonstrated in faecal extracts by immunoelectron microscopy.
3. Faecal HAV may be detected by ELISA.
4. PCR-based detection of HAV is available.

Prophylaxis

1. Proper collection and disposal of sewage
2. Passive immunization with normal immunoglobulin, which gives protection for 4–6 months

17.4 RHABDOVIRIDAE

The family Rhabdoviridae contains two genera, namely, *Vesiculovirus* and *Lyssavirus*. They infect vertebrates, invertebrates, and plants. The members of the genus *Vesiculovirus* cause vesicular stomatitis in horses, cattle, and pigs, and few serologically distinct viruses of this genus cause human infections. The genus *Lyssavirus* comprises the rabies virus and five rabies-like viruses (Molola, Lagos bat, Kotonkan, Obodhiang, and Duvenhage viruses).

17.4.1 Rabies Virus

Morphology

The rabies virus is a bullet-shaped virus with a dimension of 180×175 nm and with one end rounded or conical and the other plane or concave (Figure 17.4). The core of the virion consists of a minus-sense, 11- to 12-kilobase single-stranded RNA enclosed in a helically wound nucleocapsid. RNA-dependent RNA polymerase enzyme, which is essential for the initiation of replication of the virus, is enclosed within the virion in association with the ribonucleoprotein core. The latter is surrounded by the matrix protein (viral membrane), which may be invaginated at the plane end. The matrix protein in turn is surrounded by a lipoprotein envelope that contains glycoprotein peplomers (spikes).

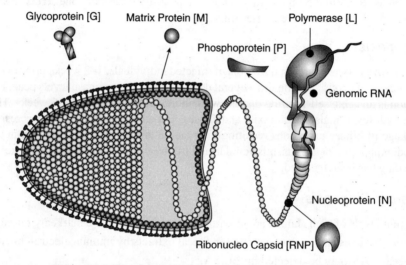

Figure 17.4 Rhabdoviridae (See page 360 for the colour image)

Resistance

The rabies virus is sensitive to alcohol and lipid solvents (ether, chloroform, and acetone) and quaternary ammonium compounds. It can be inactivated by phenol, formalin, beta-propiolactone, ultraviolet light, and heat. The virus remains stable for several days at 0°C–4°C and indefinitely at −70°C and when freeze-dried.

17.4.2 Host Range and Growth Characteristics

Animals

All warm-blooded animals (rabbits, guinea pigs, rats, and mice) are useful for primary isolation and are susceptible to rabies infection. Freshly isolated virus from natural human or animal infections is called street virus. It produces fatal encephalitis in laboratory animals, irrespective of the route of inoculation, after a long and variable incubation period of 1–12 weeks. Intracytoplasmic inclusion (Negri bodies) can be demonstrated in the brains of animals dying of street virus infection. After several serial intracerebral passages in rabbit, the virus undergoes certain changes and becomes a fixed virus. It is more neurotropic and much less infective by other routes. After intracerebral inoculation, it produces fatal encephalitis after a short and fixed incubation period of 5–8 days, and Negri bodies are usually not demonstrable in the brains of animals dying of fixed virus infection.

Chick Embryo and Tissue Culture

The rabies virus can be grown in the yolk sac of chick embryo and in several cell cultures including baby hamster kidney, mouse neuroblastoma, human diploid lung fibroblasts, and chick embryo. Most of these are used for vaccine production.

17.4.3 Pathogenesis

Rabies is a natural infection of dogs, foxes, wolves, cats, and bats. The rabies virus is excreted in the saliva of affected animals. Humans acquire the infection by the bite of rabid dogs or other animals. The infection caused by the bite of rabid animals results in the deposition of rabies-infected saliva deep in the striated muscles. The virus replicates in the muscle cells or cells of the subepithelial tissues. After it reaches a sufficient concentration, it infects the peripheral nerves in the muscle or skin. Once within the nerve fibres, it is out of reach of any circulating antibody and it travels along the axon towards the central nervous system, where it multiplies and produces encephalitis. The virus then spreads outwards along the nerve trunks to various parts of the body including salivary glands. It multiplies in the salivary glands and is shed in the saliva.

The presence of the virus in the saliva and the irritability and aggression brought on by the encephalitis ensure its transmission and survival in nature. Ultimately, the virus reaches every tissue in the body and is almost invariably present in the cornea, skin of face, and nape of the neck because of their proximity to the brain. This provides a method for the antemortem diagnosis of rabies. The virus may also be shed in the milk and urine.

17.4.4 Clinical Features

Following the bite of a rabid animal, the incubation period is usually between 1 and 2 months. After a prodromal phase of malaise, headache, and fever, the muscles become hypertonic and the patient becomes anxious with episodes of hyperactivity, aggression, and convulsions. The patient develops difficulty in drinking together with intense thirst. Attempts to drink bring painful spasm of pharynx and larynx producing choking and gagging. Thereafter, the mere sight or sound of water precipitates distressing muscular spasm leading to hydrophobia.

The furious form of rabies gradually subsides into delirium, convulsions, coma, and death. The disease once developed is almost fatal.

17.4.5 Immune Response

The rabies virus ascends to the brain along the nerves and does not come in contact with the immune system of the body. It is only after the virus spreads from the CNS along the nerve trunks to different parts of the body that antibodies are produced. By this time, irreversible damage of the neurons has already taken place, and the patient dies of respiratory paralysis. However, antibody is certainly protective when present before exposure and prompt vaccination after exposure induces resistance that is associated with antibody production.

17.4.6 Laboratory Diagnosis

1. Sections or impression smears of the brain stained by Seller's technique may reveal inclusion bodies known as Negri bodies. These are intracytoplasmic, round or oval, and

eosinophilic with basophilic inner granules. Negri bodies vary in size from 3 to 27 μm in diameter and are seen mainly in the pyramidal cells of Ammon's horn, Purkinje cells of the hippocampus, brain stem, and cerebellum.

2. The demonstration of rabies antigen by direct immunofluorescence can be done either antemortem or postmortem.
   (a) **Antemortem:** It can be done in the salivary, corneal, or conjunctival smears or skin biopsy from the nape of the neck.
   (b) **Postmortem:** It can be done in the impression smears of the cut surface of the salivary gland, hippocampus, brain stem, or cerebellum.
3. Detection of genomic RNA and viral mRNA using PCR assay and dot blot hybridization assay on skin biopsy, corneal impression, or saliva can be carried out.
4. Virus isolation can be done by inoculation in mice or cell lines (baby hamster kidney, mouse neuroblastoma, and chick embryo fibroblast). The viral genome thus isolated can be amplified by PCR, and the sequencing of the amplified products allows the identification of the infecting virus strain.
5. Rabies antibodies can be detected in the serum and CSF of the patient by ELISA.

17.4.7 Post-exposure Treatment and Prophylaxis

Post-exposure treatment in a previously unvaccinated person consists of the following steps:

1. First-aid treatment should be given to eliminate the virus from the site of infection by cleaning the wound with plenty of soap and water.
2. Human rabies immunoglobulin or heterologous (equine) antirabies serum should be taken; about half of it should be administered intramuscularly in the gluteal region and the remaining half should be infiltrated around the bite. Passive immunization is essential in the case of severe bites notably on the face, neck, and thorax. In such cases, if the antibody is not given, the virus reaches the CNS and gets fixed to neurons before the antibody is produced and the patient dies.
3. A potent cell culture vaccine should be given intramuscularly on days 0, 3, 7, 14, 30, and 90 in the deltoid region.
4. Adequate tetanus prophylaxis should be given.

17.4.8 Rabies Vaccine

Rabies is the only human disease that can be prevented by active immunization after infection because the long incubation period of the disease allows time for the immunity to develop before the onset of the disease. The types of rabies vaccine are listed below:

1. **Neural vaccines:** These vaccines include Pasteur vaccine, Fermi vaccine, Semple vaccine, beta-propiolactone vaccine, and suckling mouse brain vaccine.
2. **Non-neural vaccines:** These vaccines include duck egg vaccine and cell culture vaccines.

17.5 RETROVIRIDAE

Viruses of the family Retroviridae possess reverse transcriptase enzyme, hence the name Retroviridae (Re indicates reverse and tr transcriptase). This family has been divided into three subfamilies (Table 17.4).

Table 17.4 Human Retroviruses

Subfamily	Genus	Virus	Disease
Oncovirinae	Retrovirus	Human T Lymphotropic Virus-1 (HTLV-1)	Adult T cell leukaemia/lymphoma
		Human T Lymphotropic Virus-2 (HTLV-2)	Prevalent in intravenous drug users
Lentivirinae	Lentivirus	HIV-1	AIDS
		HIV-2	AIDS
Spumavirinae	Spumavirus	Human foamy virus	Nil

17.5.1 Human Immunodeficiency Viruses

Acquired immunodeficiency syndrome (AIDS) is a life-threatening disease caused by human immunodeficiency virus (HIV) characterized by HIV encephalopathy, HIV wasting syndrome, or certain diseases due to the immunodeficiency in a person. The emergence and pandemic spread of AIDS has threatened the economic, developmental, social welfare, and public health programmes worldwide.

Morphology

HIV is a spherical enveloped virus about 90–120 nm in diameter with a three-layer structure (Figure 17.5). There are two identical copies of ssRNA (9.2 kb each) associated with reverse transcriptase and surrounded by an icosahedral capsid, which in turn is surrounded by a matrix protein followed by a host cell membrane-derived lipid bilayer envelope that projects 72 glycoprotein peplomers.

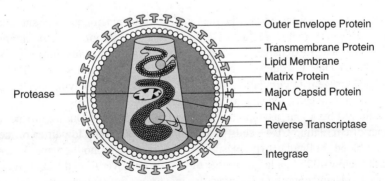

Figure 17.5 Retroviridae (See page 361 for the colour image)

The genome of HIV consists of three major genes, each coding for two or more polypeptides. The gag (group-specific antigen) gene encodes the core or capsid and matrix proteins, the pol gene encodes reverse transcriptase (polymerase), and the env gene encodes the virion envelope peplomer protein and transmembrane protein.

HIV is a delicate virus. It is thermolabile, inactivated in 10 minutes at 50°C and in seconds at 100°C. It cannot survive outside the living host; however, it can live in the blood for up to 8 days. It is susceptible to disinfectants and detergents.

Mode of Transmission of the Virus

There are three modes of transmission of HIV: sexual, parenteral, and perinatal. Of these, sexual mode of transmission is the most important. During sexual intercourse, the virus gets transmitted between and within sexes (heterosexual and homosexual, respectively). Parenteral transmission may occur through blood transfusion. The infection can also be transmitted by blood products such as plasma, serum, and cells from HIV-positive individuals. It can also be transmitted by sharing blood-contaminated syringes. Perinatal transmission indicates vertical transmission from the mother to the baby. The infection may be transmitted across the placenta before birth.

Replication of HIV

HIV gets attached via its gp120 envelope glycoprotein to the CD4 antigen complex, which is the primary HIV receptor on CD4+ (helper) T lymphocytes and the cells of the macrophage lineage. The CD4 molecule has binding avidity for gp120 of HIV. By pinocytosis, the nucleocapsid of the virion enters into the host cell. After entry into the cell, the nucleocapsid releases its RNA into the cytoplasm. The viral reverse transcriptase makes a copy of the genomic RNA. The ssDNA is made double stranded by the same enzyme. This dsDNA moves to the nucleus, and several such molecules become integrated as provirus at random sites in the host cell chromosome causing a latent infection. The integrated provirus is transcribed by cellular RNA polymerase II either for the production of mRNAs, which are translated into proteins, or for the production of genomic RNA for insertion into progeny virions.

Pathogenesis

The first cells to become infected may be the resident tissue macrophages or submucosal lymphocytes in the genital tract or rectum. The virus is then transported to the draining lymph nodes, where it replicates extensively. Two or three weeks after infection, most patients develop viraemia, a fall in CD4+ T lymphocytes, and glandular fever-like illness. As HIV infects cells expressing CD4 antigen, in circulation the virus is found in CD4+ lymphocytes and also in monocyte–macrophage cells, which may act as a reservoir for the virus.

This is followed by a long asymptomatic period of 1–15 years. During this period, only a small number of circulating CD4+ cells produce the virus and only low titres of the virus are present in the blood. Follicular hyperplasia develops in the lymphoid organs. When CD4+ T cell count falls below 400/μL, a large number of virions spill over from the degenerating lymph nodes into the blood and opportunistic infections by various microorganisms may develop. The cause of death includes opportunistic infections, malignancies, and cachexia-like state.

Clinical Features

Table 17.5 classifies HIV infection.

Table 17.5 Classification of HIV Infection

Group I	Acute HIV infection
Group II	Asymptomatic infection
Group III	Persistent generalized lymphadenopathy
Group IV	
A	Constitutional disease
B	Neurological disease
C	Secondary infectious disease
D	Secondary cancer
E	Other conditions

Acute HIV Infection

Two to six weeks after the infection, most patients develop acute onset fever with or without night sweats, malaise, headache, myalgia, arthralgia, lethargy, diarrhoea, depression, sore throat, and skin rash. Spontaneous ulceration occurs within 1 month.

Tests for HIV antibodies are usually negative at the onset of the illness but become positive during its course. Therefore, acute HIV infection is also known as seroconversion illness.

Asymptomatic Infection

All persons affected with HIV pass through a long asymptomatic period. They show positive HIV antibody tests during this phase and are infectious. However, autoimmune diseases such as Guillain–Barre syndrome, chronic demyelinating neuropathy, Reiter's syndrome, and cranial nerve palsy may occur during this period.

Persistent Generalized Lymphadenopathy

About 25–30% of patients who are otherwise asymptomatic develop enlarged lymph nodes, which are at least 1 cm in diameter, in two or more non-contagious extrainguinal sites, which persist for at least 3 months. Persistent generalized lymphadenopathy must be distinguished from other causes of lymphadenopathy such as lymphomas.

Symptomatic HIV Infection

The following features indicate disease progression:

1. The downward trend of CD4+ T cells in successive samples
2. The ease of virus culture
3. The presence of p24 antigen in the plasma
4. The loss of antibody to p24 antigen

When CD4+ T cell count falls below 400/µL, the patient develops constitutional symptoms such as fever, night sweats, diarrhoea, weight loss, and opportunistic infections. The latter includes infections of the skin and mucous membranes. The list of opportunistic infections and malignancies commonly associated with HIV infection are given in Table 17.6.

Table 17.6 Opportunistic Infections and Malignancies Commonly Associated with HIV Infection

1. Bacterial	3. Fungal
(a) *Mycobacterium avium* complex	(a) Candidiasis
(b) *Mycobacterium tuberculosis*	(b) *Cryptococcus*
(c) *Salmonella* (recurrent septicemia)	(c) Aspergillosis
2. Viral	(d) Histoplasmosis
(a) Cytomegalovirus	4. Parasitic
(b) Herpes simplex virus	(a) Toxoplasmosis
(c) Varicella zoster virus	(b) Cryptosporidiosis
(d) Epstein–Barr virus	5. Malignancies
(e) Human herpesvirus 6	(a) Kaposi's sarcoma
(f) Human herpesvirus 8	(b) Non-Hodgkin's lymphoma
	6. Slim disease (HIV wasting syndrome)

17.5.2 Laboratory Diagnosis

Table 17.7 lists the various laboratory tests for the diagnosis of HIV Infection.

Table 17.7 Laboratory Tests for the Diagnosis of HIV Infection

1. Screening (E/R/S) tests	2. Supplemental tests
(a) ELISA	(a) Western blot assay
(b) Rapid tests	(b) Immunofluorescence test
(i) Dot blot assays	3. Confirmatory tests
(ii) Particle agglutination (gelatin, latex, and microbeads)	(a) Virus isolation
(iii) HIV spot and comb tests	(b) Detection of p24 antigen
(iv) Fluorometric microparticle technologies	(c) Detection of viral nucleic acid using *in situ* hybridization and PCR
(c) Simple test based on ELISA principle, which takes less time (1–2 hours)	

Basic and expanded regimens of post-exposure prophylaxis are given below.

Basic Regimen

1. Zidovudine 300 mg BD (Twice a day) or 200 mg TDS (Three times a day) for 4 weeks
2. Lamivudine 150 mg BD for 4 weeks

Expanded Regimen

1. Basic regimen + indinavir 800 mg TDS for 4 weeks

MULTIPLE CHOICE QUESTIONS

1. Which of the following are referred to as intracellular obligatory parasites?
   (a) Bacteria   (b) Viruses
   (c) Parasites   (d) Fungi
   Ans. b

2. Which of the following symmetry has 20 equilateral triangular faces and 12 corners or apices?
   (a) Icosahedral (b) Helical
   (c) Complex   (d) All the above
   Ans. a

3. Poxviruses show _____.
   (a) Icosahedral symmetry
   (b) Helical symmetry
   (c) Complex symmetry
   (d) None of the above
   Ans. c

4. The projecting spike on the outer surface of the envelope is known as _____.
   (a) Capsid   (b) Nucleocapsid
   (c) Peplomer   (d) Virion
   Ans. c

5. Virion assembly for herpesviruses and adenoviruses takes place in the _____.
   (a) Nucleus   (b) Cytoplasm
   (c) Ribosomes (d) Mitochondria
   Ans. a

6. Virion assembly for picornaviruses and poxviruses takes place in the _____.
   (a) Nucleus   (b) Cytoplasm
   (c) Ribosomes (d) Mitochondria
   Ans. b

7. Which of the following is an RNA virus?
   (a) Poxviridae   (b) Herpesviridae
   (c) Adenoviridae   (d) Rhabdoviridae
   Ans. d

8. Which of the following is a DNA virus?
   (a) Poxviridae
   (b) Orthomyxoviridae
   (c) Paramyxoviridae
   (d) Rhabdoviridae
   Ans. a

9. Which of the following infections is caused by herpes simplex virus type 1?
   (a) Acute gingivostomatitis
   (b) Keratoconjunctivitis
   (c) Encephalitis
   (d) All the above
   Ans. d

10. Which of the following methods can be used for the diagnosis of infections caused by herpes simplex virus type 1?
    (a) Fluorescence microscopy
    (b) Production of intranuclear inclusions

(c) Production of characteristic cytopathic changes
(d) All the above

Ans. d

11. Which of the following drugs can be used for the treatment of herpes simplex virus infections?
    (a) Acyclovir
    (b) Azidothymidine
    (c) Ribavirin
    (d) All the above

Ans. a

12. Shingles is caused by _____.
    (a) *Varicella zoster virus*
    (b) *Epstein–Barr virus*
    (c) *Cytomegalovirus*
    (d) *Herpes simplex virus type 1*

Ans. a

13. Which of the following secretions of infected individuals may carry cytomegalovirus?
    (a) Saliva   (b) Semen
    (c) Urine   (d) All the above

Ans. d

14. Which of the following viruses belongs to gammaherpesvirinae?
    (a) Varicella zoster virus
    (b) Epstein–Barr virus
    (c) Cytomegalovirus
    (d) Herpes simplex virus type 1

Ans. b

15. Which of the following genera is included in the family Picornaviridae?
    (a) *Enterovirus*   (b) *Rhinovirus*
    (c) *Hepatovirus*   (d) All the above

Ans. d

16. Which of the following viruses can cause aseptic meningitis?
    (a) Polioviruses
    (b) Coxsackieviruses
    (c) Echoviruses
    (d) All the above

Ans. d

17. Which of the coxsackieviruses can cause herpangina?
    (a) Group A
    (b) Group B
    (c) Both (a) and (b)
    (d) None of the above

Ans. a

18. Which of the following viruses are acid labile?
    (a) Rhinoviruses
    (b) Echoviruses
    (c) Polioviruses
    (d) Coxsackieviruses

Ans. a

19. How many serotypes of rhinoviruses are there?
    (a) 10   (b) 20
    (c) 60   (d) More than 100

Ans. d

20. Which of the following is not a human picornavirus?
    (a) *Enterovirus*   (b) *Rhinovirus*
    (c) *Hepatovirus*   (d) *Cardiovirus*

Ans. d

21. Poliovirus is transmitted through _____.
    (a) Faecal–oral route
    (b) Ingestion
    (c) Droplet infection
    (d) Physical contact

Ans. a

22. Which of the following viruses cause common cold?
    (a) Rhinoviruses
    (b) Echoviruses
    (c) Polioviruses
    (d) Coxsackieviruses

Ans. a

23. The shape of rabies virus is _____.
    (a) Spherical   (b) Polygonal
    (c) Bullet shaped (d) Tubular
    Ans. c

24. Which of the following animals is most susceptible to rabies infection?
    (a) Skunk   (b) Dog
    (c) Cat   (d) Fowl
    Ans. c

25. Negri bodies can be demonstrated in infections caused by _____.
    (a) Fixed rabies virus
    (b) Street rabies virus
    (c) Both (a) and (b)
    (d) None of the above
    Ans. b

26. Which of the following viral infections is associated with the development of hydrophobia?
    (a) Influenza (b) Polio
    (c) Rabies   (d) Hepatitis
    Ans. c

27. Which of the following clinical specimens can be used for the demonstration of rabies antigen by direct immunofluorescence antemortem?
    (a) Salivary smears
    (b) Corneal smears
    (c) Conjunctival smears
    (d) All the above
    Ans. d

28. Which is the route of administration of purified chick embryo vaccine?
    (a) Subcutaneous
    (b) Intramuscular
    (c) Intravenous
    (d) None of the above
    Ans. b

29. Cell culture vaccines for rabies should be administered intramuscularly in the _____.
    (a) Deltoid region
    (b) Gluteal region
    (c) Anterior abdominal wall
    (d) Ventral aspect of forearm
    Ans. a

30. HIV belongs to the family Retroviridae and subfamily _____.
    (a) Oncovirinae
    (b) Lentivirinae
    (c) Spumavirinae
    (d) None of the above
    Ans. b

31. Which is the spike antigen of HIV-1?
    (a) gp120   (b) gp140
    (c) gp41   (d) gp36
    Ans. a

32. Which is the commonest mode of transmission of HIV?
    (a) Sexual   (b) Parenteral
    (c) Perinatal   (d) Oral
    Ans. a

33. What is the cause of death in AIDS patients?
    (a) Opportunistic infections
    (b) Malignancies
    (c) Cachexia-like state
    (d) All the above
    Ans. d

34. Which cells are most often infected by HIV?
    (a) CD4+ T cells   (b) CD8+ T cells
    (c) Null cells   (d) B cells
    Ans. a

35. Which is the commonest bacterial infection in HIV disease?
    (a) Mycobacterial infection
    (b) *Salmonella* infection

(c) *Bartonella* infection
(d) *Klebsiella* infection

Ans. a

36. Which of the following tests is the screening test for the diagnosis of HIV infection?
    (a) ELISA
    (b) Latex agglutination
    (c) Dot blot assay
    (d) All the above

Ans. d

37. Which of the following tests is the confirmatory test for the diagnosis of HIV infection?
    (a) Virus isolation
    (b) Detection of p24 antigen
    (c) Detection of viral nucleic acid
    (d) All the above

Ans. d

SHORT NOTES

1. General properties of viruses
2. Structure of viruses
3. Functions of capsid
4. Viral replication
5. Classification of viruses with examples
6. Cytomegalovirus
7. Varicella zoster virus
8. Epstein–Barr virus
9. Infections caused by herpes simplex virus type 1 and type 2
10. Coxsackieviruses
11. Polioviruses
12. *Rhinovirus*
13. Hepatitis A virus
14. Pathogenesis of rabies
15. Laboratory diagnosis of rabies
16. Prophylaxis of rabies
17. Morphology of rabies virus
18. Morphology of HIV virus
19. Mode of transmission of HIV infection
20. Pathogenesis of HIV infection

21. Clinical features of HIV infection
22. Laboratory diagnosis of HIV infection

ESSAYS

1. Explain in detail the structure and replication of viruses.
2. Name the viruses of the family Herpesviridae and discuss the various infections caused by herpes simplex virus type 1 and type 2.
3. Discuss the general properties and pathogenesis of enteroviruses. Add a note on their laboratory diagnosis.
4. Explain in detail the human pathogenic diseases caused by the members of Picornaviridae.
5. Explain in detail the structure of rabies virus with a diagram. Add a note on the clinical features, pathogenesis, laboratory diagnosis, and prophylaxis of rabies.
6. Explain in detail the structure of HIV virus. Describe the mode of transmission, pathogenesis, and clinical features of HIV infection. Add a note on its laboratory diagnosis as well.

18 Fungi

CHAPTER OBJECTIVES

18.1 Introduction
18.2 Morphological Classification
18.3 Taxonomical Classification
18.4 Cell Wall of Fungi
18.5 Fungal Reproduction
18.6 Mycoses

18.1 INTRODUCTION

Fungi are microscopic eukaryotic organisms that were first classified under kingdom Plantae and later separately classified under kingdom Fungi according to Robert H. Whittaker's five-kingdom system. They were classified accordingly due to the presence of unique rigid cell wall, which is chemically different from the bacterial cell wall. The fungal cell wall is rich in carbohydrates composed of polymers of acetyl glucosamine (glucose molecules linked with amino and acetyl groups) and hence is coined as chitin, which forms a thick layer protecting the inner organelles from the adverse external environment. Fungi are classified according to morphology and taxonomy.

18.2 MORPHOLOGICAL CLASSIFICATION

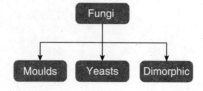

18.2.1 Moulds

Moulds are characteristic fungi that are made up of rope-like filamentous structure called hyphae. The hyphae help in the interexchange of cytosol and organelles between adjacent cells (Figure 18.1). These hyphal filaments group together to form a mess-like appearance called the mycelium.

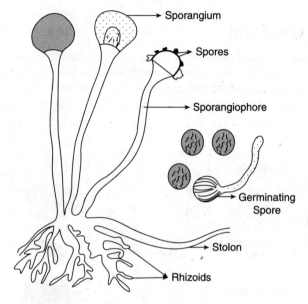

Figure 18.1 Moulds

18.2.2 Yeasts

Yeasts are unicellular eukaryotic organisms, and they resemble bacterial colonies as they appear smooth and mucoid on the media (Figure 18.2). They are aerobic organisms but their growth is enhanced in anaerobic conditions. They acquire energy from an organic compound by oxidation.

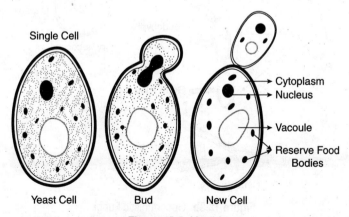

Figure 18.2 Yeasts

Yeast-like Fungi

They partly resemble yeast cells and also develop pseudohyphae resembling hyphal filaments.

18.2.3 Dimorphic Fungi

They exist in both mycelial and yeast forms in varying temperatures (Figure 18.3a and b). They exist as yeast-like colonies at 37°C and transform themselves into mould-like colonies at 25°C especially when they are found in the soil; hence, the mould forms are saprophytic, and the yeast forms are considered as pathogenic. Systemic infections are mostly caused by dimorphic fungi.

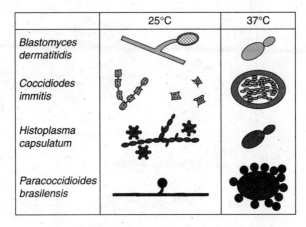

Figure 18.3a Dimorphic Fungi

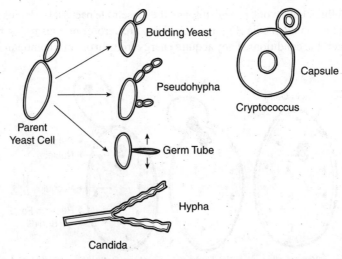

Figure 18.3b Opportunistic Fungi

18.3 TAXONOMICAL CLASSIFICATION

Fungi are classified into four different classes (Figure 18.4), which are as follows:

1. **Zygomycetes:** These are lower fungi with non-septate hyphae, and they produce sporangiospores (asexual spores).
2. **Ascomycetes:** They produce septate hyphae and ascospores. (Sexual spores are present inside the sac or ascus.)
3. **Basidiomycetes:** They produce septate hyphae and basidiospores. (Sexual spores are present in the basidium.)
4. **Deuteromycetes or Fungi imperfecti:** They produce septate hyphae and cannot be classified into sexual or asexual because their sexual state is unknown. However, most fungi appear to share common features with ascomycetes. Medically important fungi belong to this group.

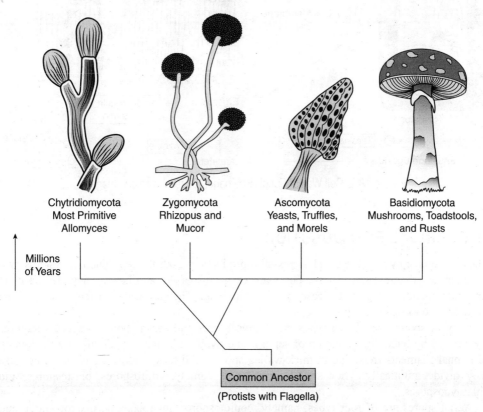

Figure 18.4 Classification of Fungi—Zygomycetes, Ascomycetes, Basidiomycetes, and Deuteromycetes (See page 361 for the colour image)

18.4 CELL WALL OF FUNGI

The major difference between the cell wall composition between fungi and plants is that; in fungi, it is made up of chitin and polysaccharide, whereas in the case of plants, it is replaced by cellulose (Figure 18.5). Even though fungi and plants share certain characteristics with each other, genetic studies have proved that the animal kingdom shares a closely related link with fungi than plants. The fungal species are ubiquitous organisms and are found in the soil, dead decaying matter, and all moist places.

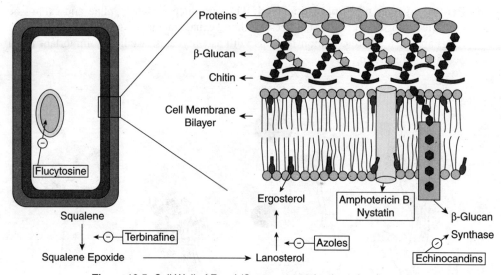

Figure 18.5 Cell Wall of Fungi (See page 362 for the colour image)

18.5 FUNGAL REPRODUCTION

They reproduce by either asexual or sexual reproduction; both the reproduction systems produce spores (Figure 18.6). These spores are easily dispersed, and hence, they are the most important infective stage in the case of infectious fungi. They are very resistant and can resist even dry and adverse conditions.

Asexual spores are of two types: vegetative spores and aerial spores. Vegetative spores are formed by budding, formation of septa in the hyphal filament, or folding and thickening of hyphal filaments (resulting in thick-walled spores). All these spores are identical because they divide mitotically. They are of three types, namely, arthrospores, blastospores, and chlamydospores.

Aerial spores are of four types, namely, conidiospores, microconidia, macroconidia, and sporangiospores. These spores are both exogenous and endogenous in nature.

Sexual spores are of three types, namely, zygospores, ascospores, and basidiospores. Haploid spores are formed inside each sporangium. Sexual reproduction is by the fusion of cells followed immediately by meiosis.

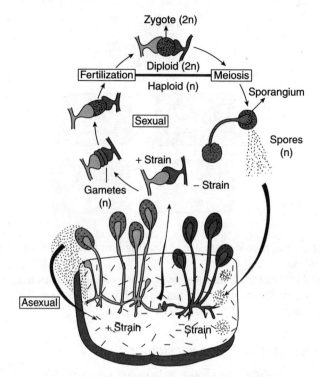

Figure 18.6 Sexual and Asexual Reproduction (See page 362 for the colour image)

18.6 MYCOSES

The diseases caused by fungi are called mycoses. These infections are caused to humans and animals through the inhalation of fungal spores (which enter the lungs, get localized, and cause respiratory infection) or inoculation of spores or pathogenic fungi through cuts (which localize in the skin). Another important cause of mycoses in humans is the alteration of the normal flora of the body due to overconsumption or heavy dose of antibiotics (which alters or kills the normal flora acting as the physical barrier and enhances the entry of pathogenic fungi). Individuals with a suppressed immune system such as those undergoing chemotherapy or in steroids due to transplantation, HIV, or diabetics are at high risk of mycoses.

These fungal infections are classified into four types, which are as follows:

1. Superficial mycosis
2. Subcutaneous mycosis
3. Deep mycosis
4. Opportunistic mycosis

18.6.1 Superficial Mycosis

Mycoses affecting the outermost layers of the skin, hair, and nail are classified under superficial mycoses. They externally localize on the layers of the skin, hair, and nail and grow well on the

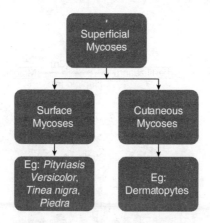

dead layers. Superficial mycoses are hence classified into two types, namely, surface mycoses and cutaneous mycoses.

Surface mycoses affect the outermost layer of the skin and do not have any kind of inciting response because they do not invade the innermost layer of the skin. Surface mycoses are mild persistent infections caused due to insanitary conditions.

Pityriasis versicolor or Tinea versicolor

This fungus causes superficial mycoses in humans. *Pityrosporum orbiculare* or *Malassezia globosa* is the fungus responsible for the surface infection.

Fungal Characteristics: They are yeasts that live on the skin in few numbers but multiply during adverse conditions leading to skin infections.

Symptoms: The common symptoms are as follows:

1. Mild patches are found on the chest, back, neck, and arms.
2. These patches lead to pigmentation.
3. This discolouration starts spreading in untreated conditions.
4. Dryness is also observed on the patches.
5. Itching is also observed.

Pathogenesis: The transmission is mainly due to contact, adherence, and invasion of the pathogenic fungi. Improper sanitation during hot weather due to over-sweating will overpopulate the yeasts present on the skin causing infection. These yeast-like fungi become pathogenic when they change their morphology from that of yeast to that of mycelium. Hence, they are dimorphic in nature. The pigmentation is mainly due to azelaic acid, which is a secondary metabolite that is produced by the fungi and acts on the melanin pigments.

Laboratory Diagnosis: The samples collected are skin scrapings of the lesions by a special technique using Sellotapes. Sellotapes are placed on the lesions and removed for further testing.

Physical examination using filtered UV light at 365 nm is done; the lesions glow golden yellow. This method of examination is called Wood's light examination.

Direct microscopic method is employed by fixing the skin scraping on a microscopic slide with 20% potassium hydroxide, covering it with a coverslip and microscopically examining it. Short unbranched hyphae with grape-like yeast cells are observed.

Isolates are cultured using Sabouraud dextrose agar (SDA) medium as the basal medium for the isolation of fungus. This medium requires an extra admission of a lipid substrate in order to enhance the growth of the pathogenic fungi; hence, the medium is overlaid with olive oil. Incubation is done at 32°C–37°C for a week's time, after which round and smooth colonies appear on the medium. These colonies are picked for further testing using a special dye, namely, lactophenol cotton blue. A wet mount is prepared for the demonstration of the yeast cells with unbranched hyphae.

Treatment: Tropical application of ointments are prescribed, which is a combination of benzoic acid, salicylic acid, ciclopirox olamine, and tincture of iodine called as Whitfield's ointment. Any sulphur-containing ointment is also prescribed. In severe cases, oral antifungals such as triazole, itraconazole, and ketoconazole are given.

Tinea nigra

Tinea nigra is the superficial mycosis caused by the fungus *Hortaea werneckii*.

Fungal Characteristics: They are dimorphic fungi that exist in both yeast and hyphal forms and are responsible for asymptomatic mycoses. These fungi are saprophytic in nature and are found on dead and decayed materials.

Symptoms: The common symptoms are as follows:

1. Mild patches are seen on palms and foot soles.
2. These patches become brownish or black in colour. They are irregular in shape.
3. Itching is also observed.
4. Skin growth like scales is seen.

Pathogenesis: The entry is mainly though the inoculation of the fungus through cuts and wounds during contact with the fungus. They enter, localize, and cause superficial infection on the palms and foot soles. They are halotolerant and hence are able to survive in human tissues by accumulating and utilizing melanin. Sometimes, this mycosis is confused with a type of skin cancer.

Laboratory Diagnosis: Skin scrapings are collected for wet mount processing (KOH mount) and viewed under the microscope demonstrating the budding yeast cells on the branched hyphal filaments. Culturing using SDA is done, and moist white-coloured colonies appear after a week's time.

Treatment: Treatment can be done through the following ways:

1. Tropical application of antifungal ointments
2. Following good hygiene
3. Avoiding the exposure to moist and dirty places

Piedra

Piedra is a disease caused in hair follicles. They are of two types, namely, white piedra and black piedra.

White Piedra: White piedra is caused by *Trichosporon beigelii*.

Fungal Characteristics: They are asexually reproductive fungi that are a part of the normal flora. These are yeast-like fungi that change to septate hyphal filaments.

Symptoms: The common symptoms are as follows:

1. Acclimation of white lump of yeast cells on the hair follicles of the head, the beard, and even pubic hair
2. Hair loss
3. Itching

Pathogenesis: The fungi are transmitted on close contact with the infected person's towel, soap, comb, and so on. The white nodules found on the hair shafts are nothing but the spores that are transmitted by contact.

Black Piedra: It is caused by *Piedraia hortae*, which causes superficial infections of the hair and scalp. It is an asexual fungi that is transmitted from one person to another through contact. Black nodules are found on the hair shafts of the beard and scalp.

Laboratory Diagnosis: Physical examination of the hair and wet mount preparation using KOH of the hair with nodules is done. The wet mount is then viewed under the microscope. The affected hairs are also collected and cultured using SDA medium.

Treatment: Treatment can be done through the following ways:

1. Applying tropical antifungal ointments (imidazoles and selenium sulphide)
2. Applying amphotericin B ointments
3. Using separate towels, soaps, and combs
4. Following proper hygiene

18.6.2 Subcutaneous Mycosis

It is a type of mycosis that causes disease in living tissues leading to tissue damage. It is of three types, namely, mycetoma, chromoblastomycosis, and rhinosporidiosis.

Mycetoma

Mycetoma is a persistent subcutaneous granulomatous infection that affects the foot. It is otherwise called as Madura foot because it was first identified in Madurai by Gill in 1842. This mycosis affects the subcutaneous tissue, and as the infection progresses, bones are also damaged.

Fungal Characteristics: Actinomycetes or filamentous fungi cause mycetoma. These fungi and actinomycetes are commonly found in the soil. Actinomycetes are higher forms of bacteria but are filamentous in nature and hence grouped under fungi. These actinomycetes are filamentous, aerobic in nature, and belong to *Mycobacterium* sp. This mycosis is grouped into two types, namely, actinomycetoma (caused by *Actinomycetes* sp.) and mycetoma (caused by the filamentous true fungi, i.e., eumycetoma).

Symptoms: The common symptoms are as follows:

1. Itching
2. Swelling in the area of itching
3. Pus formation
4. Ulceration and nodules
5. Pus expulsion
6. Irritation in the area of infection
7. Disfiguration of the leg

Pathogenesis: The mode of transmission is by the inoculation of the fungus or actinomycetes through cuts or wounds from the soil. They enter the tissue, localize, and start replicating inside live cells. The spores produced form clumps inside the cells resulting in granules. These granules vary in colour depending upon the contributing agents. The infection starts with mild itching and swelling mostly in the foot region. The swelling further enlarges due to the accumulation of dead cells with granules. The infection spreads due to the expulsion of pus as the discharge, and it affects the bones.

Laboratory Diagnosis: The samples collected are the pus exudates. These are smeared for Gram staining, and a wet mount preparation is done for the identification of actinomycetes (Gram-positive in nature) or Eumycetes. A culture method is also followed by inoculating the sample in an appropriate medium and further demonstrating the isolate.

Treatment: Treatment can be done through the following ways:

1. Surgery and removal of the abscess at the early stage should be done with proper administration of oral antifungal and antibacterial drug therapy.
2. Rifampicin, dapsone, and sulphonamides are given in the case of actinomycetoma.
3. Itraconazole and ketoconazole can be given in the case of eumycetoma.
4. The treatment should be prolonged for a certain period of time for the complete eradication of the pathogen.

Chromoblastomycosis

Chromoblastomycosis is caused by five different vegetative fungi that cause subcutaneous mycosis. They cause persistent infections that slowly progress and form granulomatous lesions ending up in the accumulation of keratinocytes in the epidermal layer resulting in the sloughing of the skin. The causative agents of chromoblastomycosis are given overleaf.

```
                          ┌─────────────────────────┐
                          │   Chromobalstomycosis   │
                          └─────────────────────────┘
   ┌──────────┬───────────┼───────────┬──────────────┐
┌──────────┐┌──────────┐┌──────────┐┌──────────┐┌──────────────┐
│Phialophora││ Fonsecaea││ Fonsecaea││Rhinocladiella││Cladophialophora│
│ verrucosa ││ compacta ││ pedrosoi ││ aquaspersa ││  carrionii   │
└──────────┘└──────────┘└──────────┘└──────────┘└──────────────┘
```

Fungal Characteristics: The morphology of the conidia differs in each type of fungi mentioned above. The conidia in *Phialophora verrucosa* are thermos-shaped phialides and mug-shaped collarettes. The conidia in *Cladophialophora carrionii* are arranged in chains and divide by budding. *Rhinocladiella aquaspersa* produces terminal conidia, which are elliptical in shape. *Fonsecaea pedrosoi* are polymorphic in nature, hence changing the shape of their conidia. The shape of the conidia produced by *Fonsecaea compacta* is spherical, and they are much smaller when compared with that of *Fonsecaea pedrosoi*.

Symptoms: The common symptoms are as follows:

1. Itching in the area of entry of the pathogen is observed.
2. Since the infection is a slow process, swelling with pus formation occurs.
3. Nodules are formed as the infection drains into the lymph.
4. Ulceration is commonly seen.

Pathogenesis: Chromoblastomycosis is a chronic subcutaneous infection that is transmitted through cuts or wounds mainly in the leg region, as it has contact with the soil or dirt more often than other parts of the body. The fungus on entry by inoculation spreads and invades the tissue very slowly because of its slow growing capacity. This fungus is very resistant due to the melanin cell wall and hence is not easily eliminated by the immune cells. Therefore, it resides, slowly localizes, and invades the tissue. The fungus affects the tissue-forming hyperplasia of the epidermis producing nodules with pus formation, which is painless. This fungus drains into the lymphatics, and in severe conditions, this might result in the damage of different organs of the body.

Laboratory Diagnosis: The pus cells or skin scrapings are used as the specimen for the identification of the causative organism by a series of tests, which are as follows:

1. KOH wet mount preparation
2. Histopathological analysis to check the multinucleated giant cells with the granules and sclerotic bodies inside and outside the cells
3. Culturing using SDA and checking for brown or black mouldy colonies

Treatment: Treatment can be done through the following ways:

1. Surgical removal of the pustule and proper oral treatment
2. Heat therapy (early stages)
3. Antifungal agents such as flucystosine, ketoconazole, and itraconazole

Rhinosporidiosis

Rhinosporidiosis, a granulomatous infection of the nose, eyes, and mouth, is caused by *Rhinosporidium seeberi*. This fungus is classified under lower aquatic protistan parasites that affect humans and amphibians.

Characterization: This species was first classified under moulds and then under protozoa; and finally, after using molecular tools, the analysis proved that it is a aquatic protistan parasite. It belongs to a group called Mesomycetozoa.

Symptoms: The accumulation of a large mass of cells that hangs out as a separate layer is seen. Pus accumulation leads to foul smell. Breathing difficulty due to the protrusion of the layer is observed.

Pathogenesis: People who are mostly affected by this infection are fishermen and washermen as they are in constant contact with water habitats facilitating the entry by the inoculation of the pathogen through cuts and wounds. The pathogen enters through the nasopharyngeal route or even eyes and external genitals. This organism enters, localizes the site, and multiplies to produce hyperplasia.

Laboratory Diagnosis: The skin biopsy is taken for histopathological analysis, and KOH wet mounting is also done for the identification of the pathogen that is in the form of endospores within the sporangium. Serological tests are also done.

Treatment: Treatment can be done through the following ways:

1. Surgery is a good option.
2. Intravenous administration of amphotericin B and dapsone is done to avoid the spread of infection.

18.6.3 Deep Mycosis

Deep mycosis is otherwise called as systemic mycosis. This infection involves both pathogenic and opportunistic fungi. The pathogenic fungi gain the entry, whereas the opportunistic fungi suppress the immune system and affect it. Histoplasmosis is a type of systemic mycosis found in humans and animals.

Histoplasmosis

Histoplasmosis is a systemic mycosis caused by *Histoplasma capsulatum*. This affects the respiratory system causing pulmonary infection in humans. It is a saprophytic fungi mostly found in the soil.

Fungal Characteristics: It is a dimorphic fungus that exists in both the mould and the yeast form. It appears to be round solo-celled microconidia (uninucleated cells). When grown in a laboratory medium, it produces mouldy white colonies.

Symptoms: As the fungus causes respiratory infection, the symptoms are as follows:

1. Dry cough
2. Body pain

3. High fever
4. Restlessness
5. Lymphadenopathy
6. Fatal in severe cases affecting the liver, eyes, and glands.

Pathogenesis: The entry of the intracellular pathogen is by inhalation, and the moulds convert themselves into the yeast form and start invading the tissues. These yeast cells are engulfed by alveolar macrophages. Consequently, the cells start replicating inside and use them as the vehicle to travel around the body and invade other parts of the body such as liver, spleen, and lymph nodes. This results in pulmonary infection mostly in men, children, and immunocompetent patients such as those with AIDS and those undergoing therapy.

Laboratory Diagnosis: The specimens collected for the isolation of *Histoplasma capsulatum* are as follows:

1. Sputum
2. Urine
3. Throat swab
4. Bone marrow aspirates

After the collection of specimens, they are processed for direct microscopy method by histopathological staining or Giemsa staining to view the intracellular yeast. Culturing using either SDA (for the isolation of mouldy colonies) or blood agar (for the isolation of yeast colonies) and isolation after a week's time are done. Serological tests such as complement fixation test are done for confirmation. Enzyme immunoassay is done to find out the circulatory antigens that are very sensitive, and it is one of the confirmatory tests done.

Treatment: Treatment can be done through the following ways:

1. Itraconazole can be administered.
2. Amphotericin B can be given.
3. Relapses may occur for immunocompromized patients; hence, prolonged treatment with itraconazole is supplemented for them to avoid cross-infection.

18.6.4 Opportunistic Mycosis

Opportunistic mycoses are infections that occur in immunocompromised patients. This type of infection depends on the load of organisms and the virulence caused by them to the host where the host defence gets suppressed leading to opportunistic mycosis.

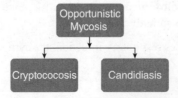

Candidiasis

Candidiasis is an opportunistic fungal infection affecting humans. Candidiasis is caused by *Candida albicans*, *Candida tropicalis*, *Candida parapsilosis*, and *Candida krusei*. These organisms are a part of the normal flora of the skin and gastrointestinal tract.

Fungal Characteristics: The cells are oval-shaped and divide by budding, which form into a pseudohyphae. Some species such as *Candida albicans* are dimorphic in nature. This fungus produces smooth white-coloured colonies on solid agar medium and appears within 24 hours.

Symptoms: This affects the skin, mouth, and genital regions causing a thrush. The common symptoms are as follows:

1. Painless lesions occur.
2. Unpleasant white patches are seen on the mouth, skin, and genital regions.
3. On eruption of the lesions, pus-like exudates with blood come out, making it very painful.
4. This eruption might lead to the spread of infection to the respiratory system and other parts of the body.
5. Haematologic diseases such as anaemia, lymphoma, and chronic granulomatous disease occur in severe cases.
6. Vulvovaginitis (irritation and vaginal discharges) is also observed.

Pathogenesis: The route of entry is airborne, or it is sexually transmitted through the genital tract. These pathogens enter the cells, suppress the immune system from acting against them, invade, and replicate in the host tissues. Airborne entry causes oral thrush infecting the tongue, lips, and palate. In untreated cases, they extend to the respiratory tract causing infection of the lungs and oesophagus. The pseudohyphae on circulation enter the bloodstream; are carried to different parts of the body; and affect various organs such as the liver, kidney, and even heart valve.

Sexually transmitted candidiasis affects the vaginal valve causing vaginal thrush called vulvovaginitis. Infections of the skin and nails, that is, cutaneous candidiasis and onychomycosis, affect the layers of the skin and nails ending up in inflammation and swelling of the nails. This type of infection is due to occupational hazards.

Laboratory Diagnosis: The samples collected for the identification of candidiasis are as follows:

1. Blood
2. Swabs (throat and vaginal)
3. Skin scrapings
4. Nails
5. Urine
6. Vaginal discharges
7. Cerebrospinal fluid

Direct microscopic method is done by histopathological staining of the tissue biopsies or Gram staining of the swabs and samples obtained to demonstrate the yeast cells and the pseudohyphae in the samples. Wet mounts of skin scrapings and nail samples are prepared using 10% KOH. Culturing is done using fungal media for performing various other tests. Serological tests are done to check the circulating cell wall antigens that are sensitive to the commercially available antibodies.

Treatment: Treatment can be done through the following ways:

1. Tropical applications are given for the infected skin and nails.
2. Ketoconazole, nystatin, and amphotericin B can be administered.
3. This type of oral treatment can be prolonged for immunosuppressive patients.
4. Less use of antibiotics is very important to maintain the balance of the microbial flora.

Cryptococcosis

Cryptococcosis refers to the opportunistic fungal infections caused by *Cryptococcus neoformans* and *Cryptococcus gattii*. These organisms are mostly found in the soil, faeces of birds, and even on trees. They are yeast-like fungi with much resistant capsules that protect them from the external environment. Hence, they are much resistant pathogenic organisms.

Fungal Characteristics: They are yeast-like fungi encapsulated by a thick cover of polysaccharide called as capsule. They can be cultured using fungal or bacteriological media by incubating for 24 hours at 37°C. They produce mucoid white colonies; this mucoid nature is mainly due to the presence of the capsule around the organisms.

Symptoms: The common symptoms are as follows:

1. Fever
2. Head ache
3. Body pain
4. Flu symptoms
5. Central nervous system affected in severe cases, causing encephalitis

Pathogenesis: The mode of transmission is by the inhalation of the fungal cells. They enter the respiratory system affecting the circulating cells and compromise the immune system by resisting the immune reactions. They multiply and infect other parts of the body, especially the central nervous system causing meningoencephalitis (inflammation of the brain tissues and meninges).

Laboratory Diagnosis: The samples collected for microscopic and culture examinations are as follows:

1. Sputum
2. Blood
3. Throat swab
4. Cerebrospinal fluid

A special staining technique is employed by using Indian ink for the demonstration of the capsule, which is the confirmatory test for cryptococcosis. The serological test for the demonstration of capsular antigen in the sample using commercial antibodies is very sensitive and effective.

Treatment: Treatment can be done through the following ways:

1. Amphotericin B and flucystosine can be administered.
2. A combination of both is prescribed in severe cases such as immunocompromized patients.
3. Prolonged treatment is provided to avoid recurrent infections.

MULTIPLE CHOICE QUESTIONS

1. The cell wall of fungi is made up of _____.
    (a) Peptidoglycan
    (b) Polysaccharide
    (c) Chitin
    (d) Lipids
    Ans. c

2. Fungi that exist in the form of both yeasts and moulds are _____.
    (a) Yeasts
    (b) Moulds
    (c) Yeast-like
    (d) Dimorphic
    Ans. d

3. Hyphal or mycelial colony is referred to as _____.
    (a) Yeasts
    (b) Moulds
    (c) Yeast-like
    (d) Dimorphic
    Ans. b

4. Chains of elongated buds are _____.
    (a) Hyphae
    (b) Yeast
    (c) Buds
    (d) Pseudohyphae
    Ans. d

5. Pityriasis versicolor is caused by _____.
    (a) Tinea nigra
    (b) *Tinea piedra*
    (c) *Candida*
    (d) *Malassezia globosa*
    Ans. d

6. Tinea nigra refers to _____.
    (a) Superficial mycoses
    (b) Cutaneous mycoses
    (c) Opportunistic mycoses
    (d) Deep mycoses
    Ans. a

7. Infections caused by fungi are termed as _____.
    (a) Mycoses
    (b) Zoonoses
    (c) Parasitic infections
    (d) Sepsis
    Ans. a

8. Mycetoma are caused by _____.
    (a) *Candida*
    (b) *Actinomycetes*
    (c) *Cryptococcus*
    (d) *Histoplasma*
    Ans. b

9. Media used for the isolation of fungal cultures is _____.
   (a) Nutrient agar
   (b) Blood agar
   (c) Sabouraud dextrose agar
   (d) Eosin methylene blue agar

   Ans. c

10. Oral thrush is caused by _____.
    (a) *Candida albicans*
    (b) *Histoplasma capsulatum*
    (c) *Cryptococcus neoformans*
    (d) *Actinomycetes*

    Ans. a

SHORT NOTES

1. Superficial mycosis
2. Opportunistic mycosis
3. Systemic mycosis
4. Classification of fungi based on reproduction
5. General laboratory diagnosis of fungi

ESSAYS

1. Brief about the pathogenesis, laboratory diagnosis, and characteristics of opportunistic fungi.
2. Brief about the pathogenesis, laboratory diagnosis, and characteristics of the fungi that cause superficial mycosis.

19 Parasites

CHAPTER OBJECTIVES

19.1 Introduction

19.2 *Entamoeba histolytica*

19.3 *Plasmodium* sp.

19.4 Parasitic Helminths

19.1 INTRODUCTION

The study of parasites is termed as parasitology. Parasites can be divided into protozoans and helminths. Protozoans are unicellular eukaryotic organisms that require a host to survive. They have two important stages in their life cycle, namely, the infective stage and the latent stage. Protozoans are classified into four groups, namely, flagellates, amoebae, sporozoa, and ciliates.

1. **Flagellates:** They are so called because of the presence of flagella, a tail-like structure that is used for locomotion. Most of the parasites belonging to this group affect the gastrointestinal region of humans, hence also termed as intestinal pathogens.
2. **Amoebae:** They are single-celled organisms that change their shape according to their body movement, which is facilitated by a locomotory organ called pseudopodia. They use humans as their hosts for survival causing parasitic infections in them. The most important parasitic infection caused in humans by amoebae is amoebic dysentery, which is caused by *Entamoeba histolytica*.
3. **Sporozoa:** They have both asexual and sexual life cycle and use two hosts for their survival; hence, they have a complex life cycle. They are most probably found in blood-sucking insects such as mosquitoes and ticks. These parasites cause infections such as malaria and filariasis.

19.2 *Entamoeba histolytica*

Entamoeba histolytica causes amoebic dysentery in humans, affecting their intestinal mucosa (Figure 19.1). They are mostly present in the lumen of humans. The pathogenic or infective stage of amoebae is the cystic stage, which is called as trophozoites. These are round cells covered by an inner and an outer membrane with a nuclear membrane covered by a chromatin granule. The nucleus divides by binary fission into four daughter cells and forms cysts.

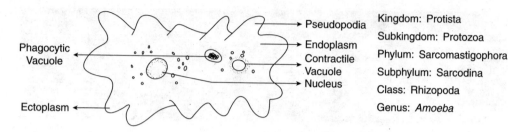

Figure 19.1 *Amoeba*

19.2.1 Life Cycle

Entamoeba histolytica exists in three forms (Figure 19.2), which are as follows:

1. Precystic stage
2. Cystic stage
3. Trophozoites

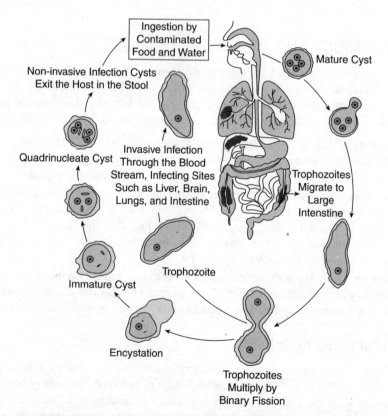

Figure 19.2 Life Cycle of Amoebiasis (See page 363 for the colour image)

The precystic and cystic stages are inactive stages. The cysts enter the body through contaminated food. As they are dormant, their outer membrane gets digested by the gastric juices in the stomach, and this loss of cell membrane ends up in the formation of an infective cystic stage called the metacyst. Metacysts are parasitic in nature and the process of formation of metacysts is called excystation. Metacysts multiply to form trophozoites, which then invade the mucosal membrane of the intestine causing the infection in the host. The trophozoites then change into cysts, which are excreted along with the faeces.

19.2.2 Pathogenesis

Transmission is through the faecal–oral route, that is, through contaminated food and water that contain the cyst. On ingestion, the cysts that enter undergo a series of attacks by the physical and chemical barriers of the human body. They survive these attacks, and trophozoites attack the mucosa of the small intestine. They localize causing ulcers and start multiplying causing lesions on the intestinal mucosa. The ulceration might lead to secondary bacterial infections; in chronic cases, the liver, lungs, and brain are also damaged due to the spread of amoebae.

Symptoms

The symptoms of amoebic dysentery are as follows:

1. Abdominal pain
2. Diarrhoea
3. Nausea
4. Malaise
5. Vomiting
6. Loss of appetite
7. Weight loss
8. Dehydration

19.2.3 Laboratory Diagnosis

The samples collected for the diagnosis are stool and blood, which are observed for the presence of cysts microscopically. Serological tests such as enzyme assay of antigen–antibody testing are done. These tests are highly effective and show positive results.

19.2.4 Treatment

In severe cases, a combination of drugs is given. The drugs commercially available for the control of the diarrhoeal disease are iodoquinol, metronidazole, diloxanide furoate, and paromomycin. The preventive measures to be taken to avoid the infection are as follows: maintaining cleanliness and keeping ourselves clean, cooking food properly, avoiding outside food, and drinking boiled water.

19.3 *Plasmodium* sp.

The *Plasmodium* species causes malaria in humans (Figure 19.3). Malaria is an infective disease transmitted to humans by the vector mosquito. The mosquitoes carry the sporozoites in their saliva and inoculate them into humans while sucking the blood. The malarial diseases are caused by four different species of *Plasmodium*, which are as follows:

1. *Plasmodium vivax*
2. *Plasmodium falciparum*
3. *Plasmodium malariae*
4. *Plasmodium ovale*

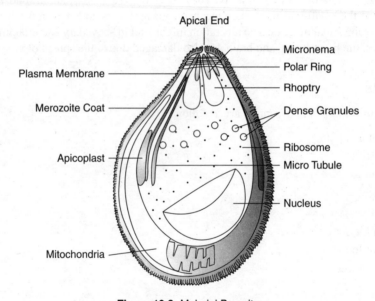

Figure 19.3 Malarial Parasite

19.3.1 Life Cycle

These protozoans are intracellular parasites that have both sexual and asexual stage (Figure 19.4). Anopheles mosquito is the vector for malarial disease. The female mosquitoes are blood suckers, whereas the male mosquitoes generally feed on fruits. The asexual stage occurs in humans. When the mosquitoes bite infected individuals, the spores enter the mosquitoes as a macrogamete, which then undergoes the sexual cycle to produce the oocyte and multiplies within the body of the mosquitoes to form sporozoites. The sporozoite is the infective stage; hence, if the mosquitoes bite during this phase, the sporozoites enter and are localized in the liver of humans. They then enter RBCs resulting in the haemolysis of RBCs.

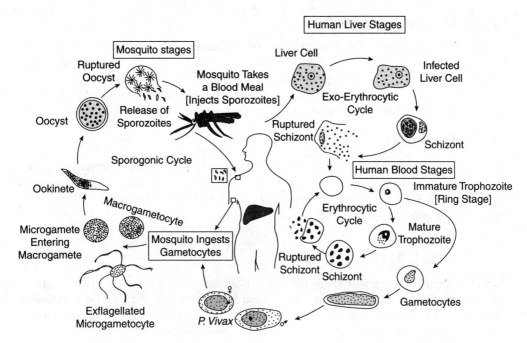

Figure 19.4 Life Cycle of Malarial Parasite (See page 364 for the colour image)

19.3.2 Pathogenesis

Transmission is through the inoculation of the pathogenic sporozoites by the biting of mosquitoes as their saliva carries the sporozoites. The malarial diseases caused by *Plasmodium vivax, Plasmodium malariae,* and *Plasmodium ovale* are less severe when compared with that by *Plasmodium falciparum,* which causes severe effects in erythrocytes leading to fatal consequences. The sporozoites invade the liver and multiply inside the liver cells forming merozoites during the endogenous phase, that is, the asexual stage. The merozoites enter the RBCs and form trophozoites, which exist in two forms, namely, ring and mature trophozoites, and the cycle goes on. This is the stage where people are clinically identified for malaria due to the presence of trophozoites in RBCs.

Symptoms

The symptoms of amoebic dysentery are as follows:

1. Chills
2. High fever
3. Vomiting
4. Nausea
5. Loss of appetite
6. High temperature ending up in chills and sweat episodes
7. Hepatomegaly and anaemia occurring in severe cases

19.3.3 Laboratory Diagnosis

Blood samples are collected for the identification of the parasite in RBCs. Thick and thin smears are prepared, stained, and viewed under the microscope to check the presence of trophozoites in RBCs. Rapid tests, such as malarial antigen detecting test using commercially available antibodies, are performed nowadays; these tests are highly specific and accurate. Molecular methods such as PCR are also being deployed for the rapid identification of the infection.

19.3.4 Treatment

The best antimalarial drugs prescribed are chloroquine and amodiaquine; in severe cases, artemisinin is given with the above-mentioned combination of drugs. The eradication of Anopheles mosquitoes by following pest control measures is very important for the eradication of malarial diseases in the suburban area.

19.4 PARASITIC HELMINTHS

Parasitic helminths are worms causing infections in humans, mainly due to the presence of helminthic eggs in uncooked food, that is, unprocessed food. These enter the intestine and develop into adult worms causing intestinal infections. Few examples of such helminthic parasites that cause infections in humans are *Taenia solium* and *Wuchereria bancrofti*.

19.4.1 *Taenia solium*

They are called tapeworms and are mostly found in pork meat as they use pig as their intermediate host, hence also called pork tapeworms (Figure 19.5).

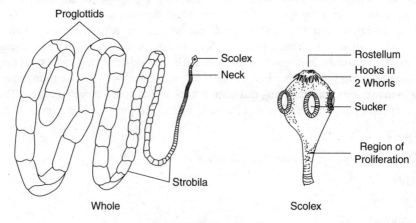

Figure 19.5 *Taenia solium*

Life Cycle

The infectious stage of the tapeworm is the cystic stage (Figure 19.6). In this stage, they enter the human intestine and hatch into adult worms that cause intestinal disorders. These adult worms can reach even metres inside the intestine. The eggs they lay stick on the walls of the intestine and are later excreted through the faeces. This faeces when consumed by pigs, transform into cysts and infect tissues like muscles. If such meats are consumed without proper cooking, they cause infections in humans and the cycle goes on.

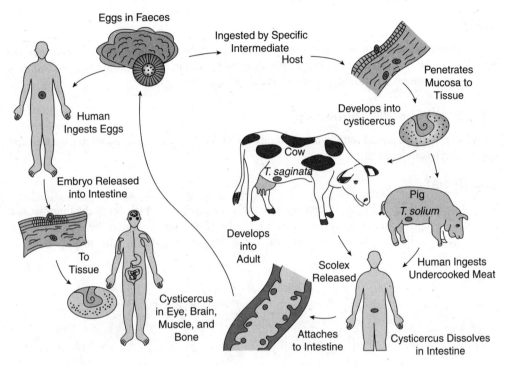

Figure 19.6 Life Cycle of *Taenia solium* (See page 364 for the colour image)

Pathogenesis

Transmission is by the ingestion of the cyst present in the pork meat, and the infection occurring is called cysticercosis. It also occurs due to the ingestion of raw or uncooked meat. The symptoms start with usual abdominal cramps, diarrhoea, vomiting, and nausea. Cysticercosis normally affects the central nervous system and is called neurocysticercosis.

Symptoms: The symptoms of *Taenia solium* infection are as follows:

1. Headache
2. Vision problem
3. Uncertainty
4. Seizures
5. Death in severe cases

Laboratory Diagnosis

The samples collected for the identification of the parasite are stool and blood. Stool samples are used for viewing the eggs laid by the adult worms in the faeces. Blood smears are prepared for the identification of the cyst in the blood.

Treatment

The commonly used antihelminthic drug is albendazole. Steroids such as corticosteroids are also being used in serious cases. Proper handling and cooking of food especially unprocessed meat before consuming should be done to avoid the infection.

19.4.2 Wuchereria bancrofti

They are otherwise called roundworms (Figure 19.7). They are transmitted by mosquitoes and cause filariasis in humans.

Life Cycle

They distribute their life cycle in two hosts, namely, mosquitoes, which act as the intermediate host, and humans, who are the definite host (Figure 19.8). The larval stage is the infective stage; the larvae are called as microfilariae, and the larvae are found in the circulation and mainly in

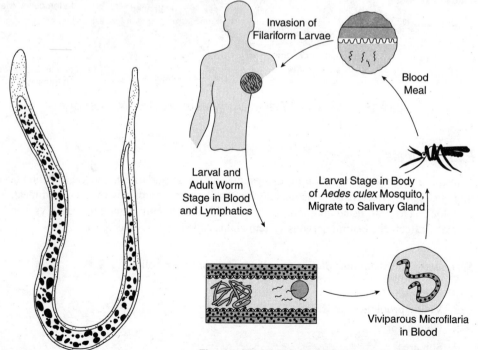

Figure 19.7 Wuchereria bancrofti

Figure 19.8 Life Cycle of Wuchereria bancrofti (See page 365 for the colour image)

the lymph affecting the lymphatics. If mosquitoes bite infected individuals, the microfilarial larvae are transmitted through these vectors to healthy individuals. The larvae enter the lymph and circulation; new worms are hatched causing the emergence of male and female worms, which mate with each other producing hundreds of microfilarial larvae.

Pathogenesis

Transmission is by the inoculation of the microfilarial larvae into the blood by vectors. Generally, the infection is asymptomatic. These worms infect the lymph and lymphatic vessels producing an inflammatory response.

Symptoms: The symptoms of filariasis are as follows:

1. Fever
2. Swelling of lymph
3. Lymphadenopathy
4. Enlargement of testes
5. Elephantiasis

Laboratory Diagnosis

Blood samples are taken for the identification of the larvae in blood circulation. Antigens are tested against adult worms. Molecular tests are performed for accurate and quick results.

Treatment

The treatment options for filariasis are as follows:

1. Surgery is the best option in the case of elephantiasis or any other inflammation.
2. Antihelminthic drugs such as albendazole can be used.
3. The infection can be prevented by controlling the pest (mosquitoes).

MULTIPLE CHOICE QUESTIONS

1. The study of parasites is termed as _____.
   - (a) Parasitology
   - (b) Mycology
   - (c) Bacteriology
   - (d) Virology

   Ans. a

2. *Entamoeba histolytica* belongs to the group _____.
   - (a) Amoebae
   - (b) Flagellates
   - (c) Sporozoa
   - (d) Ciliates

   Ans. a

3. *Entamoeba histolytica* is an _____.
   - (a) Intestinal parasite
   - (b) Blood parasite
   - (c) Intestinal Sporozoa
   - (d) Intestinal amoebae

   Ans. d

4. Malaria in humans is caused by
   _____.
   (a) *Plasmodium* sp.
   (b) *Entamoeba histolytica*
   (c) *Babesia microti*
   (d) *Taenia saginata*

   Ans. a

5. *Taenia solium* is otherwise called as
   _____.
   (a) Beef tapeworm
   (b) Pork tapeworm
   (c) Dwarf tapeworm
   (d) Dog tapeworm

   Ans. b

6. Neurocysticercosis is the symptom caused by _____.
   (a) *Taenia solium*
   (b) *Taenia saginata*
   (c) *Hymenolepis nana*
   (d) *Dipylidium caninum*

   Ans. a

7. *Wuchereria bancrofti* causes
   _____.
   (a) Gastrointestinal disorders
   (b) Elephantiasis
   (c) Neurological disorder
   (d) Cysticercosis

   Ans. b

SHORT NOTES

1. Pathogenesis of *Wuchereria bancrofti*
2. Pathogenesis of *Entamoeba histolytica*
3. Life cycle of malarial parasite
4. Life cycle of *Taenia solium*

ESSAYS

1. Brief about the life cycle, pathogenesis, and laboratory diagnosis of *Entamoeba histolytica*.
2. Brief about the life cycle, pathogenesis, and laboratory diagnosis of the malarial parasite.

20 Rodents and Vectors

CHAPTER OBJECTIVES

20.1 Rodents

20.2 Vectors

20.1 RODENTS

20.1.1 Introduction

The rodent-based infection occurs mainly due to the bites from infected animals and food contaminated with the urine and faeces of rodents. Pathogens have the tendency to spread from rodents to humans causing deadly diseases. Precautionary methods are to be followed while handling animals because if the animals are infected, the pathogens, such as viruses, might be transmitted even through air or through the cuts and wounds in the skin. The important pathogen that is transmitted from rats to humans is *Yersinia pestis*, which is responsible for the bubonic plague caused in humans.

20.1.2 *Yersinia pestis*

Morphology

Yersinia pestis is the causative agent of plague caused in rodents and humans. They were first classified under Pasteurella family but then were differentiated into *Yersinia* depending on their cultural characteristics, antigenic properties, and biochemical parameters. These are non-motile, non-spore-forming Gram-negative rods. They were first discovered by Alexandre Yersin; he revealed that the bacilli belong to Enterobacteriaceae family. They are capsulated with a slimy layer covering them, and on staining with methylene blue, they appear like safety pins (slender rods linked with each other at the two polar ends).

Cultural Characteristics

The bacilli are aerobic in nature, and they grow well at 37°C. They have a tendency to grow even in basal medium such as nutrient agar and produce no haemolytic colonies in blood agar.

They do not ferment lactose and hence are called non-lactose fermenters. *Yersinia* species grow slowly when compared with other Enterobacteriaceae species and can be differentiated when grown on selective media such as cefsulodin–Irgasan–novobiocin, which inhibits other Gram-negative bacilli and allows *Yersinia* sp. to grow well. They have a characteristic growth pattern when they grow in the broth with oil or ghee coated on top of it; due to the granular growth of the culture in addition to the oil, they become slimy and have a 'stalactites' appearance (They hang down like a jelly.).

Biochemical Properties

They do not ferment glucose, mannitol, or fructose but ferment sucrose with gas production. They are catalase positive and do not liquefy gelatin. They do not ferment glycerol but reduce nitrates.

Mode of Transmission

They are transmitted from infected rodents to humans through a vector called fleas; hence, the diseases transmitted are termed as zoonotic diseases. The cycle starts from the rodents that infect the fleas, which act as the intermediate host and transfer the pathogen to humans. After being transmitted, they take only 48 hours of incubation and the infection gets harboured in humans.

Pathogenesis

The infection is of three types, which are as follows:

1. Bubonic plague
2. Pneumonic plague
3. Septicaemic plague

Bubonic Plague: It is the most common form of plague. As the entry of the pathogen is through the bite of the flea, they enter into the circulation and affect the lymph nodes causing the swelling of the lymph nodes, which is called the buboes.

Symptoms: The symptoms of bubonic plague are as follows:

1. Chills
2. Fever
3. Malaise
4. Pain in the area of swelling
5. Meningitis in the case of untreated conditions

Pneumonic Plague: This is the secondary infection that is caused after the bubonic stage; here, the lungs are affected causing haemorrhagic lesions. This stage is infectious, and the spread of infection is by air droplets, the inhalation of which infects healthy individuals.

Symptoms: The symptoms of pneumonic plague are as follows:

1. High fever
2. Cough
3. Chills
4. Breathing difficulty

Septicaemic Plague: It is the most infective stage of all. This stage occurs mainly due to the increase of bacterial load in the blood resulting in the haemorrhage of blood vessels leading to death; hence, it is also termed as 'black death'.

Laboratory Diagnosis

Samples collected for the identification of the pathogen are as follows:

1. Blood
2. Pus
3. Lung aspirates
4. Cerebrospinal fluid

A direct microscopic examination is performed by staining the isolates with methylene blue (during which bipolar staining of the bacilli is observed). Gram staining is done, and the Gram-negative bacilli are identified. Selective medium is used for the isolation of the pathogenic bacilli, and the specific ghee broth analysis is done for the verification of the bacilli. Antigenic detection using ELISA or immunofluorescence test is done for the confirmation of the antigen present in the blood. A complement fixation serological test is done for the rapid identification of the antigen–antibody reactions.

Treatment

The drug of choice for all stages of treatment is streptomycin or else doxycycline and gentamicin. Precautionary measures should be taken for the eradication of fleas and infected rodents.

20.2 VECTORS

20.2.1 Introduction

Most vectors are insects belonging to the phylum Arthropoda. They are invertebrates having exoskeleton with their body divided into head, thorax, and abdomen. They develop into adults by undergoing a series of changes in their life cycle, which involves the egg, larval, and adult stage; this process of transformation is called metamorphosis. Arthropods such as mosquitoes, ticks, mites, lice, and fleas have the capacity to carry the infection from a primary host to humans. These arthropods are called vectors as they carry out such kind of mechanical transmission. These vectors transmit the infection mainly during their blood meal, that is, they are blood suckers and transfer the pathogen into humans while they feed on human blood by biting and sucking. There are two types of vectors, which are as follows:

1. **Mechanical vectors:** Mechanical vectors (e.g., cockroaches and fleas) are those that contaminate food materials while feeding on them by transferring the microbes that they carry on their body, wings, and legs. Individuals who consume such contaminated food get infected.
2. **Biological vectors:** Biological vectors are those that carry microbes inside their body acting as intermediate hosts and transmit them to humans. All the biological vectors are blood suckers. Few vectors that cause infections in humans are as follows:

    (a) Ticks
    (b) Mites
    (c) Lice

20.2.2 Ticks

These are eight-legged arachnids that have the ability to infect humans with a variety of pathogens, namely, bacteria, viruses, and protozoans (Figure 20.1). They are the only vector that can harbour more than one pathogen, making the diagnosis more complicated. There are two types of tick-borne infection; one is caused by hard ticks and the other by soft ticks, which are commonly found near bushes and mushy places. The ticks are differentiated from other arthropods by the presence of a shield/outer layer covering on its ventral surface.

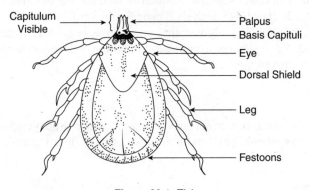

Figure 20.1 Tick

Hard Ticks

They are otherwise called Ixodid ticks, and they transmit babesiosis in humans. The infectious agent carried by the tick *Ixodes scapularis* is *Babesia microti*. Like malarial parasites, these are protozoan parasites that infect the red blood cells of humans.

Life Cycle: They are found in rodents and are transmitted by the vector. The infective stage (sporogamy) enters and divides by budding. These cells settle down on RBCs causing the lysis of the cells.

Pathogenesis: The disease is transmitted from rodents through the vector to humans. The disease shares the same symptoms as American Lyme disease, which is caused by spirochaetes.

Symptoms: The symptoms of babesiosis are as follows:

1. High fever
2. Chills
3. Headache
4. Body pain
5. Fatigue
6. Vomiting
7. Haemolytic anaemia in severe cases
8. Liver tribulations
9. Kidney failure

Laboratory Diagnosis: The sample collected for the isolation and identification of the pathogen is blood. Thin smears are prepared, and the parasite is observed by direct microscopy. Molecular detection using PCR and fluorescent in situ hybridization is done for the rapid identification of the disease.

Treatment: A combination of anti-parasitic drugs is prescribed for the eradication of the disease. The drugs are as follows:

1. Atovaquone
2. Azithromycin
3. Clarithromycin

The treatment is continued till the parasite is completely removed to avoid relapses, which might be deadly.

Soft Ticks

They belong to the family Argasidae and transmit tick-borne relapsing fever caused by *Rickettsia* and *Borrelia* belonging to the spirochaete family. Most people are infected after tick bites; the incubation period is about a week as soon as the pathogens harbour. The fever occurs at regular intervals: first onset starts with low-grade fever and chills and second onset starts with high fever, body ache, and vomiting. The fever episode subsides and starts again, hence termed after the symptoms as relapsing fever. Antibiotic treatment is given, and the treatment is continued till the bacilli are completely eradicated from the blood.

Control of Ticks

1. Ticks can be eradicated by the addition of dichlorodiphenyltrichloroethane, which is basically an insecticide and has larvicidal properties also.

2. The use of repellents and nets while sleeping is a good prophylactic method that can be followed to eliminate ticks.
3. The house must be cleaned using proper disinfectants.

20.2.3 Lice

Lice cause infections in both animals and humans. These arthropods are wingless and are of three types, namely, body louse (Figure 20.2) (*Pediculus humanus corporis*), head louse (*Pediculus humanus capitis*), and pubic louse (*Phthirus pubis*).

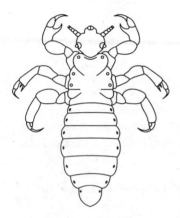

Figure 20.2 Body Louse

Body Lice

The body lice cause infection in humans, and the organisms causing infection are called *Pediculus humanus corporis*.

Life Cycle: There are three stages in the life cycle of a body louse, which are as follows:

1. Egg or nit
2. Nymph or larva
3. Adult louse

The eggs are whitish yellow in colour and are laid by the adult louse on the hair of the pubic area, body, and head. In a week's time, these eggs hatch into larvae. This stage is dormant, and the larvae hatch into adult lice in 10 days. The adult louse is an eight-legged gray-coloured insect that feeds on the blood.

Pathogenesis: Transmission is through body contact. These lice have the capacity to live on clothes for more than a month. They are very sensitive to high temperature. When transmitted to healthy individuals, these lice start laying eggs, feed on the blood, and shed the waste matter on the human skin. They cause severe itching mostly near the waist, underarms, and so on. This itching may result in skin damage.

Treatment: Anti-parasitic drugs, mainly as tropical applications on the skin, are prescribed by doctors. Permethrin and benzyl alcohol are the few drugs prescribed. Precautionary measures such as destroying the clothes infected with the lice and proper washing of the clothes with hot water might minimize louse infestation.

Head Lice

Pathogenesis: Head lice are transmitted by head-to-head contact and also by the sharing of pillows, combs, and towels. The adult louse comes in contact with the hair scalp, scrolls over, and feeds on the blood by biting the scalp. During their feeding process, when they suck the blood, they inject saliva into the scalp creating an inflammatory response. The female louse lays eggs, and the cycle continues. These lice increase in numbers causing lots of discomfort. The symptoms are itching and sores on the head due to the scratching resulting in the damage of the scalp.

Treatment: Medicines that kill lice (pediculicides) should be applied on the scalp; all the members of the family must undergo this treatment in order to eradicate the lice completely or else they tend to relapse. Cleanliness should be followed and separate towels, combs, and bed sheets should be used in order to avoid louse infestation.

Pubic Lice

Pathogenesis: They are otherwise called crab lice and are transmitted through sexual contact. The eggs attach to the pubic hair, eyelashes, and even moustache, and the adult worms crawl and bite these areas causing itching. Sores with red pustules occur, and pus cells are formed. They are diagnosed by using combs; dermatoscopy is also one of the diagnostic techniques used for the identification of the pathogen.

Treatment: Proper cleanliness should be followed. Anti-parasitic drugs prescribed for tropical application can prevent the infection.

20.2.4 Mites

Mites feed on animal blood, cause irritation in humans, and feed on plant food (Figure 20.3). These mites are of three types. Few of the medically important mites are as follows:

1. House dust mites
2. Scabies mites

House Dust Mites

They belong to the phylum Arthropoda and are scientifically called *Dermatophagoides pteronyssinus*. These mites live in house dust (in the carpets and doormats) and cause allergic reactions, such as asthma, in humans.

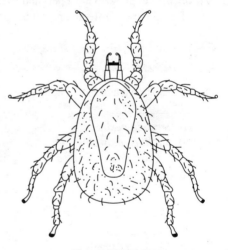

Figure 20.3 Mite

Life Cycle: They are eight-legged arthropods with the same life cycle as other arthropods. The female mates and lays more than 100 eggs every day in its lifespan of 2 weeks. Their faecal matter that contains the potent digestive enzyme is the main source of allergy to humans.

Pathogenesis: Transmission is by the ingestion of the waste or faecal matter produced by the mites. The inhalation results in asthma or wheezing due to the deposit of dust particles on the lungs, ending up in type I hypersensitivity reaction. The release of primary antibody (IgE) with the release of histamines causes asthmatic attack.

Treatment: The drug of choice is antihistamine drugs, which are given for the suppression of the hypersensitivity reaction. Salbutamol and corticosteroids are prescribed for the suppression of the reaction. A precautionary method that can be followed is covering the nose and mouth while cleaning. The mites can be eradicated from fomites by the exposure to high temperature, for example, to the Sun, or using disinfectants such as disodium octaborate tetrahydrate powder.

Scabies Mites

They are scientifically called *Sarcoptes scabiei*. They mostly infect animals and humans. They have the tendency to bore the skin and lay eggs, which is sometimes unnoticed. They are identified only after severe itching. They are contracted from animals due to close contact with them.

Pathogenesis: They cause an epizootic disease that mostly affects household pets such as cats and dogs. They cause a disease called mange in animals. The disease might be transferred to humans due to contact with infected animals. In untreated cases, it might cause even the death of the animals.

Treatment: People who have pets and veterinarians are the most affected, so care must be taken while handling pets. Infected pets must be given proper treatment.

MULTIPLE CHOICE QUESTIONS

1. The causative agent of plague is _____.
   (a) *Yersinia pestis*
   (b) *Bacillus anthracis*
   (c) *Escherichia coli*
   (d) *Rickettsia*

   Ans. a

2. *Yersinia pestis* was identified by _____.
   (a) Alexander Pestis
   (b) Michael Pestis
   (c) John Pestis
   (d) Louis Pasteur

   Ans. a

3. The vectors that carry microbes inside their body acting as intermediate hosts and transmit them to humans are termed as _____.
   (a) Chemical vectors
   (b) Biological vectors
   (c) Physical vectors
   (d) Mechanical vectors

   Ans. b

4. *Ixodes scapularis* are _____.
   (a) Hard ticks   (b) Soft ticks
   (c) Mites   (d) Lice

   Ans. a

5. The scientific name of body lice is _____.
   (a) *Pediculus humanus*
   (b) *Ixodes scapularis*
   (c) *Yersinia pestis*
   (d) *Babesia microti*

   Ans. a

6. House dust mites cause _____.
   (a) Skin disease
   (b) Gastrointestinal disorder

(c) Asthma
(d) Allergic reaction
Ans. c

7. The chemical used for the eradication of biological vectors is _____.

(a) Chlorine
(b) Dichlorodiphenyltrichloroethane
(c) Phenol
(d) Chloroform
Ans. b

SHORT NOTES

1. Hard ticks
2. Body lice
3. House dust mites
4. Pathogenesis of *Yersinia pestis*
5. Soft ticks

ESSAYS

1. Give a brief account of the cultural characteristics, pathogenesis, laboratory diagnosis, and treatment of plague.
2. Give a brief account of the infection caused by body lice.
3. Give a brief account of the infection caused by ticks in humans.

Unit 5

IMMUNOLOGY

Chapter 21	Immunity—Classification	297
Chapter 22	Antigen and Antibody Reaction	304
Chapter 23	Hypersensitivity Reaction	311
Chapter 24	Serological Tests	322
Chapter 25	Immunoprophylaxis	326

21 Immunity—Classification

CHAPTER OBJECTIVES

21.1 Introduction
21.2 Types of Immunity
21.3 Cells of the Immune System
21.4 Humoral Immunity
21.5 Cell-mediated Immunity

21.1 INTRODUCTION

The immune system is an inbuilt defensive mechanism evolved in vertebrates that protects the body from infectious organisms and neoplasia. The immune system encompasses a wide range of molecules, cells, tissues, and organs that specifically identify the foreign molecule and mount a response towards eliminating the foreign invader. All the molecules, cells, tissues, and organs work in a coherent, consistent, and dynamic manner without conflicts or overlaps.

The term '*immunis*' is of Latin origin, which means 'exempt', and the English word 'immunity' is derived from it, which means 'state of protection from infectious disease'. The science of immunology was demonstrated to the clinical world by the pioneering works of Louis Pasteur and Edward Jenner. In recognition of the contributions made by Edward Jenner, he has been honoured as the 'father of immunology'.

An immune response is classified as recognition and response (functionally).

21.1.1 Recognition

The immune system is extremely specific in terms of recognition of molecules. It can recognize the slight differences between chemical structures and sequences (DNA, RNA, or amino acids), which markedly distinguish one molecule from another at the molecular level, and the immune system is fundamentally dynamic in terms of recognizing self-cells and molecules (body's own cells and proteins) from foreign structures and molecules.

21.1.2 Response

The response phase follows the recognition phase. Once a molecule has been identified as foreign by the immune system, it ensures the participation of a variety of immune cells and organs to escalate a response, which is termed as effector response. This effector response functions towards eliminating and neutralizing the foreign body.

The cycle of recognition and effector response is unique for each pathogen, and when the system encounters the same pathogen again (later exposure), the immune system demonstrates a sensitive and sharp memory response, which is more rapid (because of the preliminary exposure).

21.2 TYPES OF IMMUNITY

On the basis of mechanisms of action, there are two types of immunity, namely, innate immunity and adaptive immunity.

21.2.1 Innate Immunity

It is also referred to as non-specific immunity, and it functions against all pathogens and foreign molecules. It is not specific to any pathogen and demonstrates external mechanisms of disease resistance. Innate immunity presents four important barriers that mediate defence. They are as follows:

1. **Anatomic barriers:** They are physical barriers that block the entry of pathogens. This is a preliminary first line of defence that the body has evolved. The skin and mucous membranes are the major anatomic barriers. The skin acts as a mechanical barrier and prevents the entry of pathogens or foreign bodies. In addition, sebum secreted by the sebaceous glands in the skin maintains the pH of the skin between 3 and 5 and this acidic environment retards the growth of microorganisms. Mucous membranes are present in the lining of alimentary, respiratory, and urogenital tracts. These tracts are rich in microbes (i.e., the normal flora), and the presence of normal flora provides competition to the pathogen for sites of attachment and nutrients. Apart from this, the mucus secreted helps in trapping the pathogen. Saliva, tears, and mucus also help in washing away the pathogen, and these secretions possess antibacterial properties as well. The cilia (hair-like protrusions of the epithelium) ensure that they propel the pathogen out of the system through their synchronous movement.

2. **Physiologic barriers:** They include temperature, pH, and chemical mediators. The normal body temperature of certain species is so high that they inhibit the growth of pathogens. The acidic pH of the stomach ensures the destruction of most of the ingested pathogens. Many chemical mediators (e.g., lysozyme, interferon, and complement) add to innate immunity. Lysozymes cleave the bacterial cell wall, interferons demonstrate antiviral properties, and complement molecules mediate the lysis of pathogens and phagocytosis.

3. **Phagocytic barriers:** Phagocytosis is a hallmark of immunity. It is a mechanism where extracellular particulate material is ingested by phagocytic cells. It is a type of endocytosis where particulate materials (whole pathogenic microorganisms) are engulfed from the external environment. It is mediated by specific cells such as monocytes, neutrophils, and macrophages, which internalize, kill, and digest whole microorganisms.

4. **Inflammatory barriers:** Inflammatory response is generated when a tissue is damaged by an invading microorganism or a wound. It is a series of complex sequences

characterized by redness (Rubor), swelling (Tumor), heat (Calor), and pain (Dolor). There are three major events of inflammation, which include the following:

(a) **Vasodilatation:** Increase in the diameter of blood vessels resulting in engorged blood capillaries, which in turn causes redness and rise in tissue temperature (heat).
(b) **Oedema:** influx of fluids from capillaries, which is facilitated by vasodilatation, resulting in the accumulation of exudates (oedema or swelling).
(c) **Capillary permeability:** Mediates the influx of phagocytes into tissues. As phagocytosis begins, the release of lytic enzymes and accumulation of dead cells result in the formation of pus.

21.2.2 Adaptive Immunity

It is otherwise known as specific immunity. It is a mechanism that demonstrates specificity and memory, which are hallmarks of an immune response. Adaptive immune response is generated within 4–7 days of the initial exposure to the pathogen. The primary response takes longer than subsequent encounters with the same antigen as memory responses are heightened in subsequent attacks. The specificity demonstrated by the adaptive response warrants the time it takes to organize itself, and hence, innate immunity provides the first line of non-specific defence to manage the pathogen. In most cases, a healthy individual eliminates the pathogen with the help of non-specific components within few days of the exposure. Failure of innate immunity to clear the pathogen results in the trigger of the adaptive immune system.

It is important to understand that both innate and adaptive immunity work together in a co-operative and consistent manner with adequate cellular and molecular interaction towards eliminating the foreign body. Their functions are not independent of each other.

Adaptive immunity demonstrates four characteristic properties, which are as follows:

1. Antigenic specificity
2. Diversity
3. Immunologic memory
4. Self and non-self recognition

The antigenic specificity allows the immune system to distinguish slight differences among antigens. The antibodies generated are extremely specific biomolecules that are capable of identifying antigens which differ by a single amino acid. The immune system is diverse in terms of generating various molecules to identify different antigens. It demonstrates immunologic memory, which can respond rapidly to a subsequent encounter. Most importantly, the immune system responds only to foreign molecules and is capable of recognizing self from non-self. Failure of this recognition leads to the immune system destroying self-cells resulting in autoimmunity.

21.3 CELLS OF THE IMMUNE SYSTEM

Immune response functions majorly with the help of lymphocytes and antigen-presenting cells (APCs). Lymphocytes are blood cells (WBCs) that are produced by haematopoiesis. Lymphocytes produce and display surface receptors for antigen binding, and they also mediate the properties of immunity. There are two major types of lymphocyte population, namely, B lymphocytes (B cells) and T lymphocytes (T cells); see Figure 21.1.

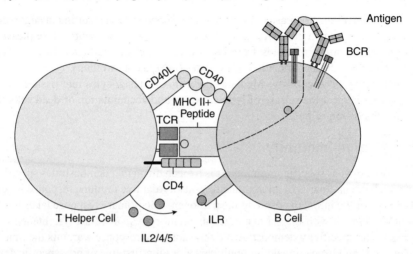

Figure 21.1 T Cells and B Cells (See page 365 for the colour image)

21.3.1 B Lymphocytes

B cells are produced and mature in the bone marrow. When they leave their site of maturation, they express a unique antigen-binding receptor on their cell surface. The receptor on the B cell is a membrane-bound antibody molecule. A naive B cell (which has not encountered any antigen) interacts with an antigen that matches and fits its membrane-bound antibody receptor, it triggers the cell to get differentiated, and it generates effector B cells (plasma cells), which secrete large amounts of specific antibodies against the antigen and memory cells (for subsequent encounters). These secreted antibodies are the effector molecules of humoral immunity.

21.3.2 T Lymphocytes

T cells are produced in the bone marrow but mature in the thymus. During their maturation, T cells express unique antigen-binding T cell receptors on their membrane. T cell receptors are different from membrane-bound B cell receptors in the sense that T cell receptors can recognize an antigen only if it is bound to cell membrane proteins called major histocompatibility complex (MHC) molecules. There are two types of T cells, namely, T helper cells (T_H) and T cytotoxic cells (T_C). T_H has CD4 and T_C has CD8 membrane glycoproteins on their surface, and they interact with class II and class I MHC molecules, respectively.

Antigen-Presenting Cells

The activation of T_H is a carefully regulated process (to prevent autoimmune consequences). T_H cells can recognize antigens only if the antigen is coupled and presented with class II MHC molecule on the surface of APCs. APCs are specialized cells (macrophages, B cells, and dendritic cells) that can express class II MHC on their membranes and also mediate a co-stimulatory signal for T_H activation.

21.4 HUMORAL IMMUNITY

B lymphocytes committed for an antigen are generated from the bone marrow and begin to circulate in the bloodstream or lymph channel or enter lymphoid organs (Figure 21.2). When a mature B cell encounters and interacts with an antigen, it begins to proliferate and differentiate, which is referred to as the activation of the B cell. The activation of B cells starts when the membrane-bound antibody molecule on the B cell (receptor) fits itself to the antigen. Receptor-mediated endocytosis is mediated for part of the antigen (bound). Once the antigen is processed, the B cell presents the antigenic peptides in combination with class II MHC and functions as an APC. A T_H cell specific for the antigen binds to the complex, stimulates B cell division and differentiation, and results in the generation of plasma cells (antibody-secreting cells) and memory cells.

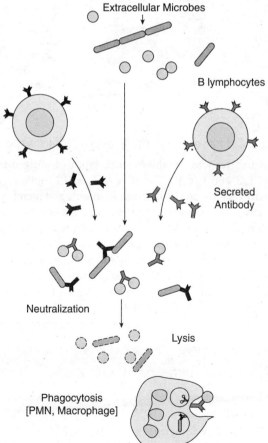

Figure 21.2 Humoral Immunity (See page 366 for the colour image)

21.5 CELL-MEDIATED IMMUNITY

Cell-mediated immunity is mediated by the subtypes of T cells (Figure 21.3). An antigen-specific T_H cell (activated) is required. These T_H cells release chemical mediators known as cytokines, which in turn activate various other effector cells important for cell-mediated response. These non-specific effector cells (natural killer cells and macrophages) are regulated only by cytokines and activated T_H cells and have no attributes of immunity.

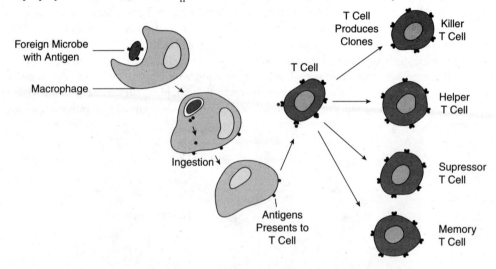

Figure 21.3 Cell-mediated Immunity (See page 367 for the colour image)

The cell-mediated immune response involves three types of antigen-specific effector cells, namely, CD4 T_H1, CD4 T_H2, and T_C cells. The effector T_H and T_C cells are efficient in activation, expression of adhesion molecules, and production of soluble and membrane effector molecules compared with their naive counterparts. T_C mediated immune reaction involves the activation and differentiation of T_C cells. When T_C cells interact with class I MHC, it results in the expression of interleukin receptors (IL-2). IL-2 mediates the proliferation and differentiation of T Cells into effector T cells. This leads to the recognition of specific target cells bearing antigen and class I MHC molecules and the subsequent cyclic process, which involves the formation of cell conjugates, orientation of cytoplasmic granules, release of granular content, formation of pore in target cell, and destruction of target cell. Cell-mediated immunity (delayed hypersensitivity) plays an important role in host defence against intracellular pathogens.

MULTIPLE CHOICE QUESTIONS

1. Which one of the following innate barriers includes chemical mediators?
   (a) Anatomic (b) Physiological
   (c) Phagocytic (d) Inflammatory

   Ans. b

2. The pH of skin is _____.
   (a) 2–5 (b) 3–5
   (c) 5–7 (d) 7

   Ans. b

3. Which one of the following cells is involved in phagocytosis?
   (a) Macrophages
   (b) Natural killer cells
   (c) Granulocytic cells
   (d) Plasma cells

   Ans. a

4. Among the four cardinal signs of inflammation, *dolor* refers to _____.
   (a) Swelling   (b) Redness
   (c) Pain   (d) Heat

   Ans. c

5. Which one of the following is not an attribute of adaptive immunity?
   (a) Recognition between self and non-self
   (b) Specificity
   (c) Cross-reactivity
   (d) Immunological memory

   Ans. c

6. CD4 receptor is present on _____.
   (a) T_H cell   (b) T_C cell
   (c) B cell   (d) Plasma cell

   Ans. a

7. CD8 receptor is present on _____.
   (a) T_H cell   (b) T_C cell
   (c) B cell   (d) Plasma cell

   Ans. b

SHORT NOTES

1. Innate immunity
2. Adaptive immunity
3. Cells of the immune system
4. Humoral immunity
5. Cell-mediated immunity

ESSAY

1. Explain in detail with diagrams the various components and types of immunity in vertebrates.

22 Antigen and Antibody Reaction

CHAPTER OBJECTIVES

22.1 Introduction

22.2 Antigen–antibody Interactions

22.3 Types of Antigen–antibody Interactions

22.1 INTRODUCTION

22.1.1 Antigens

Biomolecules, particulate material, and related substances that are capable of triggering the immune system towards an effector response are referred to as antigens (Ags). However, an Ag that is capable of triggering an immune response is more appropriately called as an immunogen. Immunogenicity is the phenomenon that deals with the ability of a substance to induce humoral and/or cell-mediated immunity. Immunogenicity of an immunogen is determined by four properties: foreignness, molecular size, chemical composition, and complexity. The degree of foreignness is a direct measure of non-self property, which is proportional to immunogenic property. Immunogenicity is best in molecules whose molecular weight is 1,00,000 Daltons. The chemical composition and heterogeneity of the molecule in terms of the presence of heteropolymers increase the molecular weight and also contribute to the structural complexity of the molecule.

22.1.2 Antibodies

These are Ag-binding proteins secreted by plasma cells present on the B cell membrane, which are also referred to as immunoglobulins. Antibody (Ab) molecules are Y-shaped structures that contain four peptide chains [two identical light chains (25,000 Da) and two heavy chains (>50,000 Da)]. Each light chain is linked with a heavy chain by disulphide bonds, and strong covalent bonds and non-covalent interactions such as hydrogen bonds and hydrophobic bonds stabilize the structure. Antibodies are classified into five subclasses based on their heavy chain differences as IgG, IgM, IgA, IgE, and IgD (Figure 22.1).

22.2 ANTIGEN–ANTIBODY INTERACTIONS

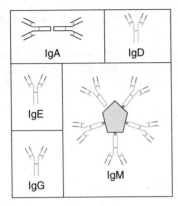

Figure 22.1 Structure and Classification of Antibodies (five subclasses) (See page 367 for the colour image)

The Ag–Ab interaction is exquisitely specific at the molecular level. It is mostly compared with the enzyme–substrate reaction. However, it is important to understand that Ag–Ab interaction does not lead to chemical alteration in either the Ag or the Ab and hence is a reversible reaction. The interactions involve various non-covalent interactions such as hydrogen bonds, ionic bonds, hydrophobic interactions, and van der Waals' interactions. The interactions are influenced by two important properties, which are commonly referred to as Ab affinity and Ab avidity.

Affinity is defined as the phenomenon that influences the interaction between an Ag and an Ab. The power of all the non-covalent interactions between a single epitope and a single Ag-binding site on an Ab (paratope) is referred to as affinity. The affinity at one site does not contribute to the total strength of the Ag–Ab interaction. Interactions between complex Ags (which contain many epitopes) and Abs (with many binding sites) mediate and influence the probability of multiple interactions. The total strength of such multiple interactions between complex Ags and multivalent Abs is referred to as avidity. IgM, which is a pentameric Ab, has a preferential low affinity, but due to its high valency (5), its avidity is higher and hence binds to the Ag more effectively.

22.3 TYPES OF ANTIGEN–ANTIBODY INTERACTIONS

22.3.1 Precipitation Reactions

Interactions between Abs and soluble Ags result in the aggregation of soluble Ags (lattice formation) in the form of a precipitate. The lattice formation (Figure 22.2) takes place immediately; however, visibility of the precipitate takes a few hours. The lattice formation between Ag and Ab depends on the higher valency of both Ag and Ab. The Ag must be bivalent or polyvalent, and the Ab must be bivalent.

Precipitation reactions can be performed and observed in fluids and gels. Quantitative reactions can be performed in fluids by testing the concentration of either Ag or Ab by keeping one of them constant and adding increasing concentrations of the other and measuring the amount of precipitate. It is important to understand that the concentrations of the Ag and Ab have to be optimum to observe maximum reaction,

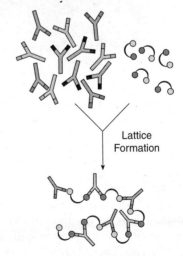

Figure 22.2 Lattice Formation (See page 368 for the colour image)

and this is referred to as the equivalence zone. Excess of either the Ag or the Ab greatly influences the significant precipitation. These precipitate matrices can be observed in gels (agarose) as well. This involves the diffusion of immune molecules (Ag and Ab) towards each other from adjacent wells through the agar matrix to form a thin white line of precipitation. Again, the precipitin line is formed at the zone of equivalence (as in fluids). Precipitation reactions in gels are of two types, which are as follows:

1. Double immunodiffusion (Ouchterlony technique)
2. Radial immunodiffusion (Mancini technique)

Double Immunodiffusion (Ouchterlony Technique)

This is a qualitative technique that establishes the presence of a given Ag–Ab (Figure 22.3). Here, the Ag and Ab are loaded on wells placed adjacent to each other on an agarose (inert) matrix. The distance between the wells should be less than 1 cm. The reacting molecules (Ag and Ab) are allowed to diffuse towards each other, and a line of precipitation is formed as they reach equivalence. The relationship between Ag–Ab systems can be identified by this technique to see if the Ag shares epitopes. The patterns of the precipitin line formed demonstrate identity, non-identity, and partial identity between epitopes of different Ags. The sensitivity of this technique is 20-200 µg Ab/ml.

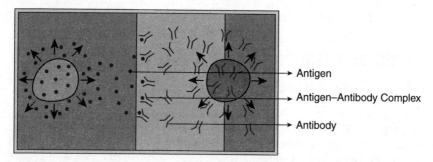

Figure 22.3 Ouchterlony Double Immunodiffusion (See page 368 for the colour image)

Radial Immunodiffusion

This is a simple quantitative assay with a sensitivity of 10–50 µg Ab/ml (Figure 22.4). This technique is used to determine the relative concentrations of the Ag. Here, the antiserum containing the Abs is mixed with the agarose matrix (at optimal temperatures to ensure that the biomolecule is not inactivated) and layered on the glass slide. The Ag samples (of unknown concentrations) are loaded in wells in the matrix and are allowed to diffuse radially (hence the name) and interact with the Ab molecules in the matrix. This results in the formation of a precipitin ring, and the diameter of the precipitin ring is directly proportional to the concentration of the Ag. A standard curve is constructed with known concentrations of the Ag, and the concentration of the unknown is determined.

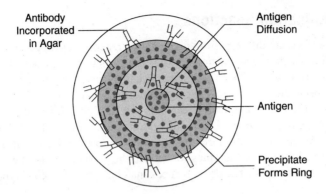

Figure 22.4 Single Radial Immunodiffusion (SRID) (See page 368 for the colour image)

Counter-Current Immunoelectrophoresis

This technique is very similar to immunodiffusion but varies with the fact that the reacting Ag and Ab molecules are forced to move towards each other under an electric field (Figure 22.5). The Ag is conferred a negative charge for ease of electrophoresis. The advantage of this technique is that the line of precipitation can be observed in less than 1 hour in contrast to diffusion techniques, which require an overnight incubation for diffusion.

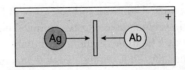

Figure 22.5 Counter-Current Immunoelectrophoresis (See page 369 for the colour image)

Immunoelectrophoresis

This technique is a combination of immunodiffusion and electrophoresis (Figure 22.6). Here, a complex Ag is electrophoresed under an electric field to separate its individual components. Following electrophoresis, two troughs are created parallel to the direction of electrophoresis and are loaded with the antiserum. The Abs present in the antiserum diffuse through the agar and interact with individual separated components of the Ag and form precipitin lines at their zones of equivalence. This is a qualitative tool that has clinical application towards identifying the presence or absence of certain proteins in serum samples. The sensitivity of this technique is 20-200 µg Ab/ml.

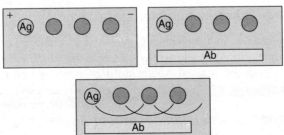

Figure 22.6 Immunoelectrophoresis (See page 369 for the colour image)

22.3.2 Agglutination Reactions

Agglutination reactions are different from precipitation reactions as they result in the formation of visible clumping due to the interaction between the particulate Ags and Ab (In precipitation reactions, the Ags are soluble.). The Abs that mediate this reaction are referred to as agglutinins. Their underlying principle is cross-linking polyvalent Ags, and it is important to note that excess of Abs interfere greatly with agglutination reactions as well. This is referred to as prozone effect.

ABO blood typing is a common example of agglutination reaction. Here, the RBCs in blood express the Ags on their membrane surface. The blood sample is mixed with anti-A and anti-B antisera and observed for visible clumping to confirm the presence or absence of a given Ag in the blood sample. It is a routine test done to check matching betweenblood types.

22.3.3 Enzyme-Linked Immunosorbent Assay

Enzyme-linked immunosorbent assay (ELISA) is a very versatile tool used in immunology and various biochemical tests in clinical practice. Enzyme-linked immunosorbent assay is used as both as a qualitative and a quantitative tool for the measurement of Ag and/or Ab. In this technique, the Ab is conjugated with an enzyme that reacts with a substrate (colourless initially and specific for the enzyme) and generates a coloured product. The presence or intensity of the colour is a measure of the amount of the Ab in the sample. Most common enzymes employed in ELISA include alkaline phosphatase, horseradish peroxidase, and beta-galactosidase. Enzyme-linked immunosorbent assay was developed to minimize radioactive exposures in radioimmunoassay (RIA). The sensitivity of ELISA is close to RIA, and it is safer and less expensive.

Many types of ELISA have been evolved based on qualitative or quantitative endpoints. Few are discussed below.

Indirect Enzyme-Linked Immunosorbent Assay

Here, the Ab is detected and quantitated. The Ag is coated on the microtitre plate and the serum containing the primary Ab is loaded to the plate and allowed to react with the Ag. Washing is done to remove unbound primary Ab, and the Ag–Ab interaction is detected by adding an enzyme-linked secondary Ab that binds to the primary Ab. Washing is done again to remove unbound secondary Ab. Addition of chromogenic substrate results in its reaction with the enzyme (linked to secondary Ab) to form a colour. The intensity of the colour is measured spectrophotometrically to quantify the amount of primary Ab.

Sandwich Enzyme-Linked Immunosorbent Assay

Here, the Ag can be detected. In this technique, the Ab is immobilized on the microtitre well and the sample containing the Ag is allowed to react with the Ab. Washing is done to remove unbound molecules. An enzyme-linked Ab for a different epitope on the Ag is added and allowed to react. The substrate is then added, and the colour is measured.

Competitive Enzyme-Linked Immunosorbent Assay

Here, the Ab is mixed with the sample containing the Ag and incubated. This Ag–Ab mixture is coated on the microtitre plate (which is coated with the same Ag). The principle of competition is exploited, that is, if more Ag is present in the mixture, then less Ab is available for binding with the microtitre plate and if less Ag is available in the mixture, then more Ab is available for binding with the microtitre plate. The addition of enzyme-linked secondary Ab quantifies the reaction.

MULTIPLE CHOICE QUESTIONS

1. Which one of the following is not a factor that determines antigenicity?
   (a) Molecular size
   (b) Foreignness
   (c) Chemical composition
   (d) Density and surface tension
   Ans. d

2. Which one of the following is not a precipitation reaction?
   (a) Mancini method
   (b) Double diffusion
   (c) ABO blood grouping
   (d) Immunoelectrophoresis
   Ans. c

3. Precipitation reaction in gels is a _____.
   (a) Quantitative tool
   (b) Qualitative tool
   (a) Both qualitative and quantitative
   (b) Neither qualitative nor quantitative
   Ans. b

4. Mancini method is a _____.
   (a) Quantitative method
   (b) Qualitative method
   (c) Both (a) and (b)
   (d) None of the above
   Ans. a

5. Double diffusion is also known as _____.
   (a) Mancini method
   (b) Ouchterlony method
   (c) Prozone effect
   (d) Zone of equivalence
   Ans. b

6. Which of the following enzymes is routinely used in enzyme-linked immunosorbent assay?
   (a) Carbonic anhydrase
   (b) Glucokinase
   (c) Horseradish peroxidase
   (d) Alpha-galactosidase
   Ans. c

7. The antigens in agglutination reactions are _____.
   (a) Soluble
   (b) Particulate
   (c) Both (a) and (b)
   (d) None of the above
   Ans. b

8. The antigens in precipitation reactions are _____.
   (a) Soluble
   (b) Particulate
   (c) Both (a) and (b)
   (d) None of the above
   Ans. a

9. The valency of IgM is _____.
   (a) 1   (b) 3
   (c) 5   (d) 6
   Ans. c

10. Which of the following determines the strength of antigen–antibody interactions?
    (a) Affinity
    (b) Avidity
    (c) Both (a) and (b)
    (d) None of the above
    Ans. c

SHORT NOTES

1. Antigen–antibody interactions
2. Precipitation reactions
3. Types of enzyme-linked immunosorbent assay
4. Mancini method
5. Immunoelectrophoresis
6. Double diffusion

ESSAYS

1. Explain in detail with diagrams the various antigen–antibody interactions and their applications.
2. Explain precipitation reaction. Add a note on its types with relevant examples.
3. Explain the principle of enzyme-linked immunosorbent assay. Add a note on its types with their applications.

23 Hypersensitivity Reaction

CHAPTER OBJECTIVES

23.1 Introduction

23.2 Classification of Hypersensitivity Reactions

23.1 INTRODUCTION

The immune response caused due to incompatible conditions resulting in allergic reactions is termed as hypersensitivity. The effector response triggered by the entry of pathogens will result in the destruction of pathogens without harming self cells. In the case of hypersensitivity reaction, the entry of antigens will trigger the immunological reaction causing the inflammatory reaction against our own cells resulting in allergic reactions leading to tissue injury and apoptosis of the cell. Such types of reactions are both antibody-mediated (humoral) response and cell-mediated response. French scientists Paul Porter and Richet were awarded the Nobel Prize for the identification of the anaphylaxis reaction in 1913.

23.1.1 Allergens

They are otherwise termed as antigens and produce immunological changes (allergies) causing harmful effects in body cells. These allergens enter through various routes, namely, ingestion, inhalation, or inoculation and mostly cause hypersensitivity reactions. Few examples of allergens are pollen, poison ivy, egg white, milk, nut varieties, seafood, animal wastes, dust, and mites. The above-mentioned intricate organic substances cause antibody-mediated reactions, whereas simple organic compounds and chemicals cause cell-mediated response (late reactions).

23.1.2 Inclination to Allergic Reaction

Allergic reactions occur when an individual comes in contact with allergens. The allergens enter through inhalation (pollen grains), ingestion (peanuts), or inoculation (any medication or vaccine), and as they come in contact with the blood circulation, they encounter immune cells and localize, resulting in the allergic reaction. The reaction occurs in two phases, which are as follows:

1. Sensitization phase
2. Effector phase

The sensitization phase is the first-degree exposure where the allergen comes in contact with the antigen-presenting cell and is processed and presented to the T helper cell, which then elicits an immune response by presenting them to B cells and T cells.

The effector phase is the second-degree exposure where the allergen re-enters and the specific antibody reacts causing an immunological reaction resulting in allergic response. The antibody-mediated response includes IgE, IgM, and IgG antibodies and also complements, which play a major role in initiating the complement pathway for triggering the allergic reaction. The cellular-mediated response is initiated by the T cell-mediated response, which has a delayed reaction and takes time to initiate the response after exposure.

23.2 CLASSIFICATION OF HYPERSENSITIVITY REACTION

Hypersensitivity reactions have been classified by two scientists, namely, P. G. H. Gell and R. R. A. Coombs into four types, which are as follows:

1. Anaphylactic hypersensitivity (type I)
2. Antibody-dependent cytotoxic hypersensitivity (type II)
3. Immune complex-mediated hypersensitivity (type III)
4. Delayed or cell-mediated hypersensitivity (type IV)

Type I, II, and III hypersensitivity are antibody- or humoral-mediated response, whereas type IV or delayed hypersensitivity is cell-mediated response. The hypersensitivity state is caused by antigens, which are called as allergens in hypersensitivity conditions. These allergens are foreign substances that evoke an immunological response resulting in inflammatory responses. These allergens may be pollen grains; poison ivy; skin or fur of animals; insect bites; and food products such as peanuts, lactose in milk, mushrooms, and few vegetables. These allergic reactions are influenced by various factors, which are as follows:

1. Age
2. Race
3. Genetic predisposition
4. Gender
5. Climate

23.2.1 Anaphylactic Hypersensitivity

This is type I-mediated hypersensitivity and is induced by the antigens that cause antibody-mediated response (Figure 23.1). The antibody that is responsible for type I-mediated response is IgE antibody, which has an exclusive region known as the Fc portion, that binds itself to the surface receptor of mast cells, thereby causing the intracellular events inside the cell releasing inflammatory mediators resulting in anaphylactic shock. The allergic reaction that is IgE mediated is mostly due to parasitic infestation and will last long after the complete elimination of the parasite from the system.

Mode of Action

The entry of the allergen into the system will activate a first-degree response referred to as sensitization, which evokes the antibody-mediated immune response. In Type I hypersensitivity IgE is secreted in the serum and Fc region of the antibody gets bound to the surface receptor of the mast cells. During its second exposure or re-entry it causes effector responses like intercellular changes inside the mast cells allowing it in the release of chemokines, namely, histamines, leukotrienes, protease, and platelet-activating factor. This process of release of inflammatory mediators is called as degranulation.

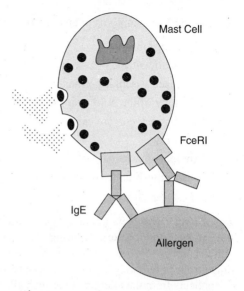

Figure 23.1 Type I Hypersensitivity (See page 369 for the colour image)

Symptoms

The symptoms are as follows:

1. Increased vascular permeability
2. Vasodilatation
3. Smooth muscle contraction
4. Cardiac attack resulting in death (serious condition)

Anaphylactic Reactions

Anaphylaxis is a severe allergic reaction that affects various systems in the human body, namely, the cutaneous, respiratory, cardiovascular, and gastrointestinal system. The reaction occurs only when the allergen comes in contact during the re-exposure to the host that has already been sensitized. The above-mentioned symptoms constitute the anaphylactic shock.

Pathogenesis: These are IgE-mediated response affecting multiple organs. Transmission is by the entry of the allergen through inoculation in the form of drugs, insect bites (venom), or ingestion of unsuitable food. The entry of the allergen triggers the immune system causing its sensitization. The IgE antibodies binds on the body cells causing an anaphylactic shock.

Symptoms: The symptoms of anaphylaxis are as follows:

1. Hypotension
2. Asthmatic attack
3. Increased vascular permeability
4. Blocking of the pulmonary region by mucus causing difficulty in breathing
5. Heart attack (severe cases)
6. Urticaria
7. Angioedema
8. Gastrointestinal involvement resulting in abdominal cramp, pain, and diarrhoea
9. Thrombocytopaenia

Laboratory Diagnosis: Laboratory tests are done for detecting the source of allergen because the patients cannot be tested during the reaction as immediate action has to be taken by giving antihistamines to avoid the anaphylactic shock. Immunological diagnosis such as skin testing done to check for the IgE immune response to an allergen should be performed for safeguarding the patients from the allergen. One such allergen skin test is wheal and flare skin test, which is done by cutaneous prick method using the allergen. The allergens are diluted and then only used because if individuals are hypersensitive, then they might start producing severe reactions. If they do not form any induration, the skin test is negative, and if they do, it is a positive reaction.

Prevention: The following preventive measures can be undertaken:

1. Avoiding the allergen that causes the reaction
2. Carrying along antihistamines such as epinephrine in the form of either injection or inhaler
3. Desensitizing the venom due to insect bites

Atopy

It is an allergic disease that is caused due to genetic predisposition. This occurs due to the entry of common allergens through the inhalation or ingestion route. The mast cells are activated due to the increase in the production of IgE antibodies, which are secreted due to the exposure to allergens. This allergen–antibody complex fixes itself to the cell causing intracellular changes resulting in the release of chemical mediators.

Types of Allergens: Few examples of allergens are as follows:

1. Pollen grains
2. Fungal spores
3. Dust mites
4. Dust
5. Animal fur/hair
6. Food allergens such as seafood, peanuts, milk, egg white and Gluten

Symptoms: The most common symptoms are allergic rhinitis and allergic asthma.

Allergic Rhinitis

Allergic rhinitis is an inflammatory atopic reaction that occurs due to the inhalation of pollen grains, dried grass, or spores present in hay. The reaction occurring due to the inhalation of pollen grains is called pollinosis, and the reaction occurring due to the inhalation of dried waste or hay is called hay fever. When the allergen enters through inhalation, it travels through the mucosal area and binds with circulating IgE antibodies (bound to mast cells/basophils) and releases histamines upon activation and results in flu-like symptoms:

1. Sneezing due to rhinorrhoea (excess nasal secretion)
2. Coughing
3. Itching
4. Hives or rashes
5. Conjunctival swelling

Diagnosis: Blood and skin test are often done for the identification of allergies caused by inhalation. The blood test will indicate the presence of increased levels of IgE in the blood. Another important test followed for the identification of the reaction is the radioimmunoassay test (RAST). The RAST is performed to check the IgE level in the circulation.

Treatment: Antihistamines are given for mild symptoms, and for serious conditions, antihistamines combined with steroids are given for suppressing the effects of inflammatory reactions.

Allergic Asthma

Allergic asthma is mediated by IgE antibodies that attack the bronchial cells resulting in allergic bronchitis. The mode of entry is air, and the allergen travels through the tracheal passage attacking the trachea and reaches the bronchi causing severe symptoms as given below:

1. Continuous coughing
2. Discomfort in the chest region
3. Bronchial obstruction
4. Chest pain
5. Difficulty in breathing

Allergic asthma is of two types: extrinsic asthma and intrinsic asthma. The former is the allergic asthma that develops at a very young age. The latter is the non-allergic asthma, also called as adult-onset asthma, which occurs in an individual whose immune system is well developed, and the onset of the infection starts after a respiratory attack.

Diagnosis: The samples collected are blood and sputum, which show an increased eosinophil count during the infections. Pulmonary function test, which shows bronchial obstruction, is one of the identification tests for asthmatic attack.

Treatment: Prophylactic measures include controlling pollution and staying away from the polluted environment; maintaining dust-free zones; and keeping homes clean and free of dust, mites, and moulds. During the onset of the attack, adrenergic bronchodilator spray should be carried along to avoid wheezing. Asthalin, theophylline, and antihistamines are prescribed. In severe cases, epinephrine can be injected to control the respiratory attack.

23.2.2 Type II Hypersensitivity Reaction

Type II hypersensitivity reaction is otherwise termed as cytotoxic hypersensitivity reaction (Figure 23.2). The antibodies are directed against the self antigen due to modification of the surface antigen of the host cell. This is a consequence of an infection or self-modifications which makes the host cell being recognized as non-self. The host defences start attacking the modified host cells by producing an immunological reaction caused by B cells. These B cells start proliferating by producing antibodies (IgG and IgM) as and when they bind with the host cells. A complement cascade initiates the reaction by forming a membrane attack complex causing inflammatory responses, which are responsible for the attack and destruction of the cells.

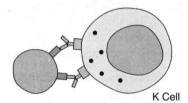

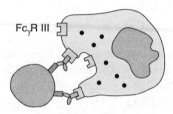

Figure 23.2 Type II Hypersensitivity (See page 370 for the colour image)

Few reactions that come under the cytotoxic hypersensitivity reaction are as follows:

1. Blood transfusion
2. Haemolytic disease of newborn (erythroblastosis fetalis)
3. Drug-induced reaction
4. Autoimmune haemolytic anaemia

Haemolytic Disease of Newborn

This disease is otherwise called as erythroblastosis fetalis, which is the lysis of erythrocytes by the mother's antibody due to Rh incompatibility. In pregnancy, if the mother is Rh negative and the foetus is Rh positive, during the process of delivery, when the placenta is separated from the baby, the foetus's blood mixes with the mother's blood producing an immunological change; then, the mother's cells start producing antibodies against the baby. As this reaction takes place only after the baby is born, the baby does not die, but the antibody produced by the B cells secrete both the plasma cells, which destroy the antigen, and memory cells, which carry out the further destruction of the antigen during re-entry. Haemolytic disease of the newborn occurs during the second consumption, where the maternal antibody crosses the placental barrier and affects the infant.

Symptoms: The symptoms are as follows:

1. Haemolytic anaemia
2. Hydrops in infants

3. Still birth
4. Hyperbilirubinemia (bilirubin getting accumulated in the plasma affecting the brain and damaging the nervous system)

Prevention: Rh-negative mothers having Rh-positive babies are given 100 μg of the commercially available concentrated anti-D (Rh) immunoglobulin within 72 hours of delivery to avoid the production of antibodies by the maternal cells.

Autoimmune Haemolytic Anaemia

This is also caused by the immunological changes due to the decrease in blood cell production resulting in aplastic anaemia. This occurs mainly due to the presence of antibodies in the serum that reacts with the bone marrow erythroblast, destroying them. The antibody responsible for causing aplastic anaemia is mostly IgG mediated. This antibody binds with the erythrocyte antigen and forms a complex with the complement cascade producing an inflammatory response.

Treatment: Individuals affected with haemolytic aplastic anaemia are often supported by total erythrocyte transformation. Splenectomy is also being done for patients with aplastic anaemia.

Drug-Induced Reaction

The most important drug that is responsible for the immunological changes causing the hypersensitivity reaction is penicillin. Penicillin and other β-lactam antibiotics are a recurrent cause of type II hypersensitivity reaction. These are IgE-mediated allergic reactions including certain reactions such as anaphylactic reaction, serum sickness, and dermatitis. Serum sickness is a severe reaction taking place in the serum when the above-mentioned protein drug comes in contact with the antibody that is complexed with the complement generating an inflammatory response that affects the skin and lymph (lymphadenopathy) and causes cardiovascular disorder and dermatitis.

Prevention: Proper skin testing before the administration of any drug is a prophylactic measure that can be followed.

23.2.3 Type III Hypersensitivity Reaction

They are otherwise known as immune complex hypersensitivity reaction or immune complex hypersensitivity reaction (Figure 23.3). These immunological reactions are mediated by the antigen–antibody complex formed by the activation of the complement causing severe immunological changes in the body. The activation of the complement leads to the release of chemotactic and vasoactive mediators, which act upon cells damaging them and also depositing the immune complex (antigen and antibody complex) in the kidney. These lead to autoimmune disorder, that is, the destruction of self antigens by host cells. The formed complexes are deposited on the organs resulting in adverse effects. The conventional examples for toxic complexed hypersensitivity reaction are as follows:

1. Arthus reaction
2. Serum sickness

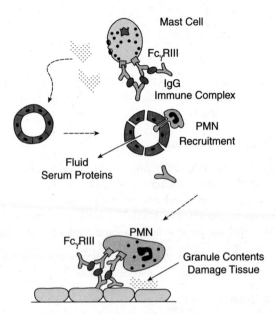

Figure 23.3 Type III Hypersensitivity (See page 370 for the colour image)

Arthus Reaction

Arthus (a Nobel laureate) in 1903, performed an experiment in a sensitized animal by injecting an antigen made of c Arthus reaction. The inflammation was the deposition of the antigen and antibody complex resulting in immune complex-mediated disease. Hence, any type of necrotic inflammation occurring due to the introduction of drugs or due to insect bites is termed as Arthus reaction.

Serum Sickness

This is caused due to the injection of foreign antigens in the form of antibiotics, vaccine preparations, and antibodies for therapeutic purpose. The pathogenesis starts as soon as the serum is injected; the antibody level increases in the circulation leading to the formation of the antigen–antibody complex initiated by a cascade of reactions carried out by the activated serum protein present in the serum. There are four stages in serum sickness, which are as follows:

1. **Phase I:** The antigen in the serum enters, and slow degradation of the antigen in the blood and tissue occurs ending up in the primary production of the antibody.
2. **Phase II:** The primary antibody in circulation combines with the antigen and forms the immune complex, which is again complexed with the circulating serum protein.
3. **Phase III:** These complexes are then deposited on blood vessels, which are then carried to different organs and deposited.
4. **Phase IV:** The antibody slowly degrades the antigen, and the complement clears the debris from the circulation but there remains a higher level of circulating antibodies.

Symptoms: The symptoms are as follows:

1. Rashes
2. Oedema
3. Lymphadenopathy
4. Vomiting

Prevention: Antihistamines are used for suppressing the reaction.

23.2.4 Cell-Mediated Hypersensitivity

Cell-mediated hypersensitivity is otherwise called as delayed type hypersensitivity (type IV; Figure 23.4). They have a cellular response against the allergens and that is why the reaction is delayed. The cells that are responsible for the delayed reaction are effector T lymphocytes. The most important and common delayed reaction is contact dermatitis, and rarely it is pneumonitis. When the sensitized host is in contact with the allergen, they end up in a cell-mediated response that can never be suppressed.

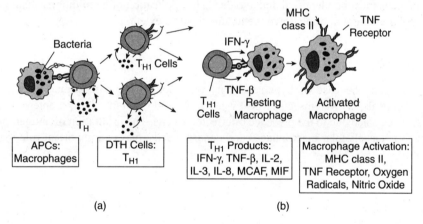

Figure 23.4 Type IV Hypersensitivity— (a) Sensitization Phase, (b) Effector Phase (See page 371 for the colour image)

Pathogenesis

Transmission is by contact or the inoculation of the antigen that has already sensitized the host. The circulating macrophages come in contact with the antigen and phagocytize them, hence called as antigen-presenting cells. These processed antigens are presented by the macrophages to T helper cells, which will be a generalized reaction, and it may take even years for a reaction to appear. The sensitization for the primary response takes few days, but the secondary response takes years to appear. The complete response may appear during the re-entry of the allergen or antigen, which on contact releases lymphokines that trigger the cells and activate the immune system to produce an immunological reaction.

Contact Dermatitis

It mainly occurs due to the skin contact, especially inflamed areas, with a variety of chemicals such as metals, dyes, drugs, and toiletries. The application of antibiotics on the affected area might also trigger the reaction. This type of hypersensitivity reactions are ruled out by doing a patch test, which is a skin test for identifying the hypersensitivity reaction. The diluted allergen is applied to a small area in the skin; if the person is sensitive to the particular allergen, he starts developing the below-mentioned symptoms within 4–5 hours.

Symptoms: The symptoms of contact dermatitis are as follows:

1. Itching
2. Swelling
3. Erythematous lesions within 24 hours
4. Lesions with pus

Prevention and Treatment: Corticosteroids are given to the patients as early as possible. The exact allergens must be identified and avoided. The application of aluminium acetate solution with a cool dressing will bring down the itching to some extent.

Pneumonitis

The allergens in the case of hypersensitivity pneumonitis are pathogenic bacteria, fungi, organic chemicals, or even insects (wheat weevil). As and when the allergen enters, a primary response takes place due to the presence of effector T cells. The infiltration of antigen-presenting cells, such as alveolar macrophages and dendritic cells, plays an important role incellular immunity.

Symptoms: The symptoms are as follows:

1. Flu-like symptoms
2. Persistent cough
3. Increase in body temperature
4. Weight loss
5. Wheezing
6. Eosinophilia
7. Granulomatous lesions in the lungs

Diagnosis: Chest X-rays show the lungs affected due to the infiltration of immune cells. Blood test indicates the presence of polymorphonuclear cells, which is high in acute conditions. Higher level of serum antibodies during the infection is also a good prognosis.

Prevention and Treatment: Corticosteroid therapy gives relief to the patients. Avoiding allergens causing occupational hazards is very important.

MULTIPLE CHOICE QUESTIONS

1. The immune response caused due to unsuitable conditions resulting in the allergic reaction is termed as _____.
   (a) Autoimmunity (b) Hypersensitivity
   (c) Hyposensitivity (d) Immunity

   Ans. b

2. French scientists Paul Portier and Richet were awarded the Nobel Prize for the identification of _____.
   (a) Delayed hypersensitivity reaction
   (b) Atopic reaction
   (c) Anaphylactic reaction
   (d) Cytotoxic reaction

   Ans. c

3. An inflammatory atopic reaction that occurs due to the inhalation of pollen grains, dried grass, or spores present in hay is _____.
   (a) Allergic rhinitis (b) Pneumonia
   (c) Flu (d) Bronchitis

   Ans. a

4. Type II hypersensitivity reaction is otherwise termed as _____.
   (a) Cytotoxic reaction
   (b) Toxic reaction
   (c) Delayed reaction
   (d) Atopic reaction

   Ans. a

5. Haemolytic disease of newborn is _____.
   (a) *Erythroblastosis fetalis*
   (b) Aplastic anaemia
   (c) Haemolysis
   (d) Cystic fibrosis

   Ans. a

6. Cell-mediated hypersensitivity is otherwise called as _____.
   (a) Immediate hypersensitivity
   (b) Delayed type hypersensitivity
   (c) Antigen–antibody-mediated hypersensitivity
   (d) Immune complex-mediated hypersensitivity

   Ans. b

7. Arthus reaction was found in _____.
   (a) 1904 (b) 1903
   (c) 1902 (d) 1919

   Ans. b

SHORT NOTES

1. Atopic reaction
2. Cytotoxic hypersensitivity
3. Type III hypersensitivity
4. Delayed type hypersensitivity
5. Erythroblastosis fetalis

ESSAY

1. Brief about the types of hypersensitivity and give a detailed account on immediate hypersensitivity reaction.

24 Serological Tests

CHAPTER OBJECTIVES

24.1 Introduction

24.2 Precipitation

24.3 Agglutination

24.1 INTRODUCTION

Serological testing is a diagnostic tool employed for the identification of the antigen and antibody interactions found in the plasma, serum, or other body fluids. The main aim of serological tests is to diagnose the circulating antibodies present in the serum due to the infection caused by pathogens. Serological tests are very reliable in terms of sensitivity and specificity and hence play an important role in the upcoming diagnostic tool for the identification and interpretation of accurate results within a short duration of time. Both qualitative and the quantitative methods are performed in serological tests. Some of the serological tests used are as follows:

1. Precipitation
2. Agglutination
3. Enzyme-linked immunosorbent assay (refer Chapter 22; Figure 24.1)

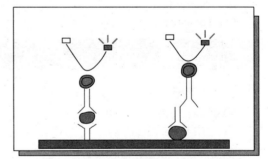

Figure 24.1 ELISA (See page 371 for the colour image)

24.2 PRECIPITATION

When a specific soluble antigen reacts with a specific soluble antibody, they form a complex substance resulting in precipitation, and this is called as precipitation reaction. The precipitation is visibly seen only at the zone of equivalence, that is, where equal amounts of antigen and antibody react with each other to form precipitin. Most of the precipitation is done by diffusing through the gel and reacting to form precipitin. There are two types of diffusion, namely, single and double diffusion, for the detection of the antibody present in the patient serum.

24.3 AGGLUTINATION

When the particulate antigen reacts with the specific soluble antibody, clumping of the cells takes place; this reaction is called as agglutination. Two types of agglutination tests are performed for qualitative and quantitative analysis, which are as follows:

1. Widal test
2. Anti-streptolysin O (ASO) test

24.3.1 Widal Test

This is the serological test done for the identification of the infection caused by *Salmonella typhi* (which causes enteric fever in humans; Figure 24.2). This is a qualitative agglutination test where the bacterium causing the fever (the antigen) is mixed with the patient's sample (serum) containing the circulating antibody to check for agglutination (based on specificity). The results are positive if there is a reaction between the antigen and the antibody. Widal test was first employed by a French physician Georges Fernand Isidore Widal, who used this test for the identification of antiserum against enteric fever.

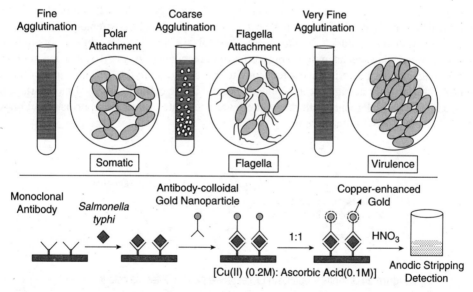

Figure 24.2 Widal Test (See page 372 for the colour image)

Methodology

Antigen Preparation: *Salmonella* H and O antigens are prepared by treating with 0.1% formalin in the overnight culture. The O antigen is grown in phenol agar for the inhibition of flagella. The bacterium that is grown is emulsified, then mixed and diluted with alcohol, and heated at 40°C for 30 minutes for inactivation. These inactivated samples are centrifuged, treated with a preservative such as chloroform, and mixed with an appropriate dye for identification.

Procedure: Widal racks containing round-bottom Felix tubes are placed on the first rack in which the patient serum is serially diluted starting from the ratio 1:10 to 1:640. Then, 0.5 ml of *Salmonella typhi* O antigen is added to the first rack. The entire set of test tubes is again double diluted in the ratio starting from 1:20 to 1:1280, which is taken as the final dilution. The same procedure is followed in the second rack containing Dreyer's tubes for H antigen (which are serially diluted). The same protocol is followed for *Salmonella* Paratyphi A and B. The whole set-up is incubated overnight at 37°C. Control tubes, which contain saline with the antigen alone, are placed for each set-up.

Interpretation

The results are interpreted by comparing the control tubes with the samples. The positive tube for antigen O will have a matt finish of agglutinated antigen and antibody complex at the bottom. In the case of H antigen, cottony or wool-like appearance will be observed. The agglutination is noted against all dilutions, and the titre value at which positive reaction is observed is also documented. The titre value indicates the severity of the infection.

24.3.2 Anti-Streptolysin O Test

Anti-streptolysin O test is a rapid latex agglutination test (Figure 24.3). It is a qualitative test done for the identification of antibodies present in the serum against streptolysin, which is a haemolytic product of group A *Streptococcus*. Streptolysin is an oxygen-labile toxin produced by *Streptococcus* that has haemolytic properties and a tendency to lyse RBCs. The antibodies produced against the bacteria have a tendency to cross-react with the self antigens present in the host. They mainly act against the collagen present in the cellular matrix of the heart, joint, brain, and skin cells. Patients with rheumatic fever and acute glomerulonephritis tend to have elevated levels of ASO in the serum.

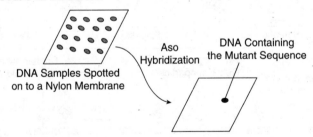

Figure 24.3 Anti-Streptolysin O Test (See page 372 for the colour image)

Methodology

This test can be done with either a slide or a card. The sample is collected from the patient, and the serum is separated. It is then added to the commercially available antigen (ASO latex reagent). This set-up is mixed well with separate applicator sticks, and the slide is moved up and down in such a way that the mixture is mixed well. Then, the results are observed. If the sample is positive, then agglutination is observed.

Interpretation

The normal serum has a value less than 200 IU/ml, and the abnormal serum has a value of 200 IU/ml. A titre level that is more than 200 IU/ml indicates streptococcal infection in the blood.

MULTIPLE CHOICE QUESTIONS

1. Widal test is a serological test used for the identification of _____.
   (a) *Salmonella*   (b) *Vibrio*
   (c) *Streptococcus*   (d) *Neisseria*
   Ans. a

2. Anti-streptolysin O test is a serological test used for the identification of _____.
   (a) *Salmonella*   (b) *Vibrio*
   (c) *Streptococcus*   (d) *Neisseria*
   Ans. c

3. Antigen O demonstrates _____.
   (a) Matt-finish agglutination
   (b) Cottony agglutination
   (c) Both (a) and (b)
   (d) None of the above
   Ans. a

4. Antigen H demonstrates _____.
   (a) Matt-finish agglutination
   (b) Cottony agglutination
   (c) Both (a) and (b)
   (d) None of the above
   Ans. b

SHORT NOTES

1. Serological testing
2. Widal test
3. Anti-streptolysin O test

ESSAY

1. Explain in detail with diagrams the importance of serological testing. Add a note on Widal test and anti-streptolysin O test.

25 Immunoprophylaxis

CHAPTER OBJECTIVES

25.1 Introduction
25.2 Active and Passive Immunity
25.3 Adjuvants
25.4 Vaccines and its Types
25.5 Vaccination Schedule
25.6 Current Approaches in Vaccines

25.1 INTRODUCTION

The history of immunoprophylaxis dates back to 400 BC. Thucydides described about the immunity gained by the individuals who were infected by plague. Those who recovered did not contract the disease and could nurse the sick. The phenomenon behind this was not understood then but was demonstrated by the famous English physician Edward Jenner in 1798. He demonstrated the process of inoculating cow pox, which actually protected the patients against small pox. He named this process as variolation, and his work was further explored by Louis Pasteur who worked on attenuated cholera bacilli, which actually provided an immune response as and when they were inoculated, and coined the name vaccine. Since then, the immune response due to the inoculation of inactivated organisms, serum (antibodies), or antibiotics, which provide an effector response to control infection, paved a new way for immunoprophylaxis.

Immunoprophylaxis is defined as the defence against infectious agents by attaining or acquiring the individual components in the form of antibodies, antibiotics, or attenuated or deactivated antigens, which elicit an immune response referred to as acquired immunity.

25.2 ACTIVE AND PASSIVE IMMUNITY

Passive immune response or passive immunity is attained by the administration of antibodies, which are transmitted either naturally or artificially. Passive immunity is further divided into the following:

1. Naturally acquired passive immunity
2. Artificially acquired passive immunity

Naturally acquired passive immunity occurs when antibodies are passively transferred from the mother to the foetus through the placenta. This immunity, which is transferred from the mother to the foetus, is temporary and exists only for few months after birth or until the immune system of the foetus is well developed.

Artificially acquired passive immunity is attained by administering readymade antibodies, which are raised against a particular antigen in another host (e.g., horse or rabbit). These antibodies are raised by inoculating the antigen that elicits an immune response. The immunity gained by administering the artificially acquired antiserum is a type of passive immunity [e.g., serum therapy for poisonous bites of spiders and snakes (antivenom)].

Active immune response is acquired by natural infections occurring in the body, which elicit an immune response due to the presence of pathogenic organisms. This active response is further classified into the following:

1. Naturally acquired active immunity
2. Artificially acquired active immunity

Naturally acquired active immunity is gained by the entry of pathogens triggering the immune system to elicit an immune response against them; it may be lifelong or for a short duration of time (e.g., immunity for chickenpox).

Artificially acquired active immunity is gained by the administration of antigens (vaccines) whose pathogenicity is removed retaining the antigenicity. The antigenic nature elicits an immune response that lasts long (e.g., oral polio vaccine).

25.3 ADJUVANTS

These are immunological agents that enhance the immune response along with antigens. They are otherwise called as activators and regulate the antigenicity for a longer period of time by retaining the antigens and elicit the maximum immune response till the antigens are being cleared by the immune cells. The commonly used adjuvants are as follows:

1. Paraffin oil
2. Aluminium hydroxide (used in toxoids)
3. Aluminium phosphate
4. Calcium phosphate
5. Freund's complete adjuvant (inactivated dried *Mycobacterium tuberculosis*) and Freund's incomplete adjuvant (water and oil emulsion)

25.3.1 Functions of Adjuvants

The functions of adjuvants are as follows:

1. They lengthen the presence of the antigen and elicit maximum immune response.
2. They activate the cells by adhering to antigen-presenting cells.
3. They are responsible for both types of immunity, namely, the humoral and cell-mediated immunity.
4. They act as a stabilizing agent in the preparation of antigens.

25.4 VACCINES AND ITS TYPES

A vaccine is a prophylactic agent that enhances the immunity for a particular infectious agent. Vaccine contains the whole organism (killed or inactivated), the inactivated toxins produced by the organism, or any product of the organism that on entry acts against the antigen resulting in the immune system recognizing it as foreign particle and eliciting a response. These immune responses are of two types, which are as follows:

1. **Effector response:** Here, plasma cells are produced; they secrete antibodies and eliminate the antigen and clear it from the site of entry.
2. **Memory response:** Memory cells are produced during a humoral response; they keep evidence of the entry of pathogens and hence produce a response as and when the pathogen re-enters the body.

There are various types of vaccine, which are as follows:

1. Live or attenuated vaccine
2. Killed vaccine
3. Recombinant subunit vaccine
4. Conjugate vaccine
5. Toxoid

25.4.1 Live or Attenuated Vaccine

A live altered bacterium or virus that retains its antigenicity constitutes a live or attenuated vaccine. The organisms are cultured and disabled in such a way that they lose their pathogenicity or virulence. The vaccine MMR (measles, mumps, and rubella) is an example of this type. All the three viruses that make up this vaccine are cultivated and grown in unfavourable conditions; hence, they tend to lose their pathogenicity and elicit an immune response. Bacillus Calmette–Guérin (BCG), a bacterial suspension, is a virulently modified vaccine for *Mycobacterium tuberculosis*.

Advantages

1. They elicit a good immune response.
2. They last long, that is, for a lifetime.

Disadvantages

1. They cannot be given to immunocompromised patients as they are live suspensions and may produce infections.
2. They also pose a risk for reversion to the virulent form.

25.4.2 Killed Vaccines

The pathogenic organisms are killed or inactivated by using heat or chemicals. Thereafter, they lose their pathogenicity and capacity to replicate but remain intact. These intact organisms are recognized as non-self by the host immune system, and it starts producing the immune response against them (e.g., viral vaccines such as polio vaccine, hepatitis A vaccine, and rabies vaccine and bacterial vaccines such as cholera vaccine).

Advantages

As the organisms are killed or inactivated, they lose their capacity to replicate and do not revert, so they are safe.

Disadvantages

The immune responses given by killed vaccines are always for a shorter duration; hence, a booster dose should be given for eliciting better immune response.

25.4.3 Recombinant Subunit Vaccine

The subunit vaccine does not contain the whole organism but has a part of the organism that can elicit the immune response. These subunits can be any of the cell organelles such as the cell wall or the protein coat of the pathogens, which are extracted or genetically engineered for providing protection against the infection. As these vaccines are prepared using molecular techniques, they are termed as recombinant vaccines (e.g., purified protein of pertussis, purified surface antigen of hepatitis B, and capsular antigens of meningococci and *Haemophilus influenzae*).

Advantages

1. They evoke a good immune response.
2. Multiple antigens (from different pathogens) can be inserted into the vector or host and provide a single-shot vaccine against various diseases. (Sometimes even without requirement of booster doses.)

Disadvantage

1. The cost of the vaccine is considerably high.

25.4.4 Conjugate Vaccine

This vaccine carries not only a single part of the organism (hence can also be grouped under recombinant vaccines) but also a specific protein in order to produce a combined immune response (e.g., pneumococcal vaccine).

Advantages

Recombinant vaccines have the ability to proliferate but do not cause any infection in the host cells, rather they elicit a good immunological response.

Disadvantage

Side effects might occur in rare cases.

25.4.5 Toxoid

Toxoids are purified proteins extracted as toxins from pathogenic organisms. These toxins are extracted, inactivated, and purified to produce immunological products called toxoids. They are inactivated by heat or chemical treatment (using formalin) and then purified to produce toxoids. For example, the neurotoxin produced by *Clostridium tetani*, which causes tetanus, is inactivated and used commercially as a tetanus toxoid. Diphtheria vaccine against *Corynebacterium diphtheriae* is given as a toxoid. It is administered in combination with pertussis and tetanus as a triple vaccine, that is, DPT.

Advantages

1. They are safe and do not cause any disease.
2. They are highly immunogenic in nature.

Disadvantage

As toxoids are inactivated toxic substances, they might cause hypersensitive reactions after inoculation; so care should be taken during and after inoculation.

25.5 VACCINATION SCHEDULE

The Centre for Disease Control and Prevention has formulated certain regulations for the administration of vaccines from birth till the age of 6 for young children in order to protect them from harmful and deadly diseases. The schedule is shown in Table 25.1.

Table 25.1 Vaccination Schedule

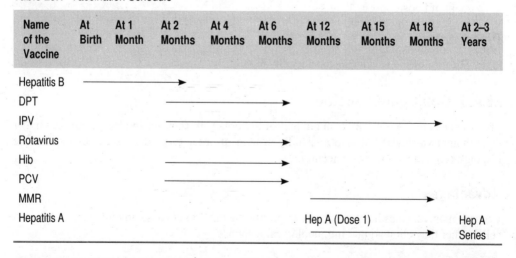

Abbreviations: IPV, inactivated polio virus; Hib, *Haemophilus influenzae* type B; PCV, pneumococcal capsular vaccine

25.6 CURRENT APPROACHES IN VACCINES

The development of new vaccines is one of the biggest challenges in clinical practice and biomedical research. Vaccines have helped eradicate many serious life-threatening diseases from various parts of the globe. However, there are many diseases for which vaccines are still unavailable. It is a challenge for researchers to evolve new vaccines for dreaded diseases and also continually improve existing vaccines by minimizing their adverse effects and lowering their cost.

The challenge and struggle to develop a vaccine for HIV has been going on since three decades. Many HIV vaccines have undergone phase I clinical trials, few of which have proceeded to phase II, and only two vaccines have reached phase III clinical trials.

Many other diseases have vaccines under trials and research and development. Examples of such vaccines include human papillomavirus, herpes simplex virus, *Chlamydia trachomatis*, *Neisseria gonorrhoeae*, *Treponema pallidum*, and *Escherichia coli*. Vaccines against Japanese encephalitis virus and yellow fever are administered in some parts of the world, and vaccines for malarial parasites, dengue virus, hookworm, and *Leishmania* species are under research.

It is important to understand that the development of new vaccines faces many barriers such as legal obstacles (flexibility of clinical trials), economic barriers (high cost involved in research), and vaccine safety issues. Researchers, scientists, and clinicians across the globe must participate in a cooperative and synchronous manner with adequate funding from national and international governmental collaborations and strive towards attaining success in vaccine development strategies.

MULTIPLE CHOICE QUESTIONS

1. An attenuated form of *Mycobacterium bovis* used as a specific vaccine for tuberculosis is _____.
   (a) BCG   (b) BGC
   (c) CBG   (d) GBC

   Ans. a

2. Edward Jenner invented the vaccine for small pox in _____.
   (a) 1778   (b) 1775
   (c) 1776   (d) 1779

   Ans. a

3. The inoculation given to stimulate immunologic memory response is known as _____.
   (a) Vaccination   (b) Booster
   (c) Variolation   (d) Inoculation

   Ans. b

4. When antibodies are passively transferred from the mother to the foetus through the placenta, then the immunity is termed as _____.
   (a) Naturally acquired passive immunity
   (b) Naturally acquired active immunity
   (c) Active immunity
   (d) Passive immunity

   Ans. a

5. An example of a live vaccine is _____.
   (a) MMR   (b) DPT
   (c) Hib   (d) IPV

   Ans. a

6. A triple vaccine given from 2 to 6 months after birth is _____.

(a) MMR   (b) DPT
(c) Hib   (d) IPV

Ans. b

7. An immunological agent that enhances the immune response along with the antigen is _____.

(a) Antigen
(b) Pathogen
(c) Toxin
(d) Adjuvant

Ans. d

SHORT NOTES

1. Types of vaccine
2. Adjuvants
3. Immunization schedule
4. Types of immune response
5. Toxoid
6. Live attenuated vaccines
7. Recombinant vaccines

ESSAY

1. Write in detail about the types of vaccine with suitable examples.

Glossary

A

ABO blood grouping – A system used to classify human blood into different groups (the four groups being A, B, AB, and O)

Acquired immunodeficiency syndrome (AIDS) – A disease caused by human immunodeficiency virus (HIV) that is characterized by the failure of the immune system followed by increased susceptibility to opportunistic infections

Aerobe – A microorganism that lives and grows in the presence of molecular oxygen

- **Facultative** – A microorganism that can grow with or without molecular oxygen
- **Obligate** – A microorganism that cannot grow without molecular oxygen

Aerobic culture – A method used to grow aerobic microorganisms from a clinical specimen

Agar agar – A gelatinous substance extracted from the seaweed red algae that is used as a solidifying agent in the preparation of culture media

Agglutination reaction – The formation of an insoluble immune complex by an antigen and an antibody

Anaerobe – A microorganism that can exist and grow only in the partial or complete absence of molecular oxygen

Anaphylaxis – An immediate hypersensitive reaction to an allergen comprising lowered blood pressure, swelling, and hives that is chiefly mediated by IgE and mast cells

Antibiotic – A substance derived from certain microorganisms that is widely used to inhibit the growth of other microorganisms

Antibiotic resistance – The ability of microorganisms to resist the effects of an antibiotic to which they were once sensitive

Antibody – A glycoprotein generally produced in response to the invaded antigen, which is a key part of the immune system, acting as 'the army of our body'

Antigen – A foreign substance that stimulates the production of antibodies

Antigen-presenting cell (APC) – A cell that displays foreign antigens to B cells and T cells in conjugation with class II MHC molecules to activate the cells (e.g., macrophages, B cells, and dendritic cells)

Antimicrobial agent – An agent that kills microorganisms or inhibits their growth

Asepsis – A technique that helps to exclude microorganisms completely and prevent contact with them

Autoclave – An apparatus employed to sterilize materials using steam under pressure

Autoimmunity – A condition characterized by the presence of serum autoantibodies and self-reactive lymphocytes that might be benign or pathogenic

B

B lymphocyte – A type of white blood cell derived from the bone marrow that produces antibodies and is one of the lymphocytes that play a major role in the body's immune response

Bacteraemia – An infection of the bloodstream caused by bacteria

Bacterial morphology – The study of form and structure of a bacterial cell with the aid of a microscope

Bacterial nutrition – The essential elements and chemicals present in the environment, whether natural niche or a laboratory, upon which the growth of bacteria is dependent

Bacterial taxonomy – The classification, nomenclature, and identification of bacteria.

Bacteriophage – A virus that infects bacteria to reproduce

Basal medium – An unsupplemented (without any special nutrient) medium that allows the growth of different types of microorganism

Biochemical reaction – A chemical reaction occurring in biological systems (plants, animals, and microorganisms) where one or more molecules interact with the aid of an enzyme and produce a product

Biomedical waste – Infectious and medical waste that is generated during diagnosis, treatment, research, and so on

Biomedical waste management – The management of hospital waste products (generated during diagnosis, treatment, and so on) to check the spread of diseases

Blood agar – A nutrient culture medium that is enriched with blood

Blood culture – A laboratory test done to detect the presence of bacteria or any other microorganism in a blood sample.

C

Capsid – The protein coat that envelopes the nucleic acid of a virion

Capsule – The mucopolysaccharide outer shell enveloping certain bacteria

Catalase test – A biochemical test used by microbiologists to identify certain bacteria (When a small amount of the bacterial isolate is added to hydrogen peroxide, O_2 bubbles can be observed if the bacteria possess catalase enzyme).

Cell culture – A complex process in which cells are grown *in vitro* under controlled conditions

Cell wall

- **Acid fast** – It contains a waxy substance called mycolic acid and a small amount of peptidoglycan.
- **Gram negative** – It consists of a thin peptidoglycan layer.
- **Gram positive** – It consists of a thick peptidoglycan layer and also contains teichoic acid.

Central dogma of molecular biology – The flow of genetic information within the biological system

Chloramphenicol – An antibiotic produced synthetically that is effective against a large number of microorganisms

Chocolate agar – Blood agar heated until the blood becomes brownish which is used to isolate especially *Haemophilus*, *Neisseria*, and other species

Coagulase – An antigenic substance produced by staphylococci

Codon – The sequence of three adjacent nucleotides that together form a unit of genetic code in DNA and RNA

Complement system – A system made up of many serum proteins that helps or 'complements' the ability of antibodies to destroy pathogens from an organism

Coombs test – A blood test performed to identify the cause of anaemia

Culture medium – A medium containing nutrients in which microorganisms and tissues are grown for scientific purposes

Cytokine – A protein that is produced by the cells of the immune system to regulate body's defence mechanism

Cytoplasm – A colourless gel-like substance composed mainly of water along with enzymes, salts, organelles, and various organic molecules

D

Delayed hypersensitivity – Type IV hypersensitivity (which is a type of cell-mediated response) in which the reaction takes 2 or 3 days to develop

Desiccation – Dehydration

Diarrhoea – A condition in which liquid faeces are discharged from the bowels more frequently than in a normal individual

Differential medium – A medium used to differentiate various types of microorganisms based on their colour and colony shape

Disinfection – The process of killing all microorganisms using a disinfectant

DNA – Deoxyribonucleic acid, which carries the genetic information in the cell and is capable of self-replication

Dye – A substance used to colour materials

E

Electron microscope – A microscope that uses a beam of electrons to produce an enlarged image of a small specimen

Endoflagella – A special type of flagella found in spirochaetes

Endospore – A resistant asexual spore produced within a bacterium during unfavourable conditions

Endotoxin – A toxic heat-stable lipopolysaccharide present inside bacteria that is released upon lysis of the cell

Enriched medium – A general-purpose medium to which special nutrients are added to encourage the growth of fastidious microbes

Enrichment medium – A medium that promotes the growth of a particular organism due to the presence of essential nutrients and certain inhibitory substances that avoid normal competitors

Enterotoxin – A bacterial toxin that targets intestinal cells and causes violent vomiting and diarrhoea

Eosinophil – A white blood cell that contains granules and is readily stained by eosin

Epitope – The part of an antigen to which an antibody gets attached

Eukaryote – A unicellular or multicellular organism with membrane-bound nucleus enclosing the genetic material

Exotoxin – A poisonous substance secreted by a microorganism and released into the medium in which it grows

Exponential phase – The period during which the cells of a defined bacterial population grow and divide continuously

F

Fever – An increase in the body temperature above normal

Fimbriae – Protein filaments (non-flagella) present on the surface of many bacteria

Flagella – A lash-like appendage arising from the cytoplasm of the cell, which is responsible for cellular motility

Fluorescence microscope – A microscope fitted with a source of ultraviolet rays to aid in the detection and examination of fluorescent specimens

Fungus – An organism that feeds on organic matter and reproduces through spores (which includes moulds, yeasts, and mushrooms)

G

Gas gangrene – The gangrene resulting from the infection of a wound by anaerobic bacteria characterized by the presence of gas in affected tissues

Genetic code – The sequence of nucleotides in DNA and RNA, which determines the specific amino acid sequence in the synthesis of proteins

Genital discharge – A fluid produced by the glands in vaginal walls or urethra that drains out from the genital opening

Graft – To implant (living tissue) surgically

Gram stain – A staining technique for the preliminary identification of bacteria that differentiates them into two large groups, namely, Gram-positive bacteria and Gram-negative bacteria

H

Heat-stable toxin – A secretory peptide produced by *Escherichia coli* that is toxic to animals

Heat sterilization – The process of destroying all forms of life in bacteriological media by means of moist or dry heat

HLA typing – A procedure done before transplantation to know the degree of tissue compatibility between a donor and a recipient

Hospital disinfection – Disinfection of medical devices undertaken to prevent the transmission of communicable diseases

Hot air oven – An instrument used for drying and sterilizing glassware

Hypersensitivity – Pathological sensitivity or extreme sensitivity

I

Immunity – The state of having sufficient biological defences to avoid infection, disease, or other unwanted biological invasion

Immunization – The process wherein a person is made immune or resistant to an infectious disease

Incineration – The process of burning something completely and eventually reducing it into ashes

Indole production – A test performed on bacterial species to determine the ability of the organism to convert tryptophan into indole

Infection – The invasion and multiplication of microorganisms that are not normally present within the body

Interferon – A protein produced by the immune system that is involved in defending the body against foreign invaders, which is also used to treat various diseases

K

Koch's phenomenon – It is immunity acquired by causing an infection, that is, if living tubercle bacilli are inoculated into tuberculous guinea pigs, then the animals become immune and reinfection does not occur.

L

Lag phase – The initial growth phase prior to exponential phase in which the cell number remains relatively constant

Light microscope – An optical microscope that uses visible light and a system of lenses to magnify images of small samples

Lipopolysaccharides (LPS) – Large macromolecules made up of lipid and polysaccharide, and the inner and outer cores are connected by a covalent bond

Log phase – Refer exponential phase

Lymphocyte – Small white blood cell that plays a role in the body's immune response

M

Major histocompatibility complex (MHC) – A group of genes that code for cell surface histocompatibility antigens, which are primarily responsible for the rapid rejection of tissue grafts between individuals

Malonate test – A biochemical test used for the identification of *Salmonella arizonae* based on the utilization of malonate in the broth as the sole source of carbon

Mantoux test – An intradermal test for the diagnosis of hypersensitivity to tuberculin in which a small amount of tuberculin is injected under the skin

Mast cell – A cell containing large basophilic granules that produces heparin, histamine, and serotonin, which is found in connective tissue and other body tissues

Medical mycology – The study of fungi that produce diseases in humans and animals and also of their ecology and epidemiology

Membrane attack complex – A complex of complement proteins that assemble to form a pore across the membrane of the cell, which allows the entry and exit of various ions and substances (resulting in the death of the cell)

Membrane filter – A filter, especially of cellulose acetate, that has pores of various diameters to check the entry of microorganisms

Methyl red test – A test to identify the ability of bacteria to perform mixed acid fermentation using methyl red as an indicator

Microaerophilic bacteria – Bacteria that live and thrive in environments with low levels of oxygen

Monoclonal antibody – An antibody produced by a single clone of cells that is also used as a therapeutic agent

Motility test – A test performed to determine the motility of microorganisms

Mould – A fungus that grows in the form of multicellular filaments called hyphae

Mutation – The process during which a permanent heritable change occurs in the nucleotide sequence of a gene in a chromosome

Mycelium – The vegetative part of a fungus consisting of mass of branching filaments (hyphae) that spread throughout the nutrient substratum

Mycosis – A fungal infection of animals, including humans

N

Negative staining – A staining technique in which the background is stained leaving the actual specimen untouched

Negri body – An eosinophilic inclusion body in the cytoplasm of certain nerve cells containing the rabies virus

Normal flora – Bacteria, yeasts, and protozoans that normally inhabit the body of healthy individuals, which are commensalists or mutualists in nature

Nosocomial infection – An infection that is acquired in a hospital

Nucleotide – The basic structural unit of nucleic acids DNA and RNA that is composed of a nucleobase, a pentose sugar, and one phosphate group

Nucleus – A membrane-bound structure enclosing the cell's hereditary information that regulates the growth and reproduction of the cell

O

Opportunistic bacterium – A bacterium that infects only when the host's resistance power is low

Opsonization – The process by which a pathogen is marked (bound with opsonin) for ingestion and destruction by a phagocyte

Oral rehydration therapy – A treatment for diarrhoea-related dehydration in which an electrolyte solution (i.e., a glucose-based salt solution) is given orally

Oxidase test – A test to identify the organisms that produce the enzyme cytochrome oxidase

P

Pasteurization – The process of partial sterilization of substances such as milk and other beverages by using high temperature to destroy microorganisms without altering the chemical make-up of the substances

Pathogen – An infectious agent (virus, bacterium, prion, or fungus) that causes disease in its host

Pathogenesis – The production and development of disease

Peptidoglycan – Also referred to as murein. It is a polymer made up of amino acids and sugars, which presents itself as a layer outside the cell membrane of bacteria.

Phagocytosis – The process by which a phagocyte engulfs a solid particle

Pilus – The hair-like structure present on the surface of bacteria

Pleomorphism – The ability of certain microorganisms to change their shape or size in response to environmental conditions

Polymerase chain reaction (PCR) – An *in vitro* molecular technology employed for amplification of DNA in few hours

Prokaryote – An organism that lacks a true nucleus (i.e., membrane-bound nucleus)

R

Radiation – The propagation and emission of energy in the form of electromagnetic waves (the types including light, heat, and sound)

Rh incompatibility – A condition developed when a pregnant woman with Rh-negative blood carries a baby with Rh-positive blood in her womb

S

Selective medium – A medium used for the growth of only selected microorganisms

Serotyping – Assigning the strains of a microorganism that have a set of antigens in common to a particular serotype

Slime layer – An easily removable, diffuse unorganized layer of extracellular material enclosing the bacterial cell

Spore – A minute, typically one-celled reproductive unit that is highly resistant to desiccation and heat and is capable of growing into a new organism when conditions are favourable

Sterilization – The total destruction of microorganisms and their spores

Syringe filter – A single-use filter cartridge used to remove particles from a sample prior to analysis by HPLC

T

Tetracycline – A broad-spectrum antibiotic used to treat a wide variety of infections

Thermophile – An extremophile that thrives at relatively high temperatures, that is, between 45°C and 80°C

Toxoid – A modified or inactivated exotoxin that has lost its toxicity but retains the ability to stimulate the production of antitoxin

Transport medium – A medium used for the safe transfer of clinical specimens to the laboratory for examination

Tumour – An uncontrolled abnormal benign or malignant growth of tissue that possesses no physiological function

Tyndallization – An old process used for sterilizing food (still used occasionally)

U

Urease test – A test employed for the identification of microorganisms such as cryptococci and *Helicobacter pylori* using their ability to produce urease

Urinary tract infection – An infection along the urinary tract

V

Vaccine – A suspension containing live, attenuated, modified, or killed microorganisms that when administered into the body stimulates the body's immune response

Vector – Carrier/transmitter of infectious agents into other living organisms

Virus – A small infectious agent that replicates only inside the living host, which can infect all types of life form

Voges–Proskauer test – A test performed to differentiate among enteric organisms

W

Widal test – A blood serum test that uses agglutination reaction to diagnose typhoid fever and *Salmonella* infections

Y

Yeast – A unicellular fungus that reproduces mainly by budding (which is used in bakeries and also in making alcoholic drinks)

Z

Zoonosis – An infectious disease that can be transmitted to humans from animals

Illustrations

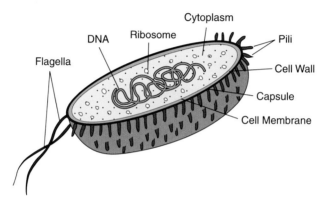

Figure 4.1 Structure and Contents of a Typical Prokaryotic (Bacterial) Cell (See also figure 4.1 on page 14)

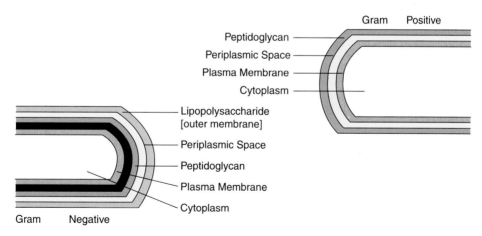

Figure 4.2 Cell Wall Differences between Gram-Positive and Gram-Negative Bacteria (See also figure 4.2 on page 15)

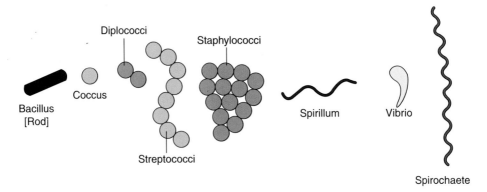

Figure 4.3 Shapes of Microbes (See also figure 4.3 on page 16)

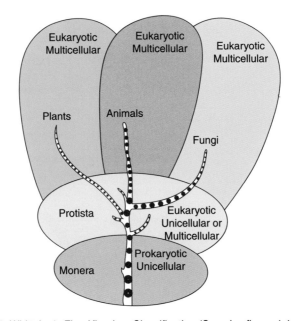

Figure 4.4 Whittaker's Five-Kingdom Classification (See also figure 4.4 on page 18)

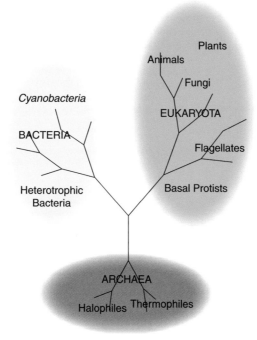

Figure 4.5 Three-Domain Concept (See also figure 4.5 on page 20)

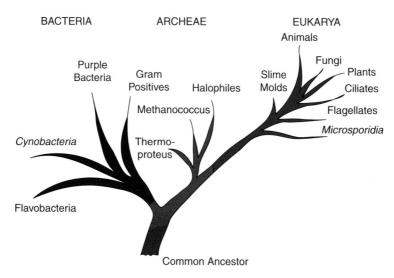

Figure 4.6 Universal Tree of Life (See also figure 4.6 on page 23)

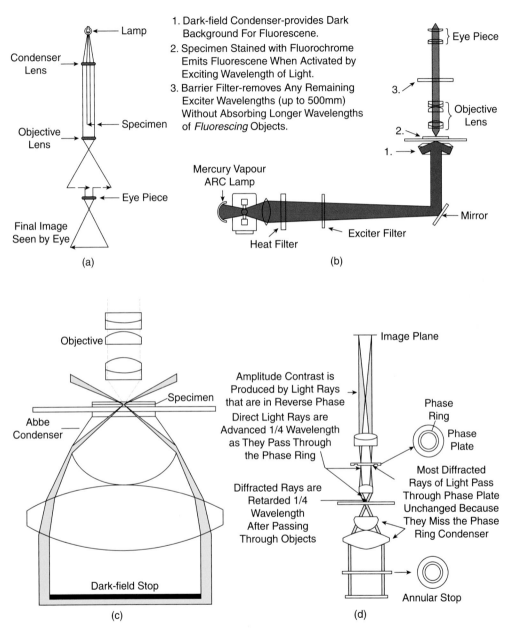

Figure 5.2 Ray Diagram of Principle of (a) Light Microscope, (b) Fluorescence Microscopy, (c) Dark-field Microscopy, (d) Phase-contrast Microscopy, (e) Transmission Electron Microscope, and (f) Scanning Electron Microscope (See also figure 5.2 on pages 31 and 32)

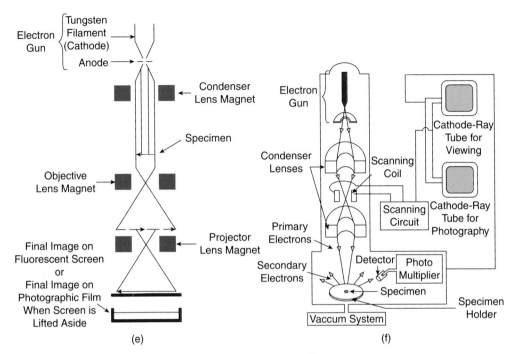

Figure 5.2 (*Continued*)

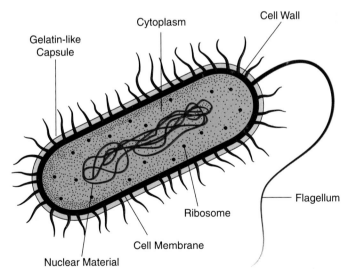

Figure 5.3 Bacterial Anatomy (See also figure 5.3 on page 34)

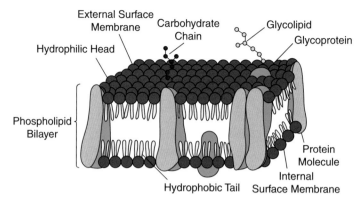

Figure 5.4 Fluid Mosaic Model of Cell Membrane (See also figure 5.4 on page 37)

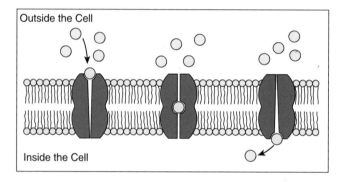

Figure 7.1 Facilitated Diffusion (See also figure 7.1 on page 58)

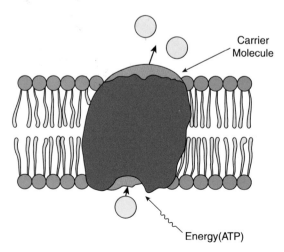

Figure 7.2 Active Transport (See also figure 7.2 on page 58)

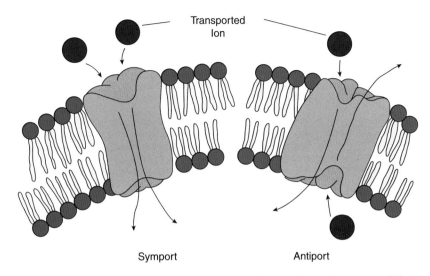

Figure 7.3 Symport and Antiport Mechanism (See also figure 7.3 on page 59)

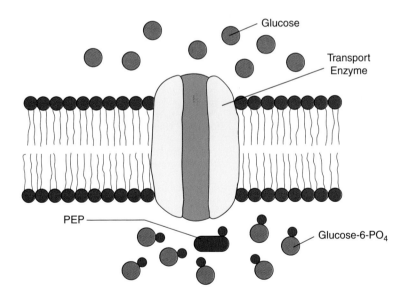

Figure 7.4 Group Translocation (See also figure 7.4 on page 60)

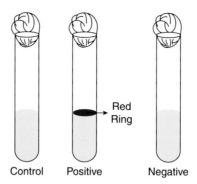

Figure 9.4 Indole Test (See also figure 9.4 on page 89)

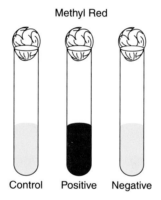

Figure 9.5 Methyl Red Test (See also figure 9.5 on page 90)

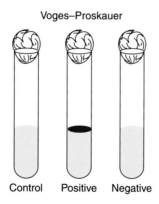

Figure 9.6 Voges–Proskauer Test (See also figure 9.6 on page 90)

Illustrations | 349

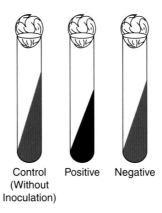

Figure 9.7 Citrate Utilization Test (See also figure 9.7 on page 90)

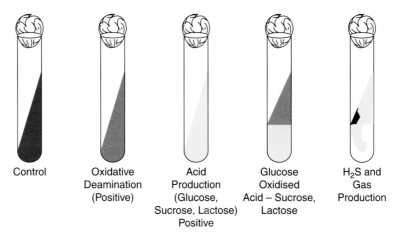

Figure 9.8 TSI Agar Test (See also figure 9.8 on page 91)

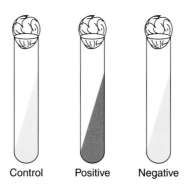

Figure 9.9 Urease Test (See also figure 9.9 on page 92)

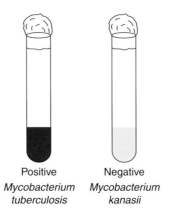

Figure 9.10 Nitrate Reduction Broth (See also figure 9.10 on page 93)

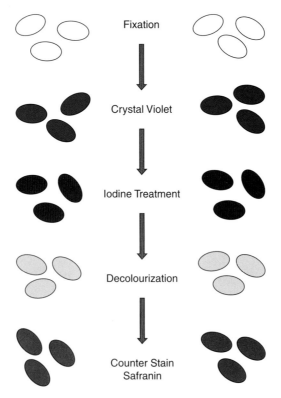

Figure 10.2a Flowchart Representation of Gram Staining (See also figure 10.2a on page 102)

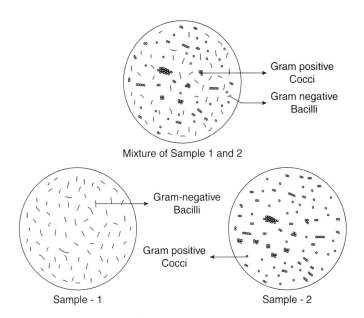

Figure 10.2b Gram Stained Organisms (See also figure 10.2b on page 103)

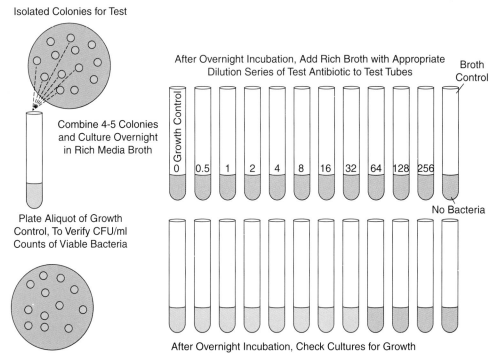

Figure 13.23 Tube Dilution Method to Determine Minimum Inhibitory Concentration (See also figure 13.23 on page 148)

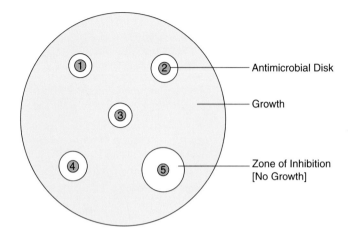

Figure 13.24 Disc Diffusion Method (See also figure 13.24 on page 149)

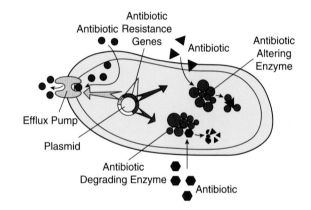

Figure 13.25 Mechanism of Antibiotic Resistance (See also figure 13.25 on page 150)

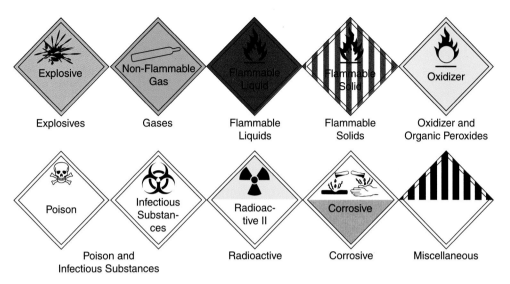

Figure 14.1 Biosafety Labels (See also figure 14.1 on page 154)

HAZARDOUS WASTE

General Information

Name_____

Address_____

_____

HANDLE WITH CARE

Figure 14.2 Container Label (See also figure 14.2 on page 155)

354 | Illustrations

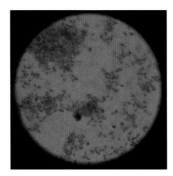

Figure 16.1 Gram-positive Cocci (See also figure 16.1 on page 177)

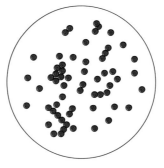

Figure 16.2 Gram-negative Cocci (See also figure 16.2 on page 184)

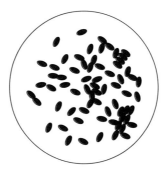

Figure 16.3 Gram-positive Bacilli (See also figure 16.3 on page 189)

Illustrations | **355**

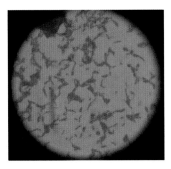

Figure 16.4 Gram-negative Bacilli (See also figure 16.4 on page 191)

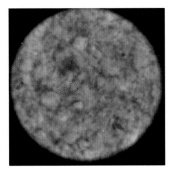

Figure 16.5 *Mycobacterium* (See also figure 16.5 on page 201)

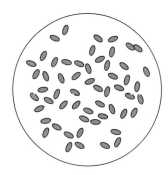

Figure 16.9 *Rickettsia* (See also figure 16.9 on page 216)

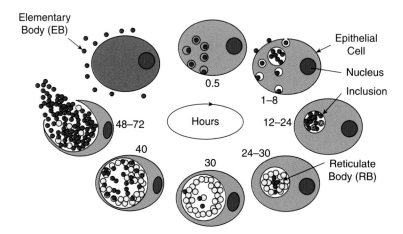

Figure 16.10 Developmental Cycle of *Chlamydia trachomatis* (See also figure 16.10 on page 218)

Characteristics	Viral Family	Important Genera
Single-Stranded DNA Non-enveloped 18–25 nm	Parvoviridae	Human parvovirus B19
Double-Stranded DNA Non-enveloped 70–90 nm	Adenoviridae 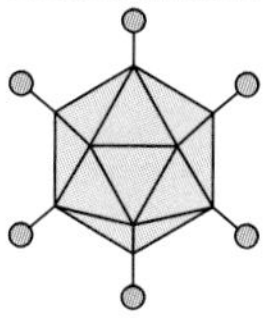	Mastadenovirus
40–57 nm	Papovaviridae 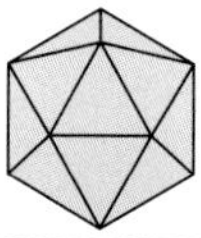	Papillomavirus [Human Wart Virus] Polyomavirus
Double-Stranded DNA enveloped 200–350 nm	Poxviridae 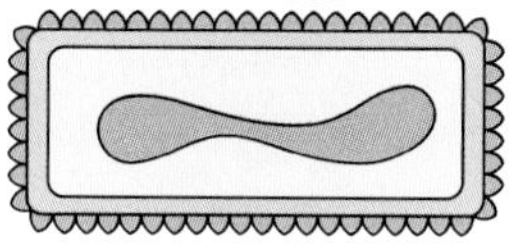	Orthopoxvirus [Vaccinia & Small Pox viruses] Molluscipoxvirus
150–200 nm	Herpesviridae 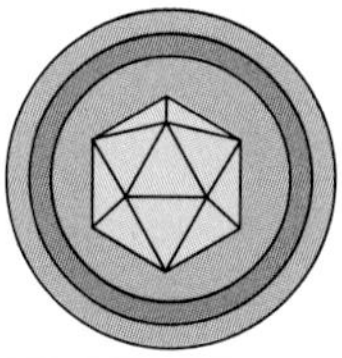	Simplexvirus [HHV-1 & 2] Varicellovirus [HHV-3] Lymphocryptovirus [HHV-4] Cytomegalovirus [HHV-5] Roseolovirus [HHV-6] HHV-7 Kaposi's Sarcoma [HHV-8]
Double-Stranded DNA enveloped 22–42 nm	Hepadnaviridae 	Hepadnavirus [Hepatitis B Virus]
Single-Stranded RNA, + Strand Non-enveloped 28–30 nm	Picornaviridae 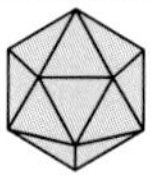	Enterovirus Rhinovirus [Common Cold Virus] Hepatitis A Virus
35–40 nm	Caliciviridae 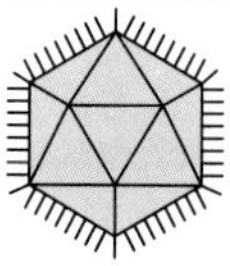	Hepatitis E Virus Norovirus

Figure 17.1 General Properties of Viruses (See also figure 17.1 on pages 233, 234, and 235)

Characteristics	Viral Family	Important Genera
Single-Stranded RNA + Strand enveloped 60–70 nm	Togaviridae 	Alphavirus Rubivirus [Rubella Virus]
40–50 nm	Flaviviridae 	Flavivirus Pestivirus Hepatitis Virus
Nidovirales 80–160 nm	Coronaviridae 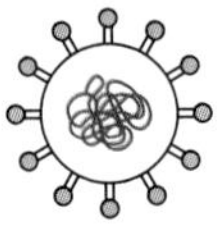	Coronavirus
Mononegavirales - Strand, One Strand of RNA 70–180 nm	Rhabdoviridae 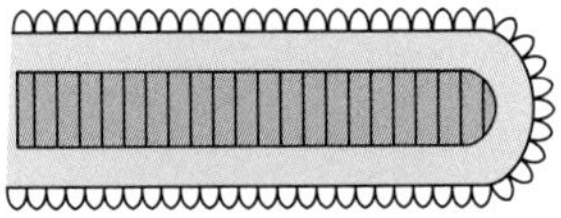	Vesiculovirus [Vesicular stomatatis Virus] Lyssavirus [Rabies Virus]
80–14,000 nm	Filoviridae 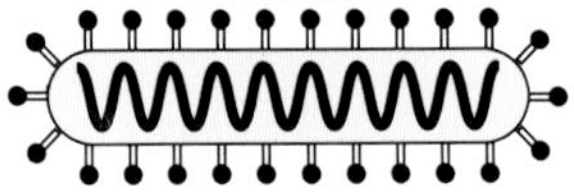	Filovirus
150–300 nm	Paramyxoviridae 	Paramyxovirus Morbillivirus [Measles-Like Virus]
- Strand one Strand of RNA 32 nm	Deltaviridae 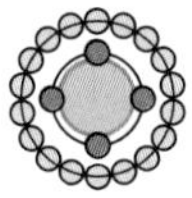	Hepatitis D

Figure 17.1 (*Continued*)

Characteristics	Viral Family	Important Genera
- Strand, Multiple Strands of RNA 80–200 nm	Orthomyxoviridae 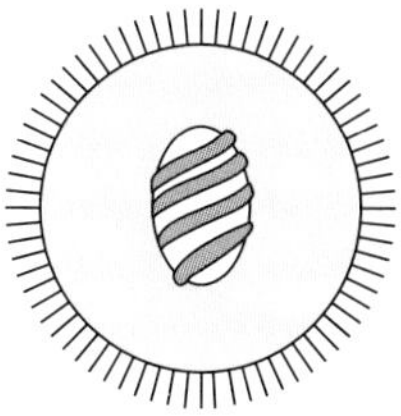	Influenza Virus A, B, and C
90–120 nm	Bunyaviridae 	Bunyavirus [Cailfornia Encephalitis Virus] Hantavirus
110–130 nm	Arenaviridae 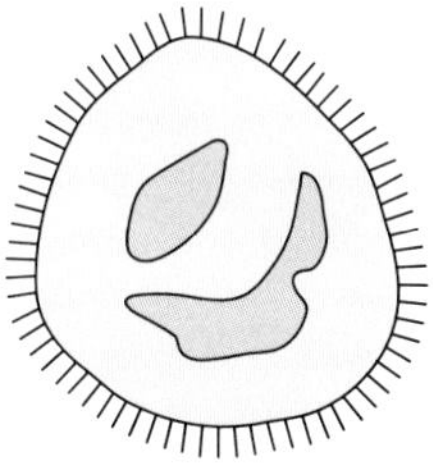	Arena Virus
Produce DNA 100–120 nm	Retroviridae 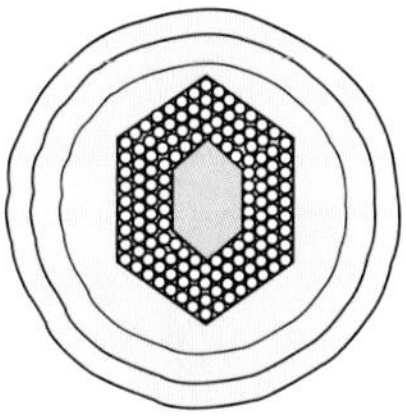	Oncoviruses Lentivirus [HIV]
Double-Stranded RNA Non-Enveloped 60–80 nm	Reoviridae 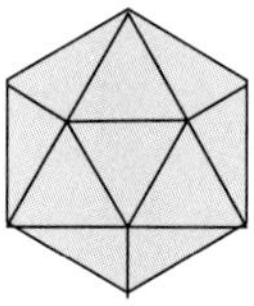	Reovirus Rotavirus

Figure 17.1 (*Continued*)

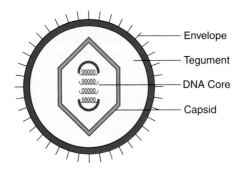

Figure 17.2 Herpesviridae (See also figure 17.2 on page 238)

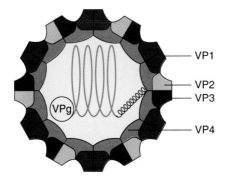

Figure 17.3 Picornaviridae (See also figure 17.3 on page 242)

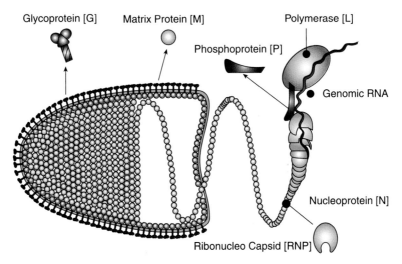

Figure 17.4 Rhabdoviridae (See also figure 17.4 on page 246)

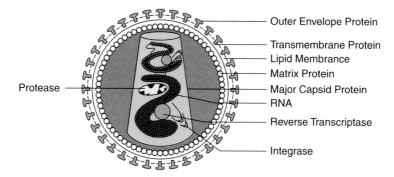

Figure 17.5 Retroviridae (See also figure 17.5 on page 249)

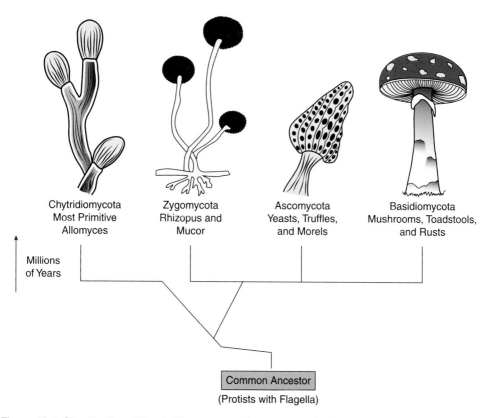

Figure 18.4 Classification of Fungi—Zygomycetes, Ascomycetes, Basidiomycetes, and Deuteromycetes (See also figure 18.4 on page 261)

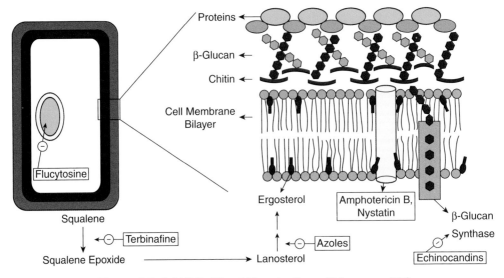

Figure 18.5 Cell Wall of Fungi (See also figure 18.5 on page 262)

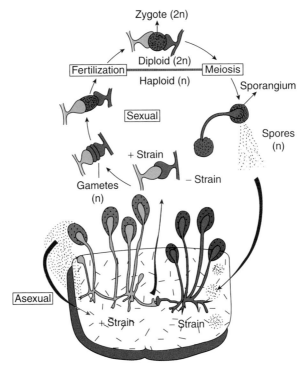

Figure 18.6 Sexual and Asexual Reproduction (See also figure 18.6 on page 263)

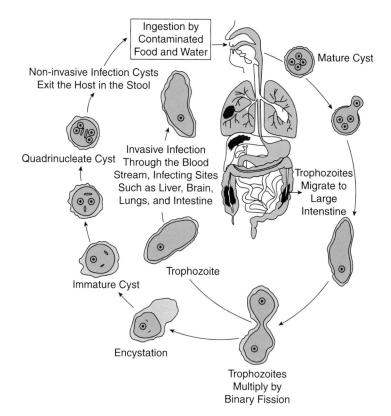

Figure 19.2 Life Cycle of Amoebiasis (See also figure 19.2 on page 276)

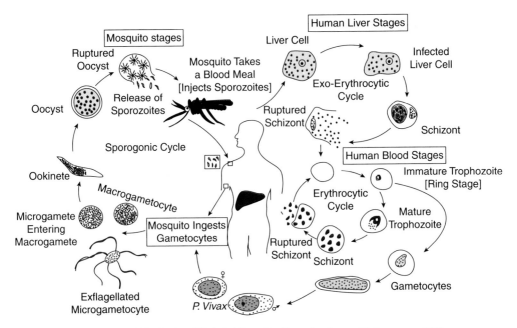

Figure 19.4 Life Cycle of Malarial Parasite (See also figure 19.4 on page 279)

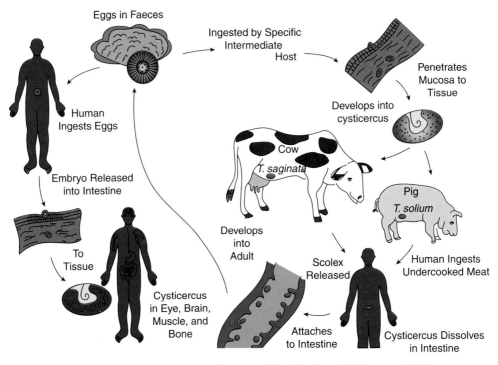

Figure 19.6 Life Cycle of *Taenia solium* (See also figure 19.6 on page 281)

Illustrations | 365

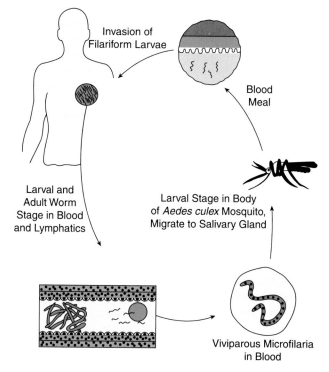

Figure 19.8 Life Cycle of *Wuchereria bancrofti* (See also figure 19.8 on page 282)

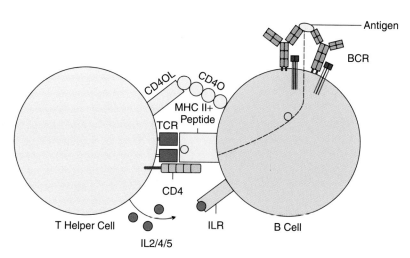

Figure 21.1 T Cells and B Cells (See also figure 21.1 on page 300)

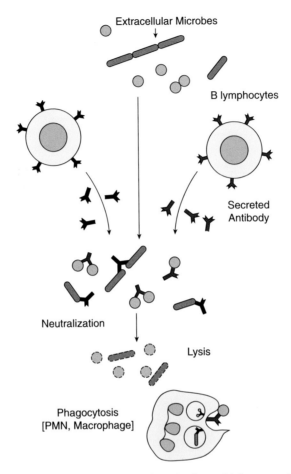

Figure 21.2 Humoral Immunity (See also figure 21.2 on page 301)

Illustrations | 367

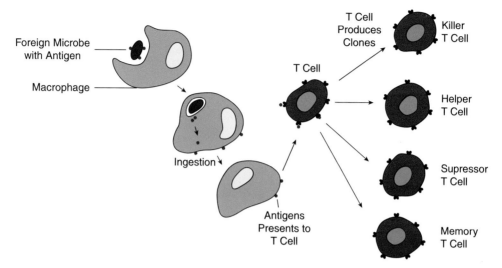

Figure 21.3 Cell-mediated Immunity (See also figure 21.3 on page 302)

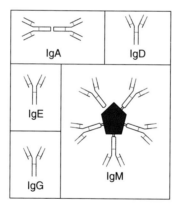

Figure 22.1 Structure and Classification of Antibodies (five subclasses) (See also figure 22.1 on page 305)

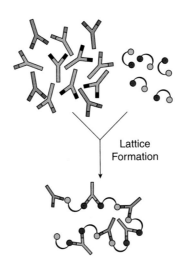

Figure 22.2 Lattice Formation (See also figure 22.2 on page 305)

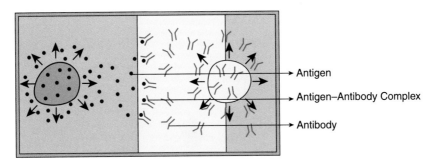

Figure 22.3 Ouchterlony Double Immunodiffusion (See also figure 22.3 on page 306)

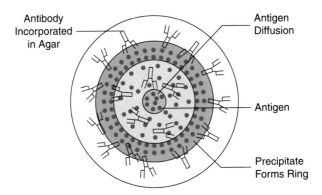

Figure 22.4 Single Radial Immunodiffusion (SRID) (See also figure 22.4 on page 307)

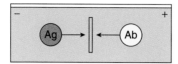

Figure 22.5 Counter-Current Immunoelectrophoresis (See also figure 22.5 on page 307)

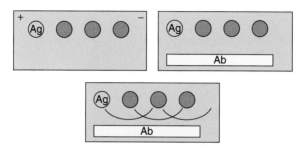

Figure 22.6 Immunoelectrophoresis (See also figure 22.6 on page 307)

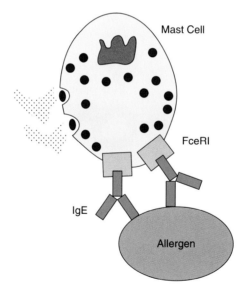

Figure 23.1 Type I Hypersensitivity (See also figure 23.1 on page 313)

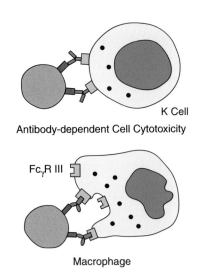

Figure 23.2 Type II Hypersensitivity (See also figure 23.2 on page 316)

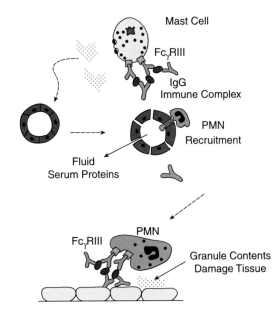

Figure 23.3 Type III Hypersensitivity (See also figure 23.3 on page 318)

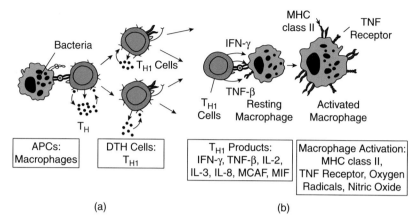

Figure 23.4 Type IV Hypersensitivity— (a) Sensitization Phase, (b) Effector Phase (See also figure 23.4 on page 319)

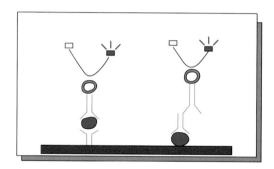

Figure 24.1 ELISA (See also figure 24.1 on page 322)

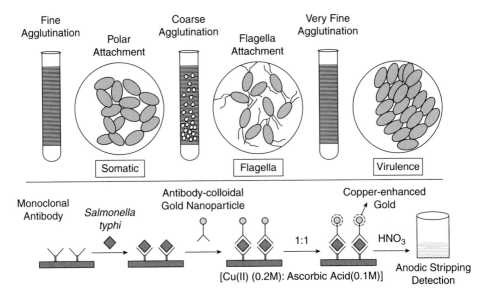

Figure 24.2 Widal Test (See also figure 24.2 on page 323)

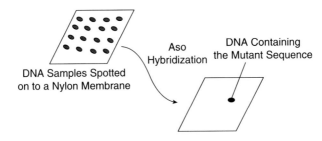

Figure 24.3 Anti-Streptolysin O Test (See also figure 24.3 on page 324)

Index

β-lactam antibiotics, 141, 317
(TCBS) agar, 78, 209
(WBCs), 299
2, 3-butanediol, 90

A

a-naphthylamine (reagent B), 93
ABO blood typing, 308
Acetoin, 90
Acetyl glucosamine, 258
Acholeplasma laidlawii, 212, 213
Acholeplasma, 212, 213
Acholeplasmataceae, 211, 212
Acid-fast bacilli, 24, 201
Acid-fast bacteria, 35, 36, 104
Acid-fast staining, 24, 103, 204, 206
Acquired Immunodeficiency Syndrome (AIDS), 249
Acridine orange, 107
Actinomycetales, 28
Actinomycetoma, 267
Active and passive immunity, 5, 326
Active transport, 57, 59
Acute gingivostomatitis, 238
Acute rheumatoid fever, 183
Acyclovir, 147, 240
Adaptive immunity, 298, 299
Adenoviridae, 233, 237
Adhesins, 41, 52
Adjuvants, 326, 327
Adsorption, 232
Aerobes, 23, 67, 93, 182, 201, 213
Aerobic, 54, 55, 184, 189, 198, 199, 201, 208, 209, 259
Aerotolerants, 67
Agar, 72, 73, 75–80, 83, 85, 91–5, 138, 148, 149, 152
Agglutination reactions, 201, 308
Agglutination, 183, 199, 201, 215, 252, 308, 322, 323
Aggression, 247

AIDS, 5, 6, 8, 33, 38, 51, 86, 88, 214, 249
Air filters, 130, 159
Albendazole, 282, 283
Albert ludwig neisser, 184
Albert stain A, 106
Albert stain B, 106
Alcohols, 121, 132
Aldehydes, 121, 131, 132
Alexandre Yersin, 285
Alimentary tract, 116
Alkaliphiles, 68
Allergic asthma, 314, 315
Allergic reactions, 5, 144, 291, 311, 312, 317
Allergic rhinitis, 314, 315
Allergy, 291
All-purpose media, 77
Alpha, 77, 90, 182
Alphaherpesvirinae, 237
Aluminium hydroxide, 327
Aluminium phosphate, 327
Amantadine, 146
American lyme disease, 289
Amies transport medium, 78
Amikacin, 198
Aminoquinolines, 147
Ammon's horn, 248
Ammonia, 53, 54, 55, 89, 92, 196, 213
Amodiaquine, 280
Amoebae, 275, 277
Amoxicillin, 165, 183, 201
Amphitrichous, 40
Amphotericin B, 142, 145, 146, 166, 262, 266, 269, 270, 272
Ampicillin, 141, 194, 196, 197, 199, 213
Amylase enzymes, 94
Anaerobes, 23, 67, 68, 71, 79, 93, 177, 191, 212

Anaerobic, 52, 54, 79, 93, 182, 184, 189, 191
Anaeroplasma, 212, 213
Anaeroplasmataceae, 212
Anaphylactic hypersensitivity (type I), 312
Anaphylactic shock, 313, 314
Anaphylaxis, 311, 313, 314
Anatomic barriers, 298
Animal waste, 152, 311
Animalia, 18, 21, 22
Antibiotic, 3, 5, 35, 38, 62, 77, 80, 87, 111
Antibody-dependent cytotoxic hypersensitivity (type II), 312
Antigenic specificity, 299
Antigen-presenting cells, 299, 300, 319, 320, 327
Antihistamines, 314–6, 319
Antiport, 37, 59
Anti-streptolysin O test, 324
Anti-streptolysin O, 183, 323–5
Anton Van Leeuwenhoek, 6, 31
Aphthovirus, 242
Aplastic anaemic, 317
Archaea, 15, 20–2, 35, 142
Arenaviridae, 235
Argasidae, 289
Arthropod vector, 216
Arthropoda, 287, 291
Arthrospores, 262
Arthus reaction, 317, 318
Arthus, 317, 318
Artificially acquired passive immunity, 326, 327
Asbestos disc filters, 129
Ascomycetes, 261
Ascospores, 261, 262
Asepsis, 3, 4, 111, 117, 119, 121, 123, 125, 127, 129
Aseptic meningitis, 239, 243, 244

Asexual, 261–3, 266, 275, 278, 279
ASO latex reagent, 324
Asparagine, 202
assay (ELISA), 308
Asthalin, 316
Atopy, 314
Atovaquone, 289
ATP-binding cassette (ABC) transporter, 58
Atypical infections, 114
Auramine–rhodamine technique, 107
Australian tick typhus, 217
Autoclave, 75, 77, 80, 123–8, 134–6, 156, 160, 161
Autoimmune disorder, 317
Autoimmune haemolytic anaemia, 316, 317
Autotrophic bacteria, 53
Autotrophs, 23, 53
Azithromycin, 187, 289

B

B cell lymphoma, 241
B cells, 203, 241, 300, 301, 312, 316
Babesia microti, 288
Babesiosis, 288, 289
Bacilli, 8, 16, 28, 29, 182, 183, 189–91, 194, 196, 198–201
Bacillus Calmette–Guerin (BCG), 328
Bacteria, 3, 5–8, 14–7, 20–6, 28–30, 33–6, 38–58, 60–5, 67–79, 82
Bacterial vaccine, 329
Bacteriocin typing, 87
Barophiles, 68
Barritt's reagent A, 90
Barritt's reagent B, 90
Basal media, 76
Basic fuchsin, 105
Basic stains, 100
Basidiomycetes, 261
Basidiospores, 261, 262
Basophilic, 218, 248
Batch culture, 61–3
Bats, 247
BCG (Bacillus Calmette–Guerin), 204
Benzoic acid salicylic acid ciclopirox olamine, 265
Benzyl alcohol, 290
Beta haemolysis, 77, 182
Beta, 77, 121, 181, 182, 246, 248, 308
Betaherpesvirinae, 237
Beta-lactamase-stable drugs, 181

Betapropiolactone, 131
Binary fission, 56, 61, 64, 212 , 216, 218, 219, 275, 276
Biochemical identification, 82, 89
Biohazardous waste, 153
Biological safety cabinets, 158, 159
Biological vectors, 115, 288
Biomedical waste, 111, 152, 153, 155, 157, 159–61, 164
Biosafety Level 1 (BSL-1), 160
Biosafety Level 2 (BSL-2), 160
Biosafety Level 3 (BSL-3), 160
Biosafety Level 4 (BSL-4), 161
Biosafety Levels, 157, 158, 160
Biotyping, 87
Biphasic media, 75, 76
Bismarck brown, 98
Bismuth subsalicylate, 194
Black piedra, 266
Blair agar, 78
Blastospores, 262
Blood agar, 77, 80, 94, 181, 182, 193, 208, 221–4, 270
Body louse, 216, 290
Body temperature, 298, 320
Boiling, 99, 124, 125
Bone marrow, 143, 201, 270, 300, 301, 317
Borrelia, 47, 209–11, 289
Boutonneuse fever, 217
Bovine heart infusion broth, 213
Brain stem, 248
Broad-spectrum antibiotics, 140, 209
Bronchitis, 166, 315
Broths, 72, 75
Bubonic plague, 285, 286
Bullet-shaped virus, 245
Bunyaviridae, 235
Burkitt's lymphoma, 240, 241

C

Calcium phosphate, 327
Calcofluor white staining, 107
Caliciviridae, 233
Candida albicans, 145, 271
Candida krusei, 271
Candida parapsilosis, 271
Candida tropicalis, 271
Candidiasis, 252, 270, 271
Candle filters, 121, 128
Capillary permeability, 299
Capsid, 231, 232, 236–8, 242, 244, 246, 249, 250

Capsomers, 231, 232, 237
Capsular K antigen, 193, 194
Capsule staining, 105, 106
Capsule, 39, 101, 105, 106, 179
Carbohydrate fermentation pattern, 91
Carbol fuchsin, 104
Carbon dioxide, 21, 23, 53, 55, 90, 131, 196, 213
Carboxysomes, 14
Cardiovirus, 242
Carl Woese's, 20
Carmine, 98
Carolus Linnaeus, 17, 18
Cary–Blair medium, 79, 80
Casein extract, 74
Castaneda, 218
Catalase test, 93, 181, 187, 221
Catalase, 67, 93, 178, 181, 182, 187, 189, 191, 196
Cationic detergents, 134
Cats, 115, 247, 292
CD4 T_H1, 302
CD4 T_H2, 302
CD4+ T lymphocytes, 250
Cefixime, 187, 201
Cefsulodin–Irgasan–novobiocin, 286
Cell wall, 5, 14, 15, 21, 22, 34–6, 43, 47
Cell-mediated hypersensitivity, 312, 319
Centre for disease control and prevention, 330
Cephalosporin, 140, 141, 187, 188, 194, 196, 198, 212, 215
Cerebellum, 248
Cervicitis, 185, 220
Cetrimide agar, 77, 80
Chalmydiaceae, 217
Chemical disinfection, 120, 156
Chemical mediators, 298, 302, 314
Chemoautotrophs, 53
Chemotactic behaviour, 49
Chemotaxis, 41, 92
Chemotherapeutic index, 138, 139, 144, 145
Chemotherapy, 111, 137, 138, 139, 141–3, 145–9, 164, 263
Chemotrophs, 23
Cherry red complex, 89
Chitin, 15, 22, 258, 262
Chlamydia pneumoniae, 217, 220
Chlamydia psittaci, 217, 220
Chlamydia trachomatis, 217–20, 331

Chlamydia, 103, 116, 140, 143, 177, 217–20, 331
Chlamydobacteriales, 28
Chlamydospores, 262
Chloramphenicol, 36, 140, 141, 143, 188, 217
Chlorine, 133
Chlorohexidine, 134
Chloroquine, 147, 280
Chloroxylenol, 134
Chlortetracycline, 142
Chocolate agar, 77, 80, 94, 187, 188
Cholera bacilli, 326
Cholera toxins, 208
Cholera vaccine, 329
Chromoblastomycosis, 266, 267, 268
Chromosomes, 14, 38, 41, 42
Chronic demyelinating neuropathy, 251
CIE, 201
Cigar bundle, 205
Ciliates, 23, 275
Ciprofloxacin, 144, 196, 201
Citrate positive, 200
Citrate, 78, 90, 198, 200, 209
Cladophialophora carrionii, 268
Clarithromycin, 289
Class, 15, 17, 155, 158, 160, 161, 211, 276, 300
Clindamycin, 143
Clofazimine, 207
Closed cabinets, 159
Clotrimazole, 145, 146
Coagulase test, 93, 181
Coagulase, 93, 166, 178, 181, 221
Coagulated, 124, 134, 202
Cocci, 16, 29, 46, 92, 177, 178, 182
Cockroaches, 288
Cold sterilization, 127
Colonization, 11, 49, 51, 52, 139, 189, 197
Colony morphology, 30, 178
Colour, 14–6, 18, 20, 23, 30, 31, 34, 37, 58–60, 77
Colour code, 153, 155, 161
Coma, 247
Comma-shaped, 29
Commensals, 77, 113, 184, 194, 211, 212
Competitive enzyme-linked immunosorbent assay, 308
Complement molecules, 298
Complex media, 76

Complex symmetry, 232
Compound microscope, 31, 33
Confirmatory tests, 182, 190, 209, 252, 270
Congenital, 117, 240, 241
Conidiospores, 262
Conjugate vaccine, 328, 329
Contact dermatitis, 319, 320
Containment, 154, 158
Convalescent carriers, 114
Convulsions, 144, 247
Coomassie blue, 98
Cord factor, 202
Corkscrew-like rotator movement, 210
Coronaviridae, 234
Corticosteroids, 282, 292, 320
Corynebacterium, 7, 23, 30, 52, 106, 177, 189
Corynebacterium diphtheriae, 7, 30, 106, 189
Corynebacterium miniutissimum, 189
Corynebacterium striatum, 189
Corynebacterium xerosis, 189
Counter-current immunoelectrophoresis, 307
Coxsackieviruses, 243, 244
Crab lice, 291
Cranial nerve, 251
Cresols, 134
Cristispira, 47, 209, 211
Cross infections, 114
Cryptococcosis, 272, 273
Cryptococcus neoformans and cryptococcus gattii, 272
Crystal violet, 36, 77, 80, 97, 98, 100–3, 105–106
Curved rods, 205, 207
Cutaneous infections, 145, 179
CV–I complex, 102
Cyanobacteria, 15, 20, 21
Cystic stage, 275–7, 280
Cysticercosis, 281
Cystitis, 192, 195, 196
Cytokines, 5, 302
Cytomegalovirus, 116, 233, 237, 241, 242, 252
Cytoplasm, 37, 38, 39, 41–3, 64, 104, 191
Cytotoxic toxin, 178
Cytotoxin (Shiga toxin), 198

D
DAPI, 98
Dapsone, 207, 267, 269
Dark-field microscope, 33
Death, 60–2, 64, 66, 69, 125, 127, 137, 144, 145, 147
Decarboxylation, 91
Deep infections, 178, 179
Deep mycosis, 263, 269
Delayed or cell-mediated hypersensitivity (type IV), 312
Delayed type hypersensitivity, 319
Delirium, 247
Deltaviridae, 234
Dendritic cells, 301, 320
Deoxycholate citrate, 198
Dermatophagoides pteronyssinus, 291
Deuteromycetes, 261
Diene's stain, 212, 213
Differential media, 78, 92, 94, 191, 193, 196
Differential staining, 97, 101, 103, 180
Diffusion method, 148, 149
Digital microscope, 33
Dimorphic fungi, 260, 265
Dimorphic leprosy, 206
Diphtheros, 189
Diplococci, 16, 184, 187, 188
Diplococcus, 29
Disc diffusion method, 148, 149
Disinfection, 3, 4, 111, 119–21, 123, 125, 127, 129, 131, 133–5
Disulphide bonds, 219, 304
Diversity, 4, 21, 299
Dogs, 115, 247, 292
Domain, 17, 20–2, 58
Dorner method, 105
Double immunodiffusion (ouchterlony technique), 306
Doxycycline, 142, 287
DPT, 330–2
Dr. Kiyoshi Shiga, 198
Dry heat sterilization, 121, 122, 134
Dyes, 36, 77, 78, 94, 97, 98, 100, 103, 104, 121
Dysentery, 79, 115, 117, 193, 198, 199, 275, 277, 279

E
Eberth, 7, 199
Eczema herpectium, 239
Edge, 30, 100
Edward Jenner, 8, 297, 326, 331

Edwin Klebs, 194
Effector phase, 312, 319
Effector response, 298, 304, 311, 313, 326, 328
Eight-legged archanids, 288
Electron microscopy (EM), 239
Elek's gel precipitation test, 190
Elementary Body (EB), 218
Elephantiasis, 283
ELISA, 201, 215, 217, 220, 239, 240, 242, 245, 248, 252
Embryonated hen's egg, 216
Encephalitis, 235, 239, 246, 247, 272, 331
Endemic, 164, 209, 242
Endocytosis, 236, 298, 301
Endoflagella, 210, 211
Endogenous infection, 164
Endospore staining, 42, 104
Endospore, 41, 42, 56, 101, 104, 105, 107
Enlargement of testes, 283
Enriched media, 76, 94, 183
Enrichment media, 77
Entamoeba histolytica, 275, 276
Enteroaggregative *Escherichia coli* (EAEC), 193
Enterobacter, 17, 77, 78, 80, 89, 91, 166, 167, 177
Enterobacteriaceae, 17, 77, 78, 89, 91, 177, 191, 194, 198, 199
Enteroinvasive *Escherichia coli* (EIEC), 193
Enteropathogenic *Escherichia coli* (EPEC), 192
Enterotoxigenic *Escherichia coli* (ETEC), 193
Enterotoxin, 178, 180, 195, 198, 221–3
Enterovirus, 233, 242–4
Envelope, 15, 34, 37, 38, 232–8, 241, 245, 246, 249, 250
Environmental infections, 164
Enzyme-linked immunosorbent, 308, 322
Eosin methylene blue, 94, 191
Eosin, 94, 99, 191
Epidemic myalgia, 244
Epidemic, 6, 8, 87, 164, 216, 242, 244
Epididymitis, 219
Epitope, 305, 306, 308
Epizootic disease, 292
Epstein–Barr virus (EBV), 238

Equivalence, 306, 307, 322
Erythematous cellulitis, 179
Erythroblastosis fetalis, 316
Erythromycin, 38, 143, 183, 190, 215, 218, 220, 221
Escherichia, 17, 38, 47, 49, 54, 57, 62, 89, 159
Ethambutol, 204
Ethidium bromide, 99
Ethyl alcohol, 102, 132
Ethylene oxide, 121, 131, 134
Eukaryotes, 14, 20–2, 35, 37, 57–9, 92, 142–5
Eukaryotic, 3, 13, 14, 18, 19, 21, 22, 29, 38
Eumycetoma, 267
Exfoliative toxin, 178, 179
Exogenous cross-infection, 164
Exotoxins, 36, 178
Exponential phase, 61, 62
Extrinsic asthma, 315
Extrinsic incubation period, 115

F

F antigen, 191
Facilitated diffusion, 57, 58
Facultative anaerobes, 67, 93, 177, 191, 213
Facultatively anaerobic, 182, 189, 191, 196, 198, 199, 207, 209
Faecal coliform, 192
Family, 17, 89, 91, 191, 198, 199, 202, 209, 212
Fastidious organisms, 74, 76, 94
Female worms, 283
Fermi vaccine, 248
Fernandez reaction, 207
Ferrous ammonium sulphate, 92
Filoviridae, 234
Fimbriae, 5, 16, 41, 75, 184, 191, 196
Firmicutes, 21
Five I's, 82, 83
Five-kingdom, 18, 19, 21, 258
Flagella staining, 106
Flagella, 11, 17, 36, 39–41, 46, 47
Flagellar H antigen, 196, 200
Flagellates, 20, 23, 275
Flagellin, 39, 40, 46–8
Flaming, 121, 122
Flaviviridae, 234
Fleas, 115, 286, 287
Flexion and extension, 210, 211
Flucystosine, 268, 273

Flucytosine, 146, 262
Fluid mosaic membrane, 37
Fluorescence microscope, 33, 100
Fluorescence microscopy, 31, 204, 239
Fluoroquinolones, 144, 187, 194, 199
Focal infection, 114, 117
Fonsecaea compacta, 268
Fonsecaea pedrosoi, 268
Formaldehyde, 97, 124, 125, 131, 132, 134
Four-kingdom, 18
Foxes, 247
Freund's complete adjuvant, 327
Freund's incomplete adjuvant, 327
Fried egg colonies, 212, 213
Friedlander, 194
Fuchsin, 99, 104, 105
Fungi imperfecti, 261
Fungi, 6, 14, 15, 18–23, 107, 115, 120
Fungus-like bacteria, 28, 201

G

Gaffky–Eberth bacillus, 199
Gamma haemolysis, 77
Gamma rays, 68, 127
Gamma-haemolytic, 182
Gammaherpesvirinae, 237
Ganciclovir, 147
Gastroenteritis, 167, 199
Gelatin, 72, 74, 78, 92, 125, 200, 208
Gelatinase, 92
Genital herpes, 239
Genital infections, 183, 221
Genital membranes, 116
Gentamicin, 142, 166, 196, 198, 287
Genus, 17, 30, 89, 92, 177, 184, 191, 196
Georg Theodor August Gaffky, 199
Georges Fernand Isidore Widal, 323
Ghon focus, 203
Giemsa, 211, 212, 216, 218–20, 270
Gill, 160, 266
Gimenez, 216
Glandular fever-like, 250
Gliding movement, 47, 92
Glomerulonephritis, 183, 324
Glucose, 60, 74, 79, 89–91, 94, 184, 187, 191, 192
Glutaraldehyde, 131, 132
Glycocalyx, 16
Glycoprotein peplomers, 237, 246, 249

Good laboratory practices, 82, 119, 120, 159
Gram staining, 9, 24, 35, 87, 89, 98, 101, 102, 106
Gram-negative infections, 165
Gram-negative pleomorphic, 196
Gram-negative, 15, 21, 23–6, 35, 36, 40, 48, 58
Gram-positive infections, 165
Gram-positive, 15, 21, 23–6, 35, 36, 40, 41, 58
Granulomatous inflammatory response, 203
Group A Streptococcus, 182, 324
Group A, 182, 244, 324
Group B, 183, 193, 244
Group translocation, 59, 60
Growth curve, 60, 61
Growth factors, 57, 74, 77
Guillain–Barre syndrome, 251

H

H antigen, 191, 193, 196, 200, 324, 327
Haeckel, 18
Haemadsorption test, 215
Haemadsorption, 215, 232
Haemagglutination test, 201
Haematopoiesis, 299
Haematoxylin, 99
Haemolytic disease of newborn, 316
Haemophilus influenzae, 329, 330
Halogens, 121, 133
Halophiles, 20–3, 65
Hand-foot-and-mouth disease, 244
Hans Christian Gram, 9, 101
Hansen's disease, 205
Hard ticks, 288
Hauch bacilli, 196
Head louse, 290
Heat (Calor), 299
Heat sterilization, 121, 122, 124, 125, 134
Heavy chain, 304
Heavy metal salts, 134
Helical symmetry, 232
Hepadnaviridae, 233
Hepatitis A vaccine, 329
Hepatitis A Virus (HAV), 245
Hepatovirus, 242, 243, 245
Herpangina, 244
Herpes simplex virus (HSV), 237

Herpes simplex virus type 2 (HSV-2), 238, 239
Herpes simplex virus type 1, 237, 238
Herpesviridae, 231, 233, 237, 238
Herpetic whitlow, 238
Heterotrophs, 23, 54
Hexachlorophene, 134
Hippocampus, 248
Histoplasma capsulatum, 285, 295, 296
Histoplasmosis, 252, 269
Hodgkin's disease, 214
Hoechst stains, 99
Horizontal evolution, 150
Hospital waste, 152
Hospital-acquired infections, 117, 163, 192, 194, 195
Hot air oven, 121–3, 134–6
House dust mites, 291–3
Human cytomegalovirus, 237
Hydrogen sulphide, 53, 91, 92
Hydrolyse geltain, 200
Hydrophobia, 247
Hydrops in infants, 316
Hyperactivity, 247
Hypersensitivity pneumonitis, 320
Hypersensitivity, 291, 292, 295, 302, 311–13, 316, 317
Hyphae, 258, 261, 265
Hypogammaglobulinemia, 214

I

Iatrogenic infections, 114, 117
Icosahedral capsid, 237, 249
Icosahedral symmetry, 232
Identification, 3, 4, 11, 17, 30, 48, 73, 82
Idoxuridine, 145, 147
IgA, 185, 304, 305
IgD, 304, 305
IgE, 304, 305, 312, 313, 314, 315, 317
IgE-mediated response, 313
IgG, 215, 239, 299, 304, 305, 312, 316, 317, 318
IgM, 215, 239, 241, 242, 304, 305, 312, 316
Imidazoles, 145–7, 266
Immune complex-mediated hypersensitivity (type III), 312
Immune complex-mediated hypersensitivity, 312
Immunis, 297

Immunity, 5, 6, 8, 9, 52, 113, 165, 189, 190, 205
Immunoelectrophoresis, 215, 307
Immunogen, 178, 196, 221, 304, 330
Immunogenicit, 304
Immunologic memory, 299, 331
Immunoprophylaxis, 295, 326
IMViC reaction, 194
Inactivated polio vaccine (Salk), 243
Inapparent infections, 114
Incineration, 121, 122, 156
Inclusion conjunctivitis, 219, 221
Inclusions, 39, 218, 219, 238, 242
Incubation, 48, 61, 62, 83–5, 92, 115, 127, 148
Indeterminate leprosy, 206
Indinavir 800 mg, 253
Indirect enzyme-linked immunosorbent assay, 308
Indole negative, 196, 200
Indole positive, 196, 208
Indole production, 92
Indole, 24, 89, 92, 192, 196, 200, 208
Indophenol blue, 93, 94
Infection control committee, 168
Infectious mononucleosis (glandular fever), 241
Infectious waste, 153
Inflammatory barriers, 298
Ingestion, 22, 117, 192, 198, 200, 221
Inhalation, 117, 182, 188, 202, 203, 205, 221
Injured skin, 116
Inner-surface protein, 37
Inoculation hoods, 159
Inoculation, 8, 48, 61, 83, 90, 92, 117, 122, 158
Inspection, 17, 83, 86
Inspissation, 124
Integral protein, 37
Interleukin receptors (IL-2), 302
Intracellular obligatory parasites, 231
Intracerebral inoculation, 219, 246,
Intranasal, 219
Intraperitoneal, 219, 244
Intrinsic asthma, 315
Invasiveness, 51
Inverted microscope, 33
Iodine, 94, 99, 102, 106–8, 133, 265
Ionizing radiation, 68, 69, 121
Iron Bacteria, 53

Isolation, 3, 9, 20, 74, 75, 83, 86, 94, 124
Itraconazole, 265, 267, 268, 270
Ixodes scapularis, 288
Ixodid ticks, 217, 288

J
John Tyndall, 7, 125
Joseph Meister, 8

K
K antigen, 191–4
Kanamycin, 142
Kaposi's sarcoma-associated herpesvirus, 237
Keratoconjunctivitis, 238
Ketoconazole, 145, 265, 267, 268, 272
Killed vaccine, 328, 329
Kingdom, 17–22, 258, 262, 276
Kirby–Bauer method, 149
Klebsiella ozaenae, 194, 195
Klebsiella pneumoniae, 167, 194, 195
Klebsiella rhinoscleroma, 194, 195
Klebsiella sp., 166, 194, 195
Klebsiella, 47–9, 89, 166, 167, 191, 194, 195
Koch's postulates, 7, 9
Kovac's reagent, 89

L
Lactose fermenter, 191, 192, 194
Lactose, 24, 78, 92, 94, 191–4, 198, 199, 208
Lag phase, 60–2
Laminar hoods, 130
Lamivudine, 253
Large colony, 213
Latent infections, 114, 203, 237
Leifson stain, 107
Leifson's, 48
Lentivirinae, 249
Lentivirus, 235, 249
Lepromatous leprosy, 206
Lepromin test, 206, 207
Leprosy, 7, 104, 205–7
Leptonema, 209
Leptospira, 29, 47, 209–11
Leptospiraceae, 209
Leucocytosis, 195
Leukaemia, 214, 249
Leukocidin, 179
L-forms, 43
Lice, 115, 211, 287, 288, 290, 291

Light chain, 304
Lipid hydrolysis test, 94
Lipopolysaccharides, 15, 185
Lipoproteins, 15, 36, 213, 246
Liquid media, 75, 80, 83, 196
Live attenuated oral polio vaccine (Sabin), 243
Live or attenuated vaccine, 328
Loffler, 189
Log phase, 61, 62
Lophotrichous, 40
Louis pasteur, 6–9, 72, 297, 326
Low-energy rays, 128
Lowenstein–Jensen medium, 76, 78, 80, 202, 204
Lxodid ticks, 217
Lymph, 51, 203, 238, 239, 243, 250, 251, 268, 270, 283
Lymphadenopathy, 251, 270, 283, 317, 319
Lymphocytes, 240, 241, 250, 299, 300, 301, 319
Lymphogranuloma venereum, 220
Lymphoma, 214, 240, 241, 249, 251, 252, 271
Lyssavirus, 234, 245

M
Macchiavello, 216, 218
MacConkey agar, 77, 78, 80, 192
Macroconidia, 262
Macroelements, 55
Macrophages, 202, 203, 250, 270, 298, 300, 302
Magnetotaxis, 41
Major Histocompatibility Complex (MHC), 300
Malachite green, 42, 78, 99, 104–106, 202
Malaria, 9, 115, 116, 138, 147, 275, 278–80
Malt Extract, 74
Mannitol, 75, 77, 94, 178, 181, 192, 198, 221
Mantoux tuberculin skin test, 204
Margin, 30
Matrix protein (viral membrane), 246
Meat extract, 72, 74
Mechanical vectors, 115, 288
media, 11, 30, 49, 53, 56, 57, 72–73, 74, 79
Mefloquine, 147
Membrane filters, 121, 129, 130, 231

Memory response, 298, 299, 328, 331
Meningitis, 179, 183, 187, 193, 195, 239, 243, 244
Mercuric chloride, 134
Merozoites, 279
Mesomycetozoa, 269
Mesophiles, 24, 65, 66
Metabolism, 37, 57, 89, 90, 139, 144
Metachromatic granule staining (Albert's Staining), 106
Metallic sheen, 191
Metamorphosis, 287
Metaphyta, 19
Metazoa, 19
Methanogens, 21
Methyl red test, 89, 90
Methylene blue, 79, 94, 97, 99, 100, 104, 107, 124, 191
Miconazole, 145
Microaerophiles, 67
Microbial susceptibility, 181
Microbicidal antibiotics, 140, 149
Microbiology, 1, 3–8, 30, 31, 82, 83, 86, 101, 103
Microbiostatic antibiotics, 140, 149
Micrococcus, 29
Microconidia, 262, 269
Microelements, 55
Microfilariae, 282
Middlebrook's medium, 202
Mineral salts, 202
Minimum inhibitory concentration (MIC), 148
Minocycline, 143
Mites, 115, 116, 217, 287, 288, 291–3, 311, 314, 316
Mitsuda reaction, 207
MMR, 328, 330–2
Moist heat sterilization, 121, 124, 125
Mollicutes, 211
Monera, 17–19
Monotrichous, 40
Monsur's tellurite taurocholate gelatin agar, 78
Mordant, 98, 99, 102
Morganella, 92, 196, 197
Mosquitoes, 115, 275, 278, 279, 280, 282, 283, 287
Motile, 48, 49, 75, 92, 106, 191, 192, 196, 199, 207
Motility, 11, 46–49, 75, 92, 196, 199, 201, 209–11
Mounting, 97, 98, 269

Mouse fibroblast cell lines, 219
MRSA strains, 181
Mucous membranes, 252, 298
Mucus, 195, 199, 298, 314
Muller–Hinton Agar, 149
Multibacillary disease, 206
Multilocus sequence typing (MLST), 88
Mupirocin, 181
Murein, 15, 35
Murine typhus, 217
Mycelium, 201, 259, 264
Mycetoma, 266, 267
Mycobacterium leprae, 205, 207
Mycobacterium tuberculosis, 9, 36, 62, 78, 80, 93, 124, 133, 159, 201
Mycobacterium, 9, 36, 62, 78, 80, 93, 104, 124
Mycolic acid, 36, 104, 202
Mycoplasma, 21, 28, 30, 103, 107, 177, 211–15
Mycoplasma fermentans, 212, 213
Mycoplasma genitalium, 212–14
Mycoplasma hominis, 212–14
Mycoplasma orale, 212, 213
Mycoplasma pneumoniae, 212–15
Mycoplasma salivarium, 212, 213
Mycoplasmataceae, 211, 212
Mycoplasmatales, 211, 213
Mycoses, 258, 263–5, 270
Mycota, 19
Myocarditis and pericarditis, 214, 244
Myxobacteriales, 28

N

N-acetylmuramic acid, 35
Nalidixic acid, 144, 196
Nalidixic, 144, 196
Narrow-spectrum Antibiotics, 140
Nasopharyngeal carcinoma, 241
Naturally acquired passive immunity, 326, 327, 331
Negative capsule staining, 106
Negri bodies, 246, 247, 248
Neisseria gonorrhoeae, 41, 77, 184, 331
Neisseria meningitidis, 187, 188
Neisseria, 41, 77, 80, 177, 184, 185, 187, 188
Neomycin, 142, 143
Neonatal infection, 239

Neural vaccines, 248
Neurocysticercosis, 281
Neurotoxin, 198, 330
Neutral red or toluylene red, 99
Niacin accumulation test, 202
Nile blue, 99
Nile red–Nile blue oxazone, 99
Nit, 290
Nitrate, 55, 93, 134, 167, 196, 202
Nitrate reduction test, 202
Nitric oxide, 93, 319
Nitrifying Bacteria, 53
Nitrofurantoin, 198
Nitrogen, 23, 39, 54, 55, 67, 92, 93, 213
Nitroso indole, 208
Nitrous oxide, 93
Non-capsulated slender, 201
Non-gonococcal urethritis, 185, 214, 219, 220
Non-haemolytic streptococci, 182
Non-ionising radiation, 128
Non-lactose fermenters, 191, 198, 199, 286
Non-motile, 46, 48, 49, 75, 92, 106, 177, 182, 184, 187
Non-neural vaccines, 248
Non-paralytic poliomyelitis, 243
Non-spore-forming, 177, 189, 191, 201, 205, 285
Norfloxacin, 144
Nosocomial bloodstream infections, 165, 166
Nosocomial infection surveillance, 163, 170
Nosocomial infections, 5, 114, 117, 163, 165–7, 169, 170, 194
Nosocomial pathogen, 197
Nosocomial pneumonia, 165, 166
Nucleocapsid, 231, 232, 236, 250
Nucleoid, 14, 35, 37, 38, 41, 210, 218
Nucleolus, 14
Nucleus, 3, 14, 18, 22, 37, 97, 216, 218, 232
Nutrients, 15, 19, 22, 37, 39, 41, 42, 49, 53, 54
Nymph, 290
Nystatin, 77, 145, 184, 262, 272

O

O antigen, 191, 192, 194, 200, 323, 324, 327
Obligate anaerobes, 67, 68
Obligate, 53, 67, 68
Ocular mucous, 116
Oedema, 299, 319
Ofloxacin, 207
Oncovirinae, 249
OPA Proteins, 185, 187
Ophthalmia neonatorum, 219, 221
Opportunistic mycoses, 270
Opportunistic mycosis, 263, 270
Optical microscope, 31
Order, 17, 22, 28, 34, 36, 47, 51–3, 59
Organ transplant, 214
Orthomyxoviridae, 235, 237
Oseltamivir, 147
Osmium tetroxide, 99
Outer-surface protein, 37
Oxaloacetate decarboxylase, 90
Oxaloacetate, 90
Oxidase negative, 93, 189, 191, 196, 198
Oxidase test, 93, 184, 187, 188
Oxidative decarboxylation, 91
Oxidizing agents, 67, 132
Oxytetracycline, 142
Ozena, 195

P

P. G. H. Gell, 312
Pain (Dolor), 299
Papovaviridae, 233
Paradoxical carriers, 114
Paraffin oil, 327
Paralytic poliomyelitis, 243
Paramyxoviridae, 234, 237
Parasitic bacteria, 54
Paratope, 305
Parvoviridae, 233, 237
Passive diffusion, 57
Passive immune response, 326
Pasteur vaccine, 248
Pasteurization, 7, 8, 124
Paucibacillary disease, 206
Paul ehrlich, 9, 103, 137, 138, 150
Paul porter, 311
Paul–Bunnell heterophile antibodies, 241
Pediculicides, 291
Pediculus humanus capitis, 290

Penicillin G, 140, 141, 181, 187, 188
Penicillin, 9, 35, 36, 138, 140, 141, 150, 181
Penicillins, 190, 212, 215
Peplomers (spikes), 246
Peplomers, 232, 236, 237, 246, 249
Peptidoglycan, 15, 21, 35, 36, 41, 101, 102, 141
Peptones, 73, 74, 92
Peracetic acid, 121, 133
Peritrichous, 40, 48, 191, 196
Permethrin, 290
Persistent generalized lymphadenopathy, 251
Personnel protective equipments, 158, 160, 161
Petroff's deposition method, 203
Petroff's method, 203
pH, 64, 68, 74, 77, 78, 79, 90, 91, 92
Phagocytic barriers, 298
Phagosomes, 236
Pharyngitis, 183, 190, 214, 220, 244
Phase contrast microscope, 33
Phase I, 318, 331
Phase II, 318, 331
Phase III, 318
Phase IV, 318
Phenols, 121, 133
Phialophora verrucosa, 268
Phototaxis, 41
Phototrophs, 23
Phthirus pubis, 290
Phylum, 17, 276, 287, 291
Physical sterilization, 120, 121
Physiologic barriers, 298
Picornaviridae, 231, 233, 237, 242
PIDs, 36, 37, 39, 55, 99, 101, 142, 185, 186
Piedra, 264, 266
Pig, 39, 53, 69, 177, 178, 182, 184, 190, 208, 215
Pike's medium, 79, 80
Pili (fimbriae), 184
Pili, 14, 16, 41, 184, 187, 188, 191
Pityriasis versicolor, 264
Plantae, 18, 21, 22, 258
Plasma membrane, 14, 15, 39, 42, 57–9, 146, 236, 278
Plasmodium, 147, 275, 278, 279
Plasmodium falciparum, 278, 279
Plasmodium malariae, 278, 279
Plasmodium ovale, 278, 279
Plasmodium vivax, 278, 279

Pleconaril, 146
Pleomorphic, 30, 189, 196, 201, 205, 212, 219
Pneumonia, 165, 166, 178, 179, 194, 195, 213, 214
Pneumonic plague, 286, 287
Pneumonitis, 220, 319, 320
Poikilothermic, 65
Polar flagella, 48, 207
Polio vaccine, 9, 243, 327, 329
Polioviruses, 236, 242, 243
Polymyxin B, 140, 142, 213
Polysaccharide, 15, 35, 39, 73, 94, 100, 187, 188, 191
POR (Porin Protein P II), 185
Pork tapeworms, 280
Positive capsule staining, 105
Potassium tellurite medium, 77, 80
Pour plate method, 85, 86
Poxviridae, 233, 237
Precipitation, 48, 190, 305, 306, 307, 308, 322
Precipitation reactions, 305, 306, 308–10
Precipitin, 306, 307, 322
Precystic stage, 276
Primaquine, 147
Primary containment, 158
Primary infection, 113, 117, 193, 203, 238, 239
Proctitis, 219, 220
Prokaryotic, 3, 13, 14, 18, 19, 21, 22, 35
Prontosil, 93, 138
Protein typing, 87
Proteus, 47, 48, 89, 92, 191, 196, 197
Proteus mirabilis, 196, 197
Proteus sp., 92, 196
Proteus vulgaris, 196, 197
Prothionamide, 207
Protista, 18, 19, 21, 269, 276
Protoplasm, 37, 56
Provindencia, 196, 197
Pseudodiptheriticum, 189
Pseudohyphae, 260, 271–3
Pseudosel agar, 77
Psittacosis, 220
Psychrophiles, 24, 66
Pubic louse, 290
Pulsed-field gel electrophoresis (PFGE), 88
Purkinje, 248

Pyrazinamide, 204
Pyruvate, 90
Pyruvic acid, 89, 90

Q

Qualitative, 306–9, 322–4
Qualitative tool, 307
Quantitative, 30, 65, 76, 305, 306, 308, 322, 323
Quantitative assay, 306
Quinine, 147
Quinolones, 144, 187, 194, 199

R

R. H. Whittaker, 18
R. R. A. Coombs, 312
Rabies vaccine, 248, 329
Rabies virus, 8, 117, 234, 245, 246, 247
Radial immunodiffusion (Mancini technique), 306
Radiation, 5, 19, 31, 41, 64, 68, 69, 120, 121, 127
Radioimmunoassay test (RAST), 315
Rapid tests, 252, 280
Rat fleas, 217
Reaction, 13, 17, 24, 36, 87, 90, 102, 142, 190, 194
Recognition, 4, 101, 297–9, 302
Recombinant subunit vaccine, 328, 329,
Red heat, 122
Red yellow blue and black containers, 153
Red-coloured smooth, 198
Redness (Rubor), 299
Reduction broth, 93
Reflection Electron Microscope (REM), 34
Reiter's syndrome, 251
Reoviridae, 235
Repetitive sequence-based PCR (Rep-PCR), 88
Reservoir hosts, 115
Respiratory infections, 117, 183, 221
Respiratory paralysis, 247
Respiratory tract, 51, 115, 116, 178, 182, 187–9, 195, 196, 206
Response, 179, 185, 189, 203, 207, 247, 264, 283, 290
Resuscitation culture medium, 79
Reticulate body (RB), 218
Retroviridae, 231, 235, 249

Retrovirus, 147, 249
Reverse transcriptase enzyme, 249
Rh incompatibility, 316
Rhabdoviridae, 231, 234, 237, 245, 246
Rhinocladiella aquaspers, 268
Rhinoscleroma, 195
Rhinosporidiosis, 266, 269
Rhinosporidium seeberi, 269
Rhinovirus, 233, 242–4
Rhodamine, 99, 107
RIA, 308
Ribavirin, 147
Ribosomes, 15, 35, 37, 38, 142, 217
Ribotyping, 88
Rice watery stools, 208
Richet, 311
Rickettsia, 177, 216, 217, 289
Rickettsiales, 28
Rickettsialpox, 217
Rifampicin, 144, 204, 207, 218, 267
Rifampin, 144
Rimantadine, 146
Ritter's disease, 179, 221
Rmp (Protein III), 185
Robert H. Whittaker's, 258
Robert Koch, 7, 9, 72, 92, 201, 207
Robertson's cooked meat medium, 79
Rocky mountain spotted fever, 217
Rodents, 285, 286, 287, 288, 289
Root colonization, 49
Roundworms, 282

S
S layer, 15
Sabouraud Dextrose Agar (SDA), 265
Sach's buffered glycerol saline, 79
Safety equipments, 158, 159
Safranin, 36, 97, 99–105, 107
Salbutamol, 292
Saliva, 164, 238, 239, 240, 241, 242, 247, 248, 278
Salmonella H and O antigens, 323
Salmonella Paratyphi A, 200, 324
Salmonella Paratyphi B, 200
Salmonella Paratyphi C, 200
Salmonella typhi, 78, 199, 200, 323, 324
Salmonella, 47, 49, 78, 80, 87, 89, 124, 160, 191, 199
Salpingitis, 185, 214, 220
salt agar, 77, 94, 181, 221

Sandwich enzyme-linked immunosorbent assay, 308
Saprophyte, 22, 54, 113
Saprophytic, 54, 260, 265, 269
Sarcina, 29
Sarcoptes scabiei, 292
Scabies mites, 291, 292
Scanning Electron Microscope (SEM), 34
Schaeffer–Fulton method, 104
Scrub typhus, 217
Sebum, 298
Secondary containment, 158
Secondary infection, 113, 117, 139, 193, 286
Secreted antibodies, 300
Selective media, 77, 80, 94, 181, 187, 188, 200, 202, 209
Selective toxicity, 138, 144, 145
Selenium sulphide, 266
Self and non-self recognition, 299
Seller's technique, 247
Semi-solid media, 49, 75
Sensitization phase, 312, 319
Septate, 261, 266
Septicaemic plague, 286, 287
Seroconversion illness, 251
Serological techniques, 215
Serological testing, 87, 322
Serratia, 191
Serum sickness, 317, 318
Sexual reproduction, 262
Shape, 6, 13, 16, 17, 29, 30, 35, 36, 98
Shigella, 48, 89, 191, 193, 198, 199
Shigella boydii, 198
Shigella dysenteriae, 198
Shigella flexneri, 198
Shigella sonnei, 198
Siberian tick typhus, 217
Siderophores, 59, 60
Simian herpes virus, 237
Simple media, 76, 208
Simple Microscope, 31
Simplex virus type 1 (HSV-1), 238
Single trilaminar cell membrane, 212
Sir Alexander Fleming, 138
Skin, 52, 115, 116, 131–3, 160, 166, 167, 178
Skin and soft tissue infections, 167
Skin infections, 133, 179, 183, 240, 264
Slide agglutination test, 183, 199

Slit skin test, 207
Sluggish, 210
Small size, 41, 231
Snakes (antivenom), 327
Sodium thiosulphate, 92
Solid media, 30, 49, 72, 83, 196, 212
Solubility, 107, 139
Somatic O antigen, 200
Specie, 17, 21, 28–30, 39, 43, 52, 56, 73
Specific immunity, 165, 298, 299
Spiders, 327
Spirilla, 16, 46
Spirochaeta, 47, 209, 211
Spirochaetaceae, 209
Spirochaetales, 28
Spirochaetes, 7, 21, 29, 33, 47, 177, 209–11, 289
Spiroplasma, 212
Spiroplasmataceae, 212
Sporan-giospores (asexual spores), 261
Sporangiospores, 261, 262
Sporozoa, 275
Sporulation, 42, 62
Spread plate method, 85
Spumavirinae, 249
Spumavirus, 249
ssRNA, 249
Stability, 88, 139, 157
Staining techniques, 6, 11, 17, 87, 67, 97, 98, 101, 103
Staphylococcal skin scalded syndrome, 178, 179, 221
Staphylococci, 7, 16, 93, 94, 166, 177, 179
Staphylococcus, 7, 29, 52, 65, 115, 124, 167, 177–9, 181, 221
Starch hydrolysis test, 94
Stationary phase, 60–63
Steam formaldehyde, 124, 125
Steam sterilizer, 124, 125
Stereo microscope, 33
Sterilization, 3, 4, 7, 111, 119–28, 131–6, 158, 168, 169
Sterol, 212
Streak plate method, 83, 84
Streptococcus, 29, 39, 177, 182, 183, 214
Streptomycin, 9, 36, 38, 138, 140, 142, 143, 287
Stuart, 75, 78, 197
Subcutaneous mycosis, 263, 266, 2
Sucrose, 78, 80, 221, 286

Sulphanilamide, 144
Sulphanilic acid (reagent A), 93
Sulphide indole motility (SIM), 92
Sulphisoxazole, 144
Sulphonamides, 138, 140, 141, 144, 147, 194, 196, 197, 267
Sulphur Bacteria, 53
Sulphur reduction, 92
Superbugs, 150
Superficial mycoses, 263, 264
Surface Vi antigen, 200
Surfactants, 134
Surgical site infections, 166
Swelling (Tumor), 299
Symbiotic bacteria, 54
Symport, 37, 59
Synthetic antibiotics, 140, 141, 145
Synthetic media, 76
Systemic mycosis, 269
Systemic, 52, 145, 157, 167, 197, 260, 269

T

T cells, 203, 251, 300, 302, 312, 320
T cytotoxic cells (T_C), 300
T helper cells (T_H), 300
Taenia solium, 280, 281
Tannic acid, 107
Tapeworms, 280
Taxonomy, 17, 101, 258
T_C cells, 302
Tears, 298
Teichoic acids, 15, 35, 36
Test, 9, 48, 65, 75, 87, 89, 90, 91, 92
Tetanus toxoid, 190, 330
Tetracycline, 140, 142, 143, 194, 196, 215, 217, 218
Tetrazolium reduction test, 215
Thayer Martin agar, 77, 184, 188
Theophylline, 316
Therapeutic dose, 139
Thermoacidophiles, 21
Thermophiles, 20, 24, 65, 66
Thioglycollate, 79, 80
ree-domain system, 21
e-kingdom, 18
, 115, 211, 217, 275, 287, 288, '89, 290, 324
e of iodine, 265
gra, 264, 265
rsicolor, 264
orphology, 97
, 324
e, 234

Toxic dose, 139
Toxic shock syndrome toxin-1, 180
Toxigenesis, 51
Toxoid, 190, 328, 330
Trachoma, 116, 219, 220
Translatory motion, 210
Translucent colonies, 198
Transmission Electron Microscope (TEM), 34
Transport media, 75, 78, 209
Traveler's diarrhoea, 193
Treponema, 7, 47, 62, 209–11, 331
Triazole, 265
Triple sugar iron (TSI) agar, 91
Trombiculid mites, 217
Trophozoites, 275–7, 279, 280
True nucleus, 14
Trybutyrin agar, 94
Tryptic soya agar, 79
Tube dilution, 148
Tuberculoid leprosy, 205, 206
Twitching motility, 49
Two-kingdom, 18
Tyndallization, 7
Type 1, 237, 238, 242
Type 2, 238, 239, 242
Type 3, 242
Type II hypersensitivity, 316, 317
Type III hypersensitivity reaction, 317
Type I-mediated hypersensitivity, 312
Typhoid-paratyphoid A and B (TAB), 201
Typing, 82, 86–88, 308

U

Universal tree of life, 13, 22, 23
Ureaplasma, 211, 212
Ureaplasma urealyticum, 212
Urease, 24, 52, 91, 92, 194, 196, 198, 200
Urease negative, 200
Urethritis, 185, 214, 219, 220
Urinary tract infection, 165, 192, 195
Utilization test, 90
UTIs, 195–7
UV rays, 128

V

Vaccination schedule, 326, 330
Vaccine bath, 124, 125
Vancomycin, 77, 141, 166, 181, 184
Varicella (chickenpox), 239
Varicella zoster virus (VZV), 239
Variolation, 8, 326, 331

Vasodilatation, 299, 313
Vectors, 115–17, 175, 283, 285, 287–9, 291–3
Venkatraman Ramakrishnan medium, 79, 80
Vertical evolution, 150
Vertical gene transmission, 22
Vesiculovirus, 234, 245
Vibrio cholerae, 7, 17, 29, 78, 115, 177, 207, 208
Vidarabine, 145, 240
Virion, 147, 232, 236, 237, 239, 240, 245, 246, 250
Viruses, 3, 6, 9, 28, 29, 34, 35, 120, 125, 128
Voges–Proskauer test, 90

W

Wassermann reaction, 214
Wax D, 202
Weil–Felix reaction, 217
Wet mount processing, 265
White piedra, 266
Whitfield's ointment, 265
Widal test, 201, 323
Wilson, 78, 80, 200
Wolves, 247
Wood's light examination, 265
Wound colonization, 52
Wuchereria bancrofti, 280, 282

X

X-rays, 68, 127, 320

Y

Yeast Extract, 74, 79, 94
Yersinia pestis, 285
Y-shaped, 304

Z

Zidovudine, 147, 253
Ziehl and Neelsen modified Koch, 201
Ziehl Neelsen Method, 104
Ziehl–Neelsen acid-fast technique, 201
Ziehl–Neelsen staining technique, 205
Zone of inhibition, 138, 149
Zoonoses, 5, 115
Zoonotic diseases, 115, 286
Zoster (Shingles), 239
Zygomycetes, 261